Springer Water

The book series Springer Water comprises a broad portfolio of multi- and interdisciplinary scientific books, aiming at researchers, students, and everyone interested in water-related science. The series includes peer-reviewed monographs, edited volumes, textbooks, and conference proceedings. Its volumes combine all kinds of water-related research areas, such as: the movement, distribution and quality of freshwater; water resources; the quality and pollution of water and its influence on health; the water industry including drinking water, wastewater, and desalination services and technologies; water history; as well as water management and the governmental, political, developmental, and ethical aspects of water.

Chaudhery Mustansar Hussain
Neha Srivastava • Ghazanfar Hussain
Editors

Fungi in Waste Water Treatment

Volume 1

Springer

Editors
Chaudhery Mustansar Hussain
Chemistry and EVSC
New Jersey Institute of Technology
Newark, NJ, USA

Neha Srivastava
Indian Institute of Technology (BHU)
Department of Chemical Engineering &Technology
Varanasi, Uttar Pradesh, India

Ghazanfar Hussain
Division of Science and Technology
Department of Education Lahore
Lahore, Pakistan

ISSN 2364-6934 ISSN 2364-8198 (electronic)
Springer Water
ISBN 978-3-032-12692-4 ISBN 978-3-032-12693-1 (eBook)
https://doi.org/10.1007/978-3-032-12693-1

This Springer imprint is published by the registered company Springer Nature Switzerland AG
The registered company address is: Gewerbestrasse 11, 6330 Cham, Switzerland

Preface

The treatment of wastewater is a critical aspect of environmental management, particularly as global urbanization and industrialization continue to place immense pressure on water resources. In recent years, the application of fungi in wastewater treatment has emerged as a promising bioremediation technique, offering an eco-friendly and sustainable solution to tackle some of the most challenging water quality issues. *Fungi in Wastewater Treatment: Volume 1* explores into the multifaceted roles of fungi in the treatment of wastewater, offering a comprehensive exploration of the subject from various perspectives.

This book begins with an introduction to the basics of mycology and wastewater chemistry, laying a strong foundation for understanding how fungi interact with wastewater systems. It further explores the critical role fungi play in nutrient cycling and their mechanisms for remediating pollutants in wastewater. The text provides a thorough examination of the adaptive mechanisms of fungi and how their remarkable resilience allows them to thrive in wastewater treatment environments, contributing to the removal of heavy metals and the degradation of organic compounds.

As the world faces increasing challenges with toxic pollutants such as per- and polyfluoroalkyl substances (PFAS), this book discusses the biodegradation and transformation of these persistent chemicals by fungi, a process crucial for addressing the contamination in various water sources. In addition to addressing the complexities of wastewater treatment, the book covers the treatment of textile wastewater, an industry known for its significant contribution to water pollution. The concept of fungal-based constructed wetlands is also introduced as a promising approach for improving wastewater treatment processes in natural and constructed environments.

In *Case Studies-I*, readers are presented with real-world examples of successful fungal-based wastewater treatment systems, providing practical insights into the implementation and challenges of these technologies. The integration of fungal and algal treatment systems is discussed as a step toward enhancing treatment efficiency, while the potential of smart materials and nanotechnology in fungal-based treatment systems is explored as a futuristic avenue for wastewater management.

The book concludes by addressing the challenges and opportunities within the field of fungal wastewater treatment. It discusses the sustainability of fungal treatment systems, emerging trends, and the potential for innovation in this area.

This volume aims to be a crucial resource for researchers, practitioners, and students interested in the application of fungi in environmental remediation and wastewater treatment. By understanding the underlying science, mechanisms, and practical applications, we hope to encourage further research and development in this promising field, contributing to a cleaner and more sustainable world.

Newark, NJ, USA Chaudhery Mustansar Hussain
Varanasi, India Neha Srivastava
Lahore, Pakistan Ghazanfar Hussain

Acknowledgments

We would like to dedicate this book to our lovely sons and daughters Sarim Hussain Chaudhery, Azlan Hussain Chaudhery, Maida Ali, Hafiza Safa Hussain Chaudhery, Hafiza Dua Hussain Chaudhery, and Aira Hussain Chaudhery for their dedicated support during compiling of this book.

(Editors: Chaudhery Mustansar Hussain and Ghazanfar Hussain)

Editor Dr. Neha Srivastava would like to acknowledge the Department of Health Research, India, for Women Scientist Project and the Department of Chemical Engineering and Technology, IIT (BHU), Varanasi-221005, India, for all technical support.

Contents

About the Editors

Chaudhery Mustansar Hussain is an Adjunct Professor and Director of Laboratories in the Department of Chemistry & Environmental Sciences at the New Jersey Institute of Technology (NJIT), Newark, New Jersey, United States. His research is focused on the applications of nanotechnology and advanced materials, environmental management, sustainability, analytical chemistry, and other various industries. Dr. Hussain is the author of numerous papers in peer-reviewed journals as well as a prolific author and editor of around two hundred (200) books, including scientific monographs and handbooks in his research areas. He has published with Elsevier, American Chemical Society, Royal Society of Chemistry, John Wiley & Sons, CRC Press, and Springer.

Neha Srivastava is currently working as Women Scientist in the Department of Chemical Engineering and Technology, IIT (BHU) Varanasi, India, since 2014. She has published over 100 research articles in Scopus-indexed, peer-reviewed journals with 3 patents and has 1 Technology Transfer. She has 9 running book series with Springer Nature, CRC Press, CABI International, and NIPA International. She is Editor in Chief of the *Journal Sustainable Processes Connect*, Scifinity International Publisher of UAE. She is Co-founder of the start-up company LCB Fertilizer which deals in the area of organic fertilizers and biomass valorization. She has a current H-index of 38. She completed her PhD from the Department of Molecular and Cellular Engineering, SHIATS, India, in 2016 in Biotechnology and Biomass-Based Biorefinaries. She has received 6 Young Scientist Awards. Presently, she is working on biofuels production (cellulase enzymes, production and enhancement, biohydrogen production from waste biomass, bioethanol production). Her fields of expertise include: Biofuels production, microbial bioprocessing, enzyme technologies, and nanomaterials and their sustainable fabrication methods.

Ghazanfar Hussain is a young Research Scholar in Science and Technology, Department of Education, Punjab, Pakistan. His key areas of research are environment, analytical chemistry, materials science and engineering, and industrial development. He is author of monographs on chemistry, sample pre-treatments, nanotechnology, and technology. He is a professional and affiliated with several organizations. He has more than 7 years of integrated teaching and research experience and published more than 12 research articles in peer-reviewed international journals along with 3 books and 19 book chapters.

Chapter 1
Introduction to Fungi in Wastewater Treatment and Fundamentals of Mycology and Wastewater Chemistry

Ntalane Sello Seroka, Evan D. Visser, Lyle A. September, Andile Mashele, Nakedi A. Mojapelo, and L. Khotseng

Abstract The rapid development of the industrial sector has contributed significantly to the global economy and the well-being of individuals, while it has consequences in terms of waste generation, consumption of resources, and energy usage, which require the development of ad hoc strategies and countermeasures to contain their impacts. Covered are microbial bioremediation, bioleaching, bioaugmentation, biostimulation, mycoremediation, and phytoremediation. Among these biological remediation approaches, mycoremediation is beneficial since it can successfully break down a wide variety of contaminants and its efficacy is not restricted to specific concentrations of pollutants. Mycoremediation is a cost-effective and environmentally benign technology that uses fungal biomass to remove hazardous chemicals from polluted water. Mushrooms are fungal fruiting structures that grow from a fibrous mass known as the mycelium. Macro fungi have demonstrated a unique capacity to break down a variety of dangerous and lignin-containing wastes produced by a variety of anthropogenic and natural activities. This chapter establishes approaches that could serve as a prospective framework for addressing the worldwide issue of wastewater.

Keywords Wastewater · Fungi · Green chemistry · Mycology

N. S. Seroka (✉)
Department of Chemistry, University of the Western Cape, Bellville, South Africa

Energy Centre, Smart Places Cluster, Council for Scientific and Industrial Research (CSIR), Pretoria, South Africa
e-mail: 3754640@myuwc.ac.za; nseroka@csir.co.za

E. D. Visser · L. A. September · A. Mashele · N. A. Mojapelo · L. Khotseng (✉)
Department of Chemistry, University of the Western Cape, Bellville, South Africa
e-mail: 3651746@myuwc.ac.za; 3727392@myuwc.ac.za; 4159390@myuwc.ac.za; 4246489@myuwc.ac.za; lkhotseng@uwc.ac.za

C. M. Hussain et al. (eds.), *Fungi in Waste Water Treatment*, Springer Water,
https://doi.org/10.1007/978-3-032-12693-1_1

1.1 Mechanism of Wastewater Treatment

Among the most devastating effects associated with human activities is the formation of large volumes of wastewater, which is rich in a wide range of contaminants and responsible for major environmental contamination and public health problems (Latif et al., 2023; Xu et al., 2022). The primary cause of industrial wastewater contamination is the leakage of untreated or insufficiently treated chemical compounds from various operations/processes. Several sectors, including paper and pulp, agrochemical, textile, and petrochemical, generate extraordinary volumes of wastewater containing a variety of harmful and resistant contaminants, including organic compounds, heavy metals, aromatic dyes, and persistent chemicals (Li et al., 2021).

Recently, several chemicals have been found in drinking, ground and surface water that have a significant impact on water characteristics (Sun et al., 2021). Untreated industrial effluents include significant amounts of carcinogenic heavy metals, and the rapid depletion of dissolved oxygen in water due to the abundance of toxic heavy metals and other industrial contaminants results in sag of oxygen (Sun et al., 2021). Hydrocarbons, such as crude oils, petroleum hydrocarbons (PHC), and polycyclic aromatic hydrocarbons (PAHs), are also a significant contributor to water pollution. These are used as energy sources in many sectors, but they contribute significantly to pollution and harm human health as carcinogens, photosensitisers, and mutagens (Ghosh et al., 2022; Patel et al., 2020). Endocrine disrupting chemicals (EDCs) are a growing concern in water and wastewater treatment research. This interest stems mainly from the various hazards that EDC can cause to the environment and living creatures. Furthermore, EDC is believed to induce neurological problems, cardiovascular diseases, various forms of cancer, and difficulties with the human reproductive system (Berger et al., 2022).

Mycoremediation is a cost-effective and environmentally benign technology that uses fungal biomass to remove hazardous chemicals from polluted water. Mushrooms are fungal fruiting structures that grow from a fibrous mass known as the mycelium. Macro fungi have demonstrated a unique capacity to break down a variety of dangerous and lignin-containing wastes produced by a range of anthropogenic and natural activities. Additionally, they have a unique method for catabolising lignin using extracellular enzymes, including copper-containing oxidases (laccase) and peroxidises. Extracellular enzymes found in many mushrooms can break down substances such as dyes, dioxins, pesticides, explosives, PAHs, and other xenobiotics by breaking down ether, amide, and ester bonds, as well as aliphatic chains or aromatic rings. Laccases are a potential class of enzymes for this use; they are mostly generated by fungi that break down wood. These fungi generate a type of extracellular ligninolytic enzyme called laccases, which break down lignin and other polyphenolic chemicals found in wood. It has previously been shown that whole fungi and pure enzymes or enzyme formulations can successfully degrade EDCs. Mycoremediation is a relatively economical, environmentally benign, and efficient technique to remove environmental contaminants. It involves the use of fungi or their derivatives (Aljuboori et al., 2021; Bala et al., 2022; Becker et al., 2017; Khan

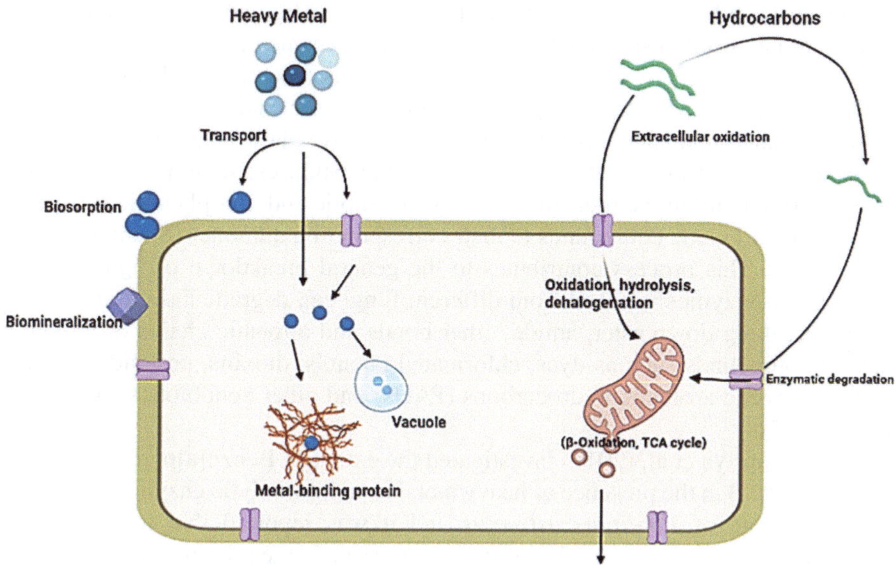

Fig. 1.1 Mechanism involved during mycoremediation by fungi

et al., 2023). Some of the mechanisms involved in the Mycoremediation of fungi are represented in Fig. 1.1.

1.1.1 Biodegradation

Microorganisms such as bacteria, fungi, and other microbes play a major role in wastewater treatment by breaking down harmful organic substances found in contaminated environments into simplified, nontoxic compounds through a process of biodegradation. Through metabolic processes, these microorganisms, which are typically native to contaminated areas, breakdown organic pollutants and use them as a source of energy (Mastropetros et al., 2022). When complex organic compounds are broken down by enzymes, simpler inorganic molecules such as nitrate, carbon dioxide, water, and other safe by-products are produced (Mastropetros et al., 2022). This process is crucial for environmental remediation because it successfully reduces the concentration, bioavailability, and toxicity of pollutants. It also helps maintain natural balance and prevents the accumulation of hazardous materials in the environment. Biodegradation provides a practical and affordable means to address environmental degradation by utilising the natural ability of microbes to break down materials (Hota et al., 2021).

Fungi are regarded effective agents for bioremediation due to their widespread distribution, ability to produce large amounts of biomass and rapid growth of hyphals, which allows them to colonise and degrade different pollutants efficiently

(Janusz et al., 2017). The production of extracellular ligninolytic enzymes gives them a unique mechanism to break down complex organic compounds such as lignin, which is a major component of plant cell walls (Jha, 2019). These enzymes include peroxidases that are very good at oxidising and cleaving the resistant lignin polymer, such as manganese peroxidase, versatile peroxidase, and lignin peroxidase (Shanmugapriya et al., 2019). Copper-containing oxidases, such as laccases produced by fungi, aid in the breakdown of both phenolic and non-phenolic substrates by oxidising aromatic compounds to their corresponding quinones (Shanmugapriya et al., 2019). This process contributes to the general breakdown of lignin. These extracellular enzymes released from different fungi can degrade hazardous chemicals by breaking down ester, amide, ether bonds and aliphatic chains or aromatic rings of compounds such as dyes, chlorinated phenols, dioxins, pesticides, explosives, polycyclic aromatic hydrocarbons (PAHs) and other xenobiotics (Narwal & Gupta, 2017).

Shanmugapriya et al. (2019) investigated the extent of Benzo[a]pyrene degradation by the PO-3 in the presence of heavy metals and ligninolytic enzyme mediators, a native isolate of *Pleurotus ostreatus* and it was reported that indicated that *P. ostreatus* PO-3 degraded 71.2% of Benzo[a]pyrene in the presence of copper ions of which a moderate degradation was observed in the presence of zinc and manganese. Furthermore, copper at a concentration of 15 mM supported the best degradation (74.2%), beyond which the degradation was progressively reduced. Among the mediators, 1 mM 2,20-azinobis-(3-ethylbenzothiazoline-6-sulfonate) supported 78.7% degradation and 83.6% degradation was observed under the influence of 5 mM vanillin. Narwal and Gupta (2017) investigated the ability of *Pleurotus dryinus* and *Trametes hirsute* to remove phenols and produce enzymes from a non-sterilized pH 7.0 biorefinery wastewater (BRW). Chemical characterisation of the BRW revealed the presence of a significant concentration of some toxicmorganic and inorganic compounds (chlorophenols, polycyclic aromatic hydrocarbons, and heavy metals). Mycoremediation was reported to have achieved high dephenolization by *Pleurotus dryinus* and *Trametes hirsute* yields (94% and 100%, respectively), and contributed to the removal of some other organic and inorganic compounds, and concomitantly produced laccase in the BRW. The use of lignocellulosic waste material as a carbon source contributed to decreasing the cost of the culture medium.

1.1.2 Biosorption

Biosorption is an efficient and passive method of removing metals and pollutants from contaminated environments using fungi. This process is particularly effective in treating industrial effluents and recovering metals (Bhattacharya et al., 2014). The biosorption mechanism can be classified according to the localisation of the pollutants: (1) extracellular accumulation or precipitation, pollutants are captured or precipitated outside the fungal cells, which means that they do not enter the cells but remain in the surrounding environment. (2) Cell surface sorption or precipitation, pollutants adhere to or precipitate on the surface of fungal cells, involving interaction with the components of the cell wall. (3) Intracellular accumulation where pollutants are taken up in fungal cells and stored or transformed inside them (Ariste et al., 2020). It can also be categorised based on its dependence on metabolism-independent cell metabolism biosorption and metabolism-dependent biosorption (Dhankhar & Hooda, 2011).

Adsorption in wastewater treatment, typically using activated carbon, has proven to be an effective technology for removing persistent pollutants that otherwise are difficult to eliminate from wastewater due to its versatility, efficiency, and low cost (Ayele et al., 2021). However, there are certain disadvantages to using this sorbent, the most significant being the requirement for chemical or thermal regeneration. This regeneration process can be costly and impractical on a large scale, generates additional pollution, and results in a significant loss of the original material (Gao et al., 2014). Due to their large cell size, a high percentage of cell wall material, simple medium / nutrients needs, high biomass harvest, and easy genetic and morphological manipulations, fungi have recently gained increased attention for their superior biosorption performance compared to other techniques (Çeçen & Aktas, 2011). They are effective in removing heavy metals and biosorption of complex pollutant mixtures. Species such as *Saccharomyces cerevisiae* and *Zygosaccharomyces rouxii* were observed to be involved in bioremediation and were involved in bioremoval of Zn (II), Fe (III), Cu (II), and Cd(II) in a high salty atmosphere (Ghosh et al., 2021).

1.1.3 Bioaccumulation

In contrast to biosorption, which involves primarily the passive binding of contaminants to the surface of cells, bioaccumulation is an active process that depends on metabolism (Katagi, 2009). In wastewater treatment, fungi have been shown to have a considerable capacity for bioaccumulation, which allows them to efficiently remove a wide range of pollutants from aqueous environments, including organic contaminants, radionuclides, and heavy metals (Hansda et al., 2016). Their robust cell wall structure, diverse enzyme and metabolite production, and adaptable metabolism all contribute to this ability. Fungi begin bioaccumulation by transporting pollutants across their cell wall to the cytoplasm (Filote et al., 2021). Depending on the type of contamination, both passive diffusion and active transport mechanisms are involved in this process. Organic contaminants can permeate through the lipid bilayer of the fungal membrane, whereas metal ions can be conveyed via ion channels. Since adenosine triphosphate (ATP) is frequently used to power active transport, fungi can ingest pollutants even in low concentrations from the surrounding environment (Newbound et al., 2010). Once inside the cell, pollutants are stored or bound to intracellular molecules, such as proteins, peptides, or organic acids. Because pollutants can be neutralised or stabilised by binding to intracellular ligands, this sequestration can help to minimise their toxic effects. Fungi produce small, cysteine-rich proteins called phytochelatins and metallothioneins that can bind and detoxify heavy metals such as mercury, zinc, and cadmium (Priyadarshini et al., 2021).

The enzymes lignin peroxidase and manganese peroxidase are produced by fungi and are reported to play a major role in the degradation of complex organic pollutants such as dyes, pesticides, and medications into simpler substances that are easier to control or remove from the cell. Pollutants that cannot be rapidly transformed or detoxified are often segregated within the fungal cell to reduce their toxicity. Typically, this is accomplished by attaching pollutants to the cell wall or encapsulating them in vacuoles (Janusz et al., 2013). Heavy metals and other toxins can be stored in vacuoles, allowing fungi to control the levels of these materials within cells and avoid interfering with essential biological functions. Ge et al. (2011) used a sequential method to investigate the tolerance of *Aspergillus foetidus* to heavy metals such as Al, Co, Cr, Cu, Fe, Mg, Mn, Ni and Zn and two mechanisms were reported. The first mechanism involves the production of extracellular metabolites that are capable of adsorbing and precipitating the metal ions on the cell surface. The second mechanism for the detoxification of metals and intracellular binding of heavy metals to thiol-containing substances and the subsequent sequestration of these metal-thiol complexes into subcellular spaces or vacuoles. These detoxification techniques produced organisms that were adapted to withstand concentrations of multiple heavy metals and had a markedly increased metal uptake because of adaptation.

1.1.4 Bioconversion

Bioconversion of biomass can be fundamentally described as a process that transforms complex organic compounds such as cellulose, hemicellulose, lignin, and other biopolymers into simpler, more useful forms of energy or chemicals (Ge et al., 2011). Typically, this transformation involves the hydrolysis of these complex biomolecules, which are then fermented by microorganisms to produce valuable biochemicals and biofuels such as ethanol or biogas. Because of the complexity of the original biomass, which gives plant materials their structural integrity, these compounds must be efficiently converted through pretreatment and the use of specialised enzymes or microorganisms. In addition to using the abundance of biomass and its renewable nature, this conversion makes it possible to produce a variety of sustainable energy and chemical products, supporting a circular economy and reducing dependence on fossil fuels (Cho et al., 2020).

Fungi can play an important role in improving the bioconversion of activated sludge, a byproduct of wastewater treatment that consists of a mixture of water, organic matter, and various microorganisms. Owing to their wide range of extracellular enzymes, fungi are much more favourable for the bioconversion of more recalcitrant compounds than bacteria. They also exhibit greater resistance to toxic compounds, such as heavy metals and xenobiotics, commonly found in industrial wastewater, which often inhibit bacterial growth (Meza & Monterrey, 2019). Integrating fungi into activated sludge systems can also enhance the settling and dewatering properties of the sludge, potentially reducing the costs associated with the handling and disposal of the sludge (Sperandio & Ferreira Filho, 2019). Moreover, fungi can be cocultured with bacteria in a synergistic relationship where fungal enzymes break down complex substrates into simpler forms, which are then further metabolised by bacteria, leading to a more complete and efficient biodegradation process. Chen et al. (2023) investigated the aerobic elimination of total inorganic nitrogen (TIN) from saline wastewater under the influence of static magnetic fields (SMFs) of varying intensities. The results demonstrated that under static magnetic field (SMF), fungal nitrogen dissimilation removal was significantly increased and aerobic total inorganic nitrogen removal efficiency reached 76.16% under 50 mT, this was attributed by the fungi-bacteria symbiosis relation. The stable operation of aerobic granular sludge is significantly threatened by structural breakdown due to the excessive growth of filamentous microorganisms in bacterial-fungal interactions. Huang et al. (2024) conducted an experiment to provide a foundation for controlling fungal growth and preserving the structural integrity of HAGS (hydrodynamically active granular sludge). The results revealed that during reactivation in the wastewater treatment system, filamentous fungi could extend their hypha from the core to the surface of the HAGS, and it was also found that filamentous fungi could break down extracellular polymeric substances secreted by bacteria and degrade organic matter in saline wastewater, functioning effectively under both anaerobic and aerobic conditions. These properties enable the fungi to thrive and gain a competitive edge, especially when bacterial activity in HAGS is suppressed due to low temperatures and anaerobic environments.

1.2 Mycoremediation of Industrial Waste

The practical application of mycoremediation research is crucial, especially given the increasing urgency of environmental contamination worldwide (Chandran et al., 2023). Conventional water treatment techniques, such as chemical coagulation, membrane filtration, and activated sludge, have long been accepted as best practices; however, their affordability, sustainability, and ability to handle a variety of contaminants are often lacking. Because these techniques are well known and frequently used, they have proven to be effective in a variety of settings. They do have certain drawbacks, though, such as high energy and chemical consumption and secondary pollution production (Mishra et al., 2018). Fungal bioremediation presents a potentially innovative solution due to its ability to degrade a variety of complex mixtures that are resistant to conventional treatments (Fig. 1.2). Fungi have special enzyme pathways that can convert complex organic molecules into simpler, non-toxic compounds, including those involving ligninolytic enzymes. Due to these characteristics, fungi are especially useful in the treatment of stubborn pollutants

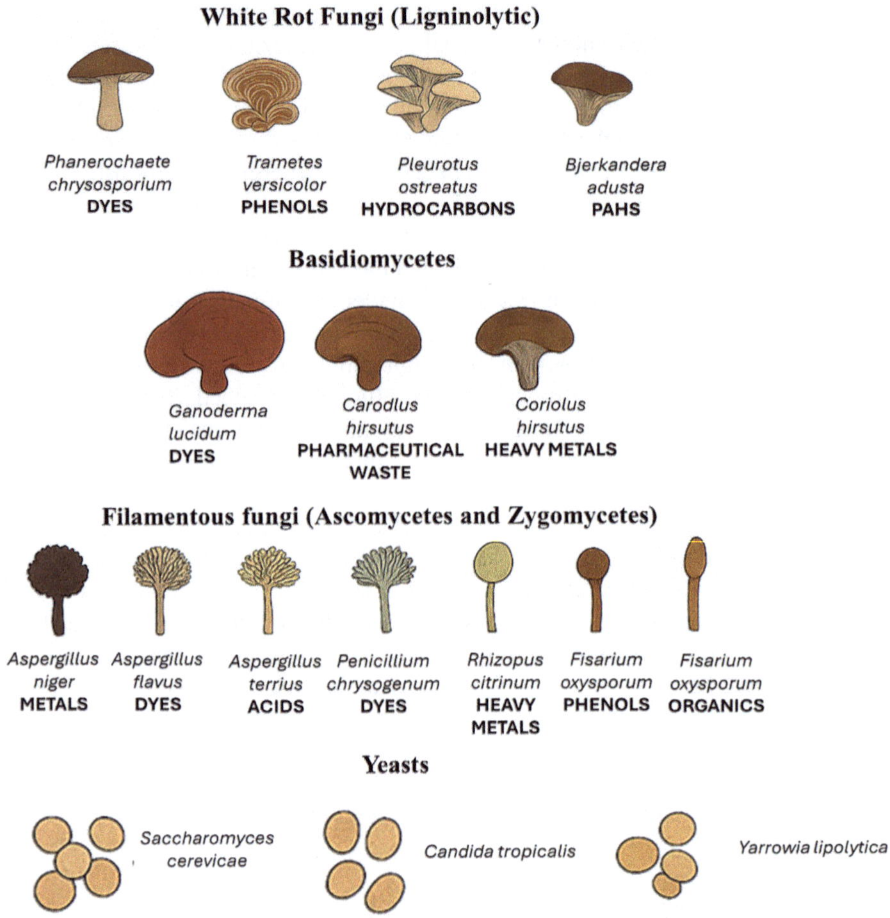

Fig. 1.2 Fungal species used in bioremediation with the specific contaminant

such as synthetic dyes, pharmaceutical residues, and polycyclic aromatic hydrocarbons (PAHs), see Fig. 1.3 (Varjani & Patel, 2018).

1.2.1 Organic Pollutants

Human activities, such as industrial processes, agricultural practices, and improper waste management, release a wide range of organic pollutants into the environment. These contaminants include polycyclic aromatic hydrocarbons (PAHs) (like petroleum products), halogenated compounds (such as chlorinated solvents), pesticides, polychlorinated biphenyls (PCBs) and chloroforms that pose serious threats to human health, wildlife and ecosystems due to their toxicity, persistence, and

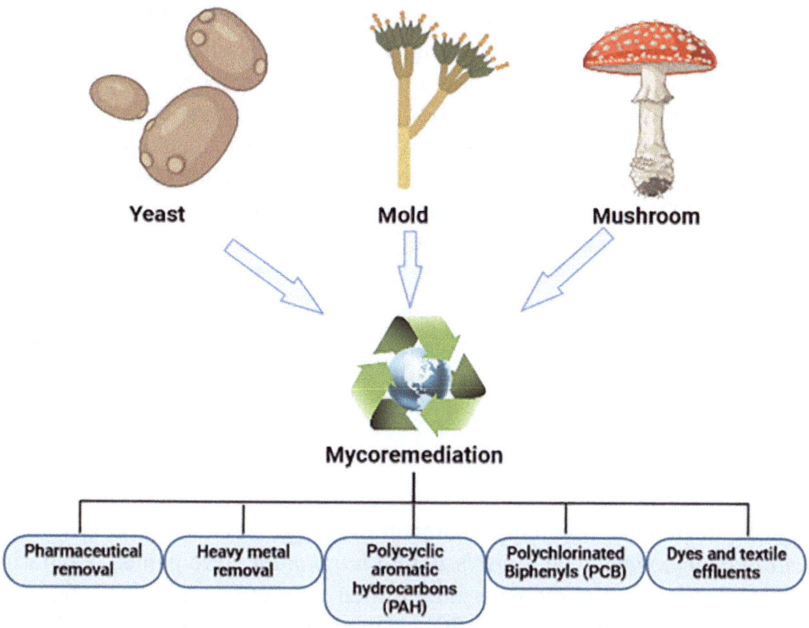

Fig. 1.3 Schematic representation of emerging applications of fungi in wastewater treatment for the removal of various pollutants

bioaccumulation ability (McCarty, 2010). Many PAHs are known for their mutagenic (DNA-damaging) and carcinogenic (cancer-causing) properties. Benzo[a]pyrene, a well-known PAH, is recognised as a potent carcinogen and, as a result, chronic exposure to PAHs can cause various health problems, including respiratory problems, skin disorders, and cancers, particularly lung, skin, and bladder cancers (Gupta & Pathak, 2020).

Fungi, especially white-rot fungi such as *Phanerochaete chrysosporium* and *Pleurotus ostreatus*, have demonstrated the ability to break down PAHs through enzymatic processes (Pozdnyakova, 2012). Giraud et al. (2001) conducted a quantitative analysis of mycoflora and a selection of 10 micromycetes strains were investigated for FA and AC degradation, toxicity assays, and detection of phenoloxidases (POx). Among them Absidia *cylindrospora*, sphaerospermum and *Ulocladium chartarum* were able to highly degrade Anthracene (AC) and Fluoranthene (FA) efficiently. Recently, Zhang et al. (2023) studied the biodegradation of benzo[a]pyrene (BaP) by the white-rot fungus *Phlebia acerina* that showed higher BaP degradation capabilities (57.7%) after 32 days of incubation under ligninolytic conditions. Three non-ionic surfactants were added to increase the biodegradation rate even further. Adding 1 or 2 $g{\cdot}L^{-1}$ of polyethylene glycol monododecyl ether (Brij 30) led to almost complete biodegradation of BaP by *P. acerina*. Anthracene and fluoranthene are among the abundant PAHs in the environment, and therefore are considered a pollution indicator (Morillo et al., 2008). Anthraquinone (AQ) is a major class of synthetic dyes used in the textile industry, known for their vivid colours and excellent resistance to light, heat, and washing. These dyes are characterised by the presence of an anthraquinone chromophore group, which is responsible for their intense and diverse colouration (Tariq et al., 2024). In addition, anthraquinone dyes are also among the most problematic environmental pollutants due to their non-biodegradable nature, high chemical stability, and toxic effects on aquatic life (Ismail et al., 2019). Melati et al. (2023) evaluated the decolorization of AQ in the presence of a dark septate endophytic (DSE) strain and the results showed that all tested DSE fungi successfully decolorised 89% of RBBR and 97% of AB 113 dyes. Their findings imply that the chosen DSE fungi could bioremediate dye wastewater, which is typically acidic in pH.

1.2.2 Heavy Metals

Biological treatments involving filamentous fungi, such as *Aspergillus niger*, *Penicillium chrysogenum*, and *Trichoderma harzianum*, have gained popularity for heavy metal removal and recovery due to their high efficiency, cost effectiveness, and wide availability (Ahmad et al., 2011; Dusengemungu et al., 2020). These fungi are known for their high biosorption capacities for various heavy metals, including lead (Pb), zinc (Zn), cadmium (Cd), copper (Cu), chromium (Cr), arsenic (As) and nickel (Ni) (Negi & Das, 2023), attributed to their unique cell wall composition that contains functional groups such as carboxyl, hydroxyl, and amine, which provide multiple binding sites for metal ions (Gadd, 2004). Their rapid biomass production on inexpensive substrates, versatility under diverse environmental conditions, and capacity for metal recovery further enhance their suitability for large-scale applications. Filamentous fungi employ several mechanisms for metal adsorption, such as biosorption (a rapid and reversible process involving surface interactions), bioaccumulation (an energy-dependent process involving intracellular sequestration), chelation and precipitation (through extracellular compounds such as organic acids), and redox transformations (altering metal oxidation states to less toxic forms) (Dinakarkumar et al., 2024). These fungi are increasingly used in wastewater treatment for industrial effluents, such as those from the mining, textile, and tannery industries, and in combination with plants for enhanced phytoremediation. Future directions include advances in genetic engineering to further boost their metal binding capacity and resilience, making filamentous fungi stand out from other microorganisms (Raj & Das, 2023).

Lead (Pb) is a persistent environmental pollutant that poses serious health risks as a result of its toxic effects on various biological systems. Its accumulation in ecosystems and contamination of water sources pose significant challenges for environmental management and public health (Thirumalaivasan et al., 2024). As a result, there has been increasing interest in innovative methods for removing lead from wastewater. Mycoremediation, which takes advantage of fungus's ability to degrade or transform pollutants, has emerged as a promising strategy for addressing

Pb contamination (Ghyadh et al., 2019). Several *Aspergillus* fungal species exhibit remarkable biosorption and bioaccumulation capabilities of heavy metals due to their cell wall components, such as chitin and glucans (Fig. 1.4), which interact with metal ions. Qader and Shekha (2023) investigated the ability of two fungal species, *Aspergillus niger* and *Rhizopus oligosporus*, used in wastewater treatment to lower the concentrations of heavy metals such as lead (Pb) and cadmium (Cd). The results indicated that *Aspergillus niger* exhibited a higher efficiency in reducing the concentrations of cadmium (Cd) and lead (Pb), achieving a 100% removal rate, which was superior to that of *Rhizopus oligosporium.* Qader et al. also investigated the bioremediation capability of *Aspergillus niger* and *Candida albicans* in the removal of heavy metals (Pb and Cd). The experiment was carried out using varying concentrations (5, 15, 35, and 50 ppm) of heavy metals, and the results showed that the lowest concentration (5 ppm) achieved the highest reduction rates. At this concentration, the two fungal strains, *Aspergillus niger* and *Candida albicans*, removed the lead by 85.6% and 84.2%, respectively, while cadmium removal was 80% and 78.4%, where *Aspergillus niger* demonstrated greater effectiveness in remediating both components at all concentrations tested compared to *Candida albicans*. This ability to tolerate and bioremediate nanometals highlights *Aspergillus* as a good candidate for bioremediation technologies aimed at treating multimetal contaminated wastewater. Shalaby et al. (2023) studied the biosorption capacity of the multimetal-tolerant fungus *Aspergillus* sp. for the simultatenous removal of various metallic nanoparticles as potential nanoscale metal pollutants from aqueous solutions. The results of this investigation showed that *Aspergillus* sp. is a good candidate for Zn, Se, Fe, and Ag NPs removal either in single or mixture state where at pH 7, the maximum removal percentages for the four studied metals were observed, with 38.8%, 68.1%, 80.4%, and 82.0% for Zn-, Fe-, Se-, and Ag-NPs, respectively.

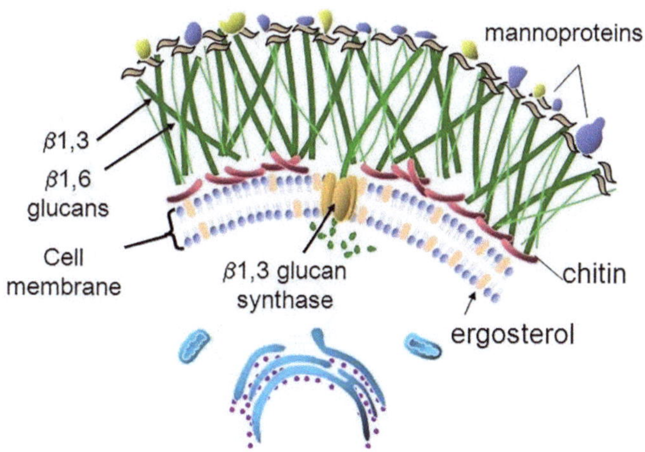

Fig. 1.4 Cell wall structure of *Aspergillus* sp

1.2.3 *Pharmaceutical Waste*

In recent years, the use of non-steroidal anti-inflammatory drugs (NSAIDs) has increased significantly, particularly among patients undergoing COVID-19 treatment (Kushner et al., 2022). Nonsteroidal anti-inflammatory drugs such as diclofenac, ibuprofen, and paracetamol (Fig. 1.5) are frequently available without restrictions and are often used without medical guidance to manage basic inflammatory symptoms (Fokunang et al., 2018). One of the most significant concerns associated with pharmaceutical pollutants in the environment is the promotion of antibiotic resistance. Antibiotic residues can exert selective pressure on bacteria, leading to the evolution of antibiotic-resistant strains. These resistant bacteria can transfer their resistance genes to other bacteria, exacerbating the spread of antibiotic resistance in both clinical and natural settings (Larsson & Flach, 2022). Consumption of water contaminated with pharmaceutical ingredients can alter reproduction, hormonal imbalance, cancer, and respiratory problems. Diclofenac has been associated with both lethal and sublethal effects in fish, including organ damage, behavioural changes, and decreased success rates during reproduction (Schwarz et al., 2017).

Despite the widespread presence of these pharmaceuticals in the environment, the toxicity, degradation pathways, microbial mechanisms, and genetic basis of these bioremediation processes are still incompletely understood. More research is required to fully elucidate these mechanisms and optimise microbial approaches for the efficient removal of these micropollutants from wastewater (Saxena et al., 2020). Edrees et al. (2017) evaluated the ability of *Aspergillus niger* and *Fusarium oxysporium to* degrade paracetamol from pharmaceutical wastewater effluent. Paracetamol biodegradation was found to be influenced by glucose concentration at optimal temperature and pH, and the results demonstrated that *Aspergillus niger* and *Fusarium oxysporium* were able to degrade 35.7% and 26.1% of 1000 and 2000 mg/L, respectively, paracetamol in 60 days. Chandrashekar et al. (2024) further investigated the

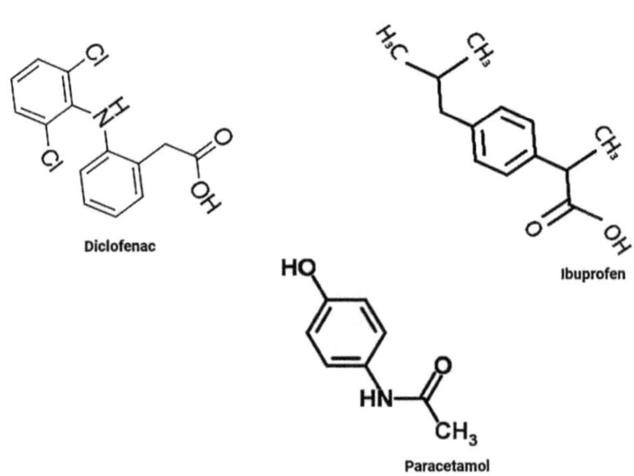

Fig. 1.5 Chemical structure of diclofenac, ibuprofen, and paracetamol

mycodegradation of paracetamol by *Aspergillus niger* KCAC. The isolated paracetamol degradation fungal strain demonstrated that a 99.6% degradation rate was achieved. White-rot fungi (WRF) are one of the most prominent biocatalysts due to their broad substrate specificity, achieved through the activity of both intracellular enzymes, such as the cytochrome P450 system, and extracellular enzymes (Lin et al., 2022). This ability to degrade a wide range of xenobiotics makes them highly suitable for bioremediation purposes. Currently, WRF have been primarily employed in the removal of endocrine-disrupting chemicals. Tran et al. (2010) investigated the degradation of 10 selected pharmaceutical active compounds by a WRF *Trametes versicolor* where a complete removal of diclofenac (DCF), naproxen (NPX), indomethacin (IDM), ibuprofen (IBP), and fenoprofen (FEP) were observed after 48 h of incubation with the 7-day-old liquid fungal culture both in the presence and absence of ABTS (2,2′-azino-bis(3-ethyl-benzothiazoline-6-sulfonic acid)) as a laccase-mediator. Cytochrome P450 (CYP) enzymes play a crucial role in the biotransformation and elimination of pharmaceuticals in wastewater (Fig. 1.5). They often work in synergy with extracellular mechanisms, such as the action of oxidative enzymes such as laccases and peroxidases, which initiate the breakdown of complex pharmaceutical molecules outside the cell. This combination enhances the efficiency of degradation processes by creating intermediates that are more accessible to CYP enzymes for further metabolism. Olicón-Hernández et al. (2019) reported the biotransformation of diclofenac by *Penicillium oxalicum* for the first time. The Pellets of *P. oxalicum* degraded 100 μM of DFC in 24 h, and the activity of the CYP450 enzymes played an important role in the elimination of the drug.

1.3 Fungal Symbiosis in Wastewater Treatment

Fungal symbiosis with bacteria or algae in wastewater treatment represents a promising strategy to enhance the efficiency of bioremediation processes (Mahesh et al., 2021). This synergistic interaction leverages the complementary metabolic capabilities of fungi, bacteria, and algae to degrade complex pollutants more effectively. In a fungal-bacterial consortium, fungi can produce highly effective enzymes that break down recalcitrant organic compounds and certain aromatic hydrocarbons into simpler molecules that bacteria can further degrade; while, bacteria can provide essential nutrients like nitrogen and phosphorus, supporting fungal growth (Mani & Hameed, 2019). In a fungal-algal symbiosis, fungi can supply carbon dioxide and other growth factors that promote algal growth, while algae, in turn, contribute oxygen through photosynthesis, which is essential for aerobic fungal and bacterial metabolism (Wang et al., 2023). The co-culturing of these microorganisms can also enhance survival rates and resilience under harsh environmental conditions, such as extreme pH, temperature, or salinity, which are common in contaminated sites. By fostering such symbiotic interactions, researchers aim to develop more effective, sustainable and versatile bioremediation strategies to address various environmental contaminants, particularly those involving heavy metals that pose significant ecological and human health risks (Kapoore et al., 2022).

The co-cultivation (Fig. 1.6) of fungi and microalgae is gaining prominence due to its highly efficient bio-flocculation of microalgal cells, which eliminates the need for external chemical additives and requires minimal energy input (Wrede et al., 2014). Certain fungal and microalgal species have demonstrated remarkable capabilities in wastewater purification, effectively removing nutrients, heavy metals, and organic pollutants. This consortium not only treats wastewater, but also produces biomass that can be harvested as a renewable and sustainable feedstock for the production of biofuels (Ambaye et al., 2021). The synergistic relationship between fungi and microalgae improves nutrient uptake and pollutant degradation, while fungi provide a structural matrix that aids in the aggregation and settlement of microalgal cells, thus simplifying harvesting processes. This integrated approach offers a cost-effective and environmentally friendly solution for wastewater management while contributing to the circular economy through biofuel generation (Egede et al., 2016).

Harvesting microbial cells, including fungi and microalgae, from growth media is technically challenging and represents a significant cost factor, accounting for more than 20% of the total cost of biomass production. This high cost is primarily due to the small size, low density, and similar buoyant properties of cells, making separation from the surrounding liquid medium difficult (Ghazvini et al., 2022). In

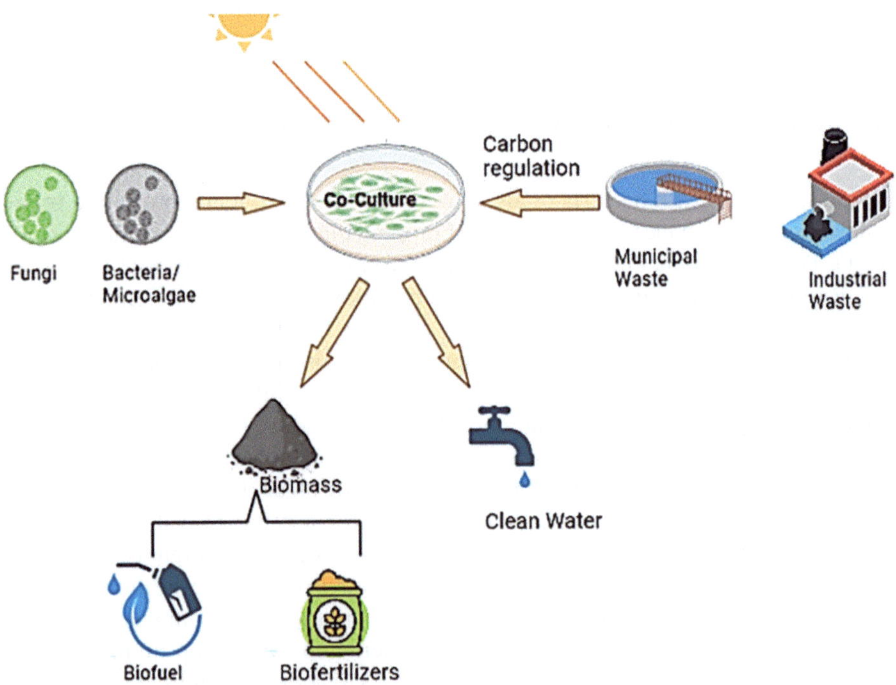

Fig. 1.6 Co-cultivation of fungi with other microbes

addition, the harvesting process typically requires energy-intensive techniques, such as centrifugation, filtration, or flocculation, each of which comes with its own limitations. Zhang and Hu described a new pelletised cell cultivation process using a combination of *Aspergillus oryzae* and *Chlorella pyrenoidosa*. Wang et al. (2022) developed a facile novel method of one-step co-culture and harvesting of microalgae and fungi, for efficient starch wastewater treatment and high-value biomass production that resulted in the conversion of wastewater nutrients into useful microbial biomass and purified water. Synergistic interaction between cultures was also reported to significantly improve the removal efficiencies of the main pollutants, reaching 92.08% for chemical oxygen demand (COD), 83.56% for total nitrogen (TN), and 96.58% for total phosphorus (TP), while also increasing the final biomass concentration, along with notable improvements in protein and lipid concentrations, reaching 1.92 g/L and 0.99 g/L, respectively. Recently, Lu et al. (2023) conducted a study in which starch processing effluent (SPE) is used as a fermentation substrate to co-culture *Chlorella vulgaris* and *Rhodotorula glutinis* for the production of biofuel feedstock. Under optimal conditions, the microorganisms achieved a lipid yield of 1.81 g/L and a carbon conversion ratio of 82.53%, while in monoculture, the lipid yield ranged from 0.79 to 0.81 g/L, and the carbon conversion ratio was between 55.93% and 62.61%.

1.4 Future Perspectives

With the increase in urbanization and industrialization the need for more sustainable and cheaper wastewater remediation techniques is evident. The theoretical application of fungi-derived solutions has great potential, but the realised application is limited due to a lack of research to support its implementation. The discovery of new fungi species will continue to unlock more potential candidates for the various field of application within wastewater remediation and will likely continue to make the case for fungi-based treatment techniques.

1.5 Conclusion

The increasing complexity and difficulty of treating paper, agrochemical, textile, heavy metal, and organic pollutants in water systems leads to a constant search for new treatment methods. Today, a great number of physical, chemical, and biological methods are available. However, biological methods such as fungal mycoremediation offer a cost-effective, innovative, and environmentally friendly way for treating water systems. Fungi have shown great promise in both understanding mechanisms and remediating contaminants. Mycoremediation in fungi provides an easy way of removing heavy metals, hydrocarbons, and pesticides from the environment. Macro fungi have demonstrated a unique capacity to break down a variety of dangerous and lignin-containing wastes produced by a variety of anthropogenic and natural activities. This chapter addressed the efficient strategies and metabolic abilities of fungi to break down complex pollutants into simpler and less harmful substances.

The incorporation of fungal species into wastewater treatment systems is especially advantageous due to its resilience and adaptability to changing environmental conditions such as pH, temperature, and chemical composition. Furthermore, the biochemical mechanisms underlying fungal pollutant degradation, such as enzymatic breakdown of organic matter and metal sequestration, are becoming better understood, opening new opportunities for optimising fungal applications in wastewater systems. As the field of fungal bioremediation expands, new fungal species with impressive remediation capabilities are discovered together with a better understanding of the complex biochemical pathways involved in pollutant breakdown. Fungal symbiosis is one example of this type of development, a promising strategy for enhancing the efficiency of bioremediation processes.

Acknowledgments This research was supported by the Sustainable and Renewable Energy Nanomaterials Research Group, the University of the Western Cape. This research was funded by the National Research Foundation: 138079 and Eskom (South Africa): 2002/015527/0 is hereby acknowledged.

Conflicts of Interest The authors declare no conflicts of interest.

Data Availability Data used is available in the manuscript.

References

Ahmad, I., Imran, M., Ansari, M. I., Malik, A., & Pichtel, J. (2011). Metal tolerance and biosorption potential of soil fungi: Applications for a green and clean water treatment technology. In *Microbes and microbial technology: Agricultural and environmental applications* (pp. 321–361). Springer.

Aljuboori, Z. M., Singh, H., Haddad, H., Al-Ramahi, N. M., & Ali, M. A. (2021). Intellectual capital and firm performance correlation: The mediation role of innovation capability in Malaysian manufacturing SMEs perspective. *Sustainability, 14*(1), 154.

Ambaye, T. G., Vaccari, M., Bonilla-Petriciolet, A., Prasad, S., van Hullebusch, E. D., & Rtimi, S. (2021). Emerging technologies for biofuel production: A critical review on recent progress, challenges and perspectives. *Journal of Environmental Management, 290*, 112627.

Ariste, A. F., Batista-García, R. A., Vaidyanathan, V. K., Raman, N., Vaithyanathan, V. K., Folch-Mallol, J. L., & Cabana, H. (2020). Mycoremediation of phenols and polycyclic aromatic hydrocarbons from a biorefinery wastewater and concomitant production of lignin modifying enzymes. *Journal of Cleaner Production, 253*, 119810.

Ayele, A., Haile, S., Alemu, D., Tesfaye, T., & Kamaraj, M. (2021). Mycoremediation: Fungal-based technology for biosorption of heavy metals–A review. In *Strategies and tools for pollutant mitigation: Avenues to a cleaner environment* (pp. 355–373). Springer.

Bala, S., Garg, D., Thirumalesh, B. V., Sharma, M., Sridhar, K., Inbaraj, B. S., & Tripathi, M. (2022). Recent strategies for bioremediation of emerging pollutants: A review for a green and sustainable environment. *Toxics, 10*(8), 484.

Becker, D., Rodriguez-Mozaz, S., Insa, S., Schoevaart, R., Barceló, D., de Cazes, M., & Wagner, M. (2017). Removal of endocrine disrupting chemicals in wastewater by enzymatic treatment with fungal laccases. *Organic Process Research & Development, 21*(4), 480–491.
Berger, R. G., Bordewick, S., Krahe, N. K., & Ersoy, F. (2022). Mycelium vs. fruiting bodies of edible fungi—A comparison of metabolites. *Microorganisms, 10*(7), 1379.
Bhattacharya, S., Das, A., Prashanthi, K., Palaniswamy, M., & Angayarkanni, J. (2014). Mycoremediation of benzo [a] pyrene by Pleurotus ostreatus in the presence of heavy metals and mediators. *3 Biotech, 4*(2), 205–211.
Çeçen, F., & Aktas, Ö. (2011). *Activated carbon for water and wastewater treatment: Integration of adsorption and biological treatment*. John Wiley & Sons.
Chandran, P., Suresh, S., Balasubramain, B., Gangwar, J., Raj, A. S., Aarathy, U. L., & Sebastian, J. K. (2023). Biological treatment solutions using bioreactors for environmental contaminants from industrial waste water. *Journal of Umm Al-Qura University for Applied Sciences, 11*, 185–207.
Chandrashekar, K., Chatterjee, A., & Shah, M. P. (2024). Paracetamol mineralization strategy in laboratory scale using aspergillus Niger KCAC. *International Microbiology, 27*(3), 907–916.
Chen, J., Cui, Y. W., Huang, M. Q., Yan, H. J., & Li, D. (2023). Static magnetic field increases aerobic nitrogen removal from hypersaline wastewater in activated sludge with coexistence of fungi and bacteria. *Bioresource Technology, 382*, 129194.
Cho, E. J., Trinh, L. T. P., Song, Y., Lee, Y. G., & Bae, H. J. (2020). Bioconversion of biomass waste into high value chemicals. *Bioresource Technology, 298*, 122386.
Dhankhar, R., & Hooda, A. (2011). Fungal biosorption–an alternative to meet the challenges of heavy metal pollution in aqueous solutions. *Environmental Technology, 32*(5), 467–491.
Dinakarkumar, Y., Ramakrishnan, G., Gujjula, K. R., Vasu, V., Balamurugan, P., & Murali, G. (2024). Fungal bioremediation: An overview of the mechanisms, applications and future perspectives. *Environmental Chemistry and Ecotoxicology, 6*, 293–302.
Dusengemungu, L., Kasali, G., Gwanama, C., & Ouma, K. O. (2020). Recent advances in biosorption of copper and cobalt by filamentous fungi. *Frontiers in Microbiology, 11*, 582016.
Edrees, W. H., Abdullah, Q. Y., Naji, K. M., & Al-Kaf, A. G. (2017). Biodegradation of paracetamol by native fungal species inhabiting wastewater of a pharmaceutical factory in Sana'a, Yemen. *Universal Journal of Pharmaceutical Research, 2*(6), 35–41.
Egede, E. J., Jones, H., Cook, B., Purchase, D., & Mouradov, A. (2016). Application of microalgae and fungal-microalgal associations for wastewater treatment. In *Fungal applications in sustainable environmental biotechnology* (pp. 143–181). Springer.
Filote, C., Roşca, M., Hlihor, R. M., Cozma, P., Simion, I. M., Apostol, M., & Gavrilescu, M. (2021). Sustainable application of biosorption and bioaccumulation of persistent pollutants in wastewater treatment: Current practice. *PRO, 9*(10), 1696.
Fokunang, C., Fokunang, E. T., Frederick, K., Ngameni, B., & Ngadjui, B. (2018). Overview of non-steroidal anti-inflammatory drugs (nsaids) in resource limited countries. *MOJ Toxicology, 4*(1), 5–13.
Gadd, G. M. (2004). Microbial influence on metal mobility and application for bioremediation. *Geoderma, 122*(2–4), 109–119.
Gao, J., Ye, J., Ma, J., Tang, L., & Huang, J. (2014). Biosorption and biodegradation of triphenyltin by Stenotrophomonas maltophilia and their influence on cellular metabolism. *Journal of Hazardous Materials, 276*, 112–119.
Ge, W., Zamri, D., Mineyama, H., & Valix, M. (2011). Bioaccumulation of heavy metals on adapted aspergillus foetidus. *Adsorption, 17*(5), 901–910.
Ghazvini, M., Kavosi, M., Sharma, R., & Kim, M. (2022). A review on mechanical-based microalgae harvesting methods for biofuel production. *Biomass and Bioenergy, 158*, 106348.
Ghosh, S., Selvakumar, G., Ajilda, A. A. K., & Webster, T. J. (2021). Microbial biosorbents for heavy metal removal. In *New trends in removal of heavy metals from industrial wastewater* (pp. 213–262). Elsevier.

Ghosh, A., Tripathy, A., & Ghosh, D. (2022). Impact of endocrine disrupting chemicals (EDCs) on reproductive health of human. *Proceedings of the Zoological Society, 75*(1), 16–30.

Ghyadh, B. A., Al-Ashoor, A., & Ibrahim, Z. M. (2019). Biotechnology of wastewater treatment with fungi (aspergillus Niger and Rhizopus oligosporium). *Plant Archives, 19*(2), 1546–1549.

Giraud, F., Guiraud, P., Kadri, M., Blake, G., & Steiman, R. (2001). Biodegradation of anthracene and fluoranthene by fungi isolated from an experimental constructed wetland for wastewater treatment. *Water Research, 35*(17), 4126–4136.

Gupta, S., & Pathak, B. (2020). Mycoremediation of polycyclic aromatic hydrocarbons. In *Abatement of environmental pollutants* (pp. 127–149). Elsevier.

Hansda, A., Kumar, V., & Anshumali. (2016). A comparative review towards potential of microbial cells for heavy metal removal with emphasis on biosorption and bioaccumulation. *World Journal of Microbiology and Biotechnology, 32*(10), 170.

Hota, S., Sharma, G. K., Subrahmanyam, G., Kumar, A., Shabnam, A. A., Baruah, P., & Yadav, A. N. (2021). Fungal communities for bioremediation of contaminated soil for sustainable environments. In *Recent trends in mycological research: Volume 2: Environmental and industrial perspective* (pp. 27–42). Springer International Publishing.

Huang, J. L., Cui, Y. W., & Chen, C. (2024). Overgrowth of filamentous fungi in halophilic activated granule sludge reactivated after refrigeration: Insights into bacterial-fungal interactions. *Chemical Engineering Journal, 482*, 148859.

Ismail, M., Akhtar, K., Khan, M. I., Kamal, T., Khan, M. A., Asiri, M., & Khan, S. B. (2019). Pollution, toxicity and carcinogenicity of organic dyes and their catalytic bio-remediation. *Current Pharmaceutical Design, 25*(34), 3645–3663.

Janusz, G., Kucharzyk, K. H., Pawlik, A., Staszczak, M., & Paszczynski, A. J. (2013). Fungal laccase, manganese peroxidase and lignin peroxidase: Gene expression and regulation. *Enzyme and Microbial Technology, 52*(1), 1–12.

Janusz, G., Pawlik, A., Sulej, J., Świderska-Burek, U., Jarosz-Wilkołazka, A., & Paszczyński, A. (2017). Lignin degradation: Microorganisms, enzymes involved, genomes analysis and evolution. *FEMS Microbiology Reviews, 41*(6), 941–962.

Jha, H. (2019). Fungal diversity and enzymes involved in lignin degradation. In *Mycodegradation of lignocelluloses* (pp. 35–49). Springer International Publishing.

Kapoore, R. V., Padmaperuma, G., Maneein, S., & Vaidyanathan, S. (2022). Co-culturing microbial consortia: Approaches for applications in biomanufacturing and bioprocessing. *Critical Reviews in Biotechnology, 42*(1), 46–72.

Katagi, T. (2009). Bioconcentration, bioaccumulation, and metabolism of pesticides in aquatic organisms. *Reviews of Environmental Contamination and Toxicology, 204*, 1–132.

Khan, M. F., Hof, C., Niemcová, P., & Murphy, C. D. (2023). Recent advances in fungal xenobiotic metabolism: Enzymes and applications. *World Journal of Microbiology and Biotechnology, 39*(11), 296.

Kushner, P., McCarberg, B. H., Grange, L., Kolosov, A., Haveric, A. L., Zucal, V., & Bissonnette, S. (2022). The use of non-steroidal anti-inflammatory drugs (NSAIDs) in COVID-19. *npj Primary Care Respiratory Medicine, 32*(1), 35.

Larsson, D. J., & Flach, C. F. (2022). Antibiotic resistance in the environment. *Nature Reviews Microbiology, 20*(5), 257–269.

Latif, W., Ciniglia, C., Iovinella, M., Shafiq, M., & Papa, S. (2023). Role of white rot fungi in industrial wastewater treatment: A review. *Applied Sciences, 13*(14), 8318.

Li, P., Karunanidhi, D., Subramani, T., & Srinivasamoorthy, K. (2021). Sources and consequences of groundwater contamination. *Archives of Environmental Contamination and Toxicology, 80*(1), 1–10.

Lin, S., Wei, J., Yang, B., Zhang, M., & Zhuo, R. (2022). Bioremediation of organic pollutants by white rot fungal cytochrome P450: The role and mechanism of CYP450 in biodegradation. *Chemosphere, 301*, 134776.

Lu, Q., Ma, C., Guo, L., Lu, Y., & Li, H. (2023). Co-fermentation of Chlorella vulgaris with oleaginous yeast in starch processing effluent as a carbon-reducing strategy for wastewater treatment and biofuel feedstock production. *Fermentation, 9*(5), 476.

Mahesh, R., Panda, S. K., Das, M., Yashavanth, P. R., Dhull, S., Negi, B. B., & Maiti, S. K. (2021). Advances in biotechnological tools for bioremediation of wastewater using bacterial–algal symbiotic system. In *Wastewater treatment* (pp. 385–411). Elsevier.

Mani, A., & Hameed, S. A. S. (2019). Improved bacterial-fungal consortium as an alternative approach for enhanced decolourisation and degradation of azo dyes: A review. *Nature Environment and Pollution Technology, 18*(1), 49–64.

Mastropetros, S. G., Pispas, K., Zagklis, D., Ali, S. S., & Kornaros, M. (2022). Biopolymers production from microalgae and cyanobacteria cultivated in wastewater: Recent advances. *Biotechnology Advances, 60*, 107999.

McCarty, P. L. (2010). Groundwater contamination by chlorinated solvents: History, remediation technologies and strategies. In *In situ remediation of chlorinated solvent plumes* (pp. 1–28). Springer.

Melati, I., Rahayu, G., Effendi, H., Henny, C., & Yanto, D. H. Y. (2023). Biodecolorization of anthraquinone and azo dyes by dark septate endophytic fungi. *Bioresource Technology Reports, 22*, 101427.

Meza, E. M., & Monterrey, N. L. (2019). Use of fungal spores to enhance the settling of activated sludge from a domestic wastewater treatment plant.

Mishra, S., Chowdhary, P., & Bharagava, R. N. (2018). Conventional methods for the removal of industrial pollutants, their merits and demerits. In *Emerging and eco-friendly approaches for waste management* (pp. 1–31). Springer.

Morillo, E., Romero, A. S., Madrid, L., Villaverde, J., & Maqueda, C. (2008). Characterization and sources of PAHs and potentially toxic metals in urban environments of Sevilla (southern Spain). *Water, Air, and Soil Pollution, 187*(1), 41–51.

Narwal, S. K., & Gupta, R. (2017). Biodegradation of xenobiotic compounds: An overview. In IGI Global (Ed.), *Handbook of research on inventive bioremediation techniques* (pp. 186–212).

Negi, B. B., & Das, C. (2023). Mycoremediation of wastewater, challenges, and current status: A review. *Bioresource Technology Reports, 22*, 101409.

Newbound, M., Mccarthy, M. A., & Lebel, T. (2010). Fungi and the urban environment: A review. *Landscape and Urban Planning, 96*(3), 138–145.

Olicón-Hernández, D. R., Camacho-Morales, R. L., Pozo, C., González-López, J., & Aranda, E. (2019). Evaluation of diclofenac biodegradation by the ascomycete fungus Penicillium oxalicum at flask and bench bioreactor scales. *Science of the Total Environment, 662*, 607–614.

Patel, A. B., Shaikh, S., Jain, K. R., Desai, C., & Madamwar, D. (2020). Polycyclic aromatic hydrocarbons: Sources, toxicity, and remediation approaches. *Frontiers in Microbiology, 11*, 562813.

Pozdnyakova, N. N. (2012). Involvement of the ligninolytic system of white-rot and litter-decomposing fungi in the degradation of polycyclic aromatic hydrocarbons. *Biotechnology Research International, 2012*(1), 243217.

Priyadarshini, E., Priyadarshini, S. S., Cousins, B. G., & Pradhan, N. (2021). Metal-fungus interaction: Review on cellular processes underlying heavy metal detoxification and synthesis of metal nanoparticles. *Chemosphere, 274*, 129976.

Qader, M. Q., & Shekha, Y. A. (2023). Bioremediation of heavy metals by using aspergillus Niger and Candida albicans. *Zanco Journal of Pure and Applied Sciences, 35*(3), 180–186.

Raj, K., & Das, A. P. (2023). Lead pollution: Impact on environment and human health and approach for a sustainable solution. *Environmental Chemistry and Ecotoxicology, 5*, 79–85.

Saxena, G., Kumar, V., & Shah, M. P. (2020). *Bioremediation for environmental sustainability: Toxicity, mechanisms of contaminants degradation, detoxification and challenges*. Elsevier.

Schwarz, S., Schmieg, H., Scheurer, M., Köhler, H. R., & Triebskorn, R. (2017). Impact of the NSAID diclofenac on survival, development, behaviour and health of embryonic and juvenile stages of brown trout, Salmo trutta f. fario. *Science of the Total Environment, 607*, 1026–1036.

Shalaby, M. A., Matter, I. A., Gharieb, M. M., & Darwesh, O. M. (2023). Biosorption performance of the multi-metal tolerant fungus aspergillus sp. for removal of some metallic nanoparticles from aqueous solutions. *Heliyon, 9*(5), e16125.

Shanmugapriya, S., Manivannan, G., Selvakumar, G., & Sivakumar, N. (2019). Extracellular fungal peroxidases and laccases for waste treatment: Recent improvement. In *Recent advance-*

ment in white biotechnology through fungi: Volume 3: Perspective for sustainable environments (pp. 153–187). Springer International Publishing.
Sperandio, G. B., & Ferreira Filho, E. X. (2019). Fungal co-cultures in the lignocellulosic biorefinery context: A review. *International Biodeterioration & Biodegradation, 142*, 109–123.
Sun, K., Song, Y., He, F., Jing, M., Tang, J., & Liu, R. (2021). A review of human and animals exposure to polycyclic aromatic hydrocarbons: Health risk and adverse effects, photo-induced toxicity and regulating effect of microplastics. *Science of the Total Environment, 773*, 145403.
Tariq, B., Mansha, A., Asim, S., & Kausar, A. (2024). Effect of substituents on solubility, medicinal, absorption, emission and cationic/anionic detection properties of anthraquinone derivatives. *Journal of Fluorescence, 34*(4), 1527–1544.
Thirumalaivasan, N., Gnanasekaran, L., Kumar, S., Durvasulu, R., Sundaram, T., Rajendran, S., & Kanagaraj, K. (2024). Utilization of fungal and bacterial bioremediation techniques for the treatment of toxic waste and biowaste. *Frontiers in Materials, 11*, 1416445.
Tran, N. H., Urase, T., & Kusakabe, O. (2010). Biodegradation characteristics of pharmaceutical substances by whole fungal culture Trametes versicolor and its laccase. *Journal of Water and Environment Technology, 8*(2), 125–140.
Varjani, S. J., & Patel, R. K. (2018). Fungi: A remedy to eliminate environmental pollutants. In *Mycoremediation and environmental sustainability* (Vol. 1, pp. 53–67). Springer.
Wang, S. K., Yang, K. X., Zhu, Y. R., Zhu, X. Y., Nie, D. F., Jiao, N., & Angelidaki, I. (2022). One-step co-cultivation and flocculation of microalgae with filamentous fungi to valorize starch wastewater into high-value biomass. *Bioresource Technology, 361*, 127625.
Wang, J., Tian, Q., Zeng, W., Qiu, G., & Shen, L. (2023). Insights about fungus-microalgae symbiotic system in microalgae harvesting and wastewater treatment: A review. *Renewable and Sustainable Energy Reviews, 182*, 113408.
Wrede, D., Taha, M., Miranda, A. F., Kadali, K., Stevenson, T., Ball, A. S., & Mouradov, A. (2014). Co-cultivation of fungal and microalgal cells as an efficient system for harvesting microalgal cells, lipid production and wastewater treatment. *PLoS One, 9*(11), e113497.
Xu, H., Jia, Y., Sun, Z., Su, J., Liu, Q. S., Zhou, Q., & Jiang, G. (2022). Environmental pollution, a hidden culprit for health issues. *Eco-Environment & Health, 1*(1), 31–45.
Zhang, W., Li, Q., Wang, J., Wang, Z., Zhan, H., Yu, X., & Zhou, L. W. (2023). Biodegradation of benzo [a] pyrene by a white-rot fungus Phlebia acerina: Surfactant-enhanced degradation and possible genes involved. *Journal of Fungi, 9*(10), 978.

Chapter 2
Nutrient Cycling and Mechanisms of Fungal Remediation for Fungi in Waste Water Treatment

Kanu Murmu, Arup Sarkar, Sidhu Murmu, Samapriya Saha, and Sramana Sen Sarma

Abstract This review explores the use of fungi for nutrient recycling and mechanism of fungal remediation for fungi in waste water treatment. Nutrient removal is connected with both direct and indirect uptake, with the former being associated with biomass concentration and growing environment. By using fungi's inherent metabolic skills to break down and purify a variety of contaminants, fungal bioremediation offers a viable and sustainable solution to environmental contamination. The methods, uses, and prospects for fungal bioremediation are all well covered in this review paper. Laccases, peroxidases, and hydrolases are just a few of the many enzymes that fungi possess that enable the breakdown of heavy metals, complex organic compounds, and xenobiotics into less toxic forms. Because of their adaptability, fungus can be used in a variety of environmental settings, such as soil, water, and air remediation. The ability of fungal bioremediation to break down persistent organic pollutants like petroleum hydrocarbons, polychlorinated biphenyls (PCBs), and polycyclic aromatic hydrocarbons (PAHs) as well as to immobilize and change heavy metals through biosorption and bioaccumulation is evidence of its effectiveness. The difficulties and restrictions related to fungal bioremediation are also covered in this paper, including the requirement for ideal environmental conditions and possible ecological effects. The integration of omics technology for the clarification of fungal metabolic pathways and the creation of biotechnological advancements to scale up fungal bioremediation processes are two examples of future research objectives that are emphasized. The importance of fungi in environmental remediation is emphasized in this review, along with the necessity of further study and technical development to fully utilize their potential in tackling environmental issues worldwide.

K. Murmu (✉) · A. Sarkar · S. Saha · S. S. Sarma
Department of Agronomy, F/Ag, Bidhan Chandra Krishi Viswavidyalaya, Krishi Viswavidyalaya, Nadia, West Bengal, India

S. Murmu
Department of Agricultural Chemistry and Soil Science, F/Ag, Bidhan Chandra Krishi Viswavidyalaya, Krishi Viswavidyalaya, Nadia, West Bengal, India

C. M. Hussain et al. (eds.), *Fungi in Waste Water Treatment*, Springer Water,
https://doi.org/10.1007/978-3-032-12693-1_2

Keywords Nutrient cycling · Fungi · Bioaccumulation · Metabolites · Remediation · Waste water treatment

2.1 Introduction

Due to urbanization and industrialization, certain pollutants and their byproducts are now discharged into water bodies, and it is estimated that waterborne illnesses claim the lives of 1.8 million people annually (Pujari & Kapoor, 2021; Roy et al., 2022). Water resources are becoming more limited, and a number of contaminants that affect aquatic habitats and human health because of their toxicity, resistance to degradation, and longevity pose a severe threat to the quality of water bodies (Netherway et al., 2021; Wagg et al., 2019). Waste water treatment plants are designed to provide clean drinking water and improved living conditions to feed the world's growing population while halting the spread of such diseases. The ecosystem and human health may suffer if wastewater is not adequately handled. Beach closures and other limits on recreational water usage, pollution of drinking water, damage to fish and wildlife populations, and oxygen depletion are a few examples of these effects. Microorganisms have an important function in detoxifying habitats and soil. A promising technique called mycoremediation employs fungus to break down or transform dangerous substances into less toxic or non-toxic forms (Zeghal et al., 2021). Mycoremediators include *Pleurotus ostreatus*, *Rhizopus arrhizus*, *Phanerochaete chrysosporium*, *Phanerochaete sordida*, *Trametes hirsuta*, *Trametes versicolor*, *Lentinus tigrinus*, and *Lentinus edodes*. Bioremediation, particularly mycoremediation, uses microorganisms' metabolic capacity to convert environmental contaminants into less dangerous or non-hazardous forms, making it a promising solution with low chemical and energy input (Goodell et al., 2020). For the benefit of society and our future, wastewater treatment is an essential undertaking that has to be given more careful consideration. Wastewater released into river systems, particularly that containing textile colors, can severely pollute the ecosystem and endanger the life of microorganisms in the water. Every year, approximately 100,000 different kinds of commercial dyes are produced, and the textile industry alone releases over 1000 tons of dyes into water systems. Organic chemical reagents known as dyes are used extensively in the printing, leather tanning, textile, paper, and plastics industries. Although a variety of materials, including cloth, copper mesh, organic and inorganic film, and aerogels, have been employed in wastewater treatment, it is challenging to degrade dye effluent using conventional waste water treatment methods (Mansouri et al., 2017; Turusov et al., 2002). Physical, chemical, and biological treatments are all part of conventional wastewater treatment. The goal of biological treatments, which use microorganisms to remove contaminants, is to provide environmentally safe and efficient wastewater treatment technologies. According to reports, fungi also improve wastewater sludge's degradability, settleability, and dewaterability, which helps with sludge management. Extracellular enzymes are typically used in fungal-mediated bioremediation to break down contaminants and lessen or eliminate hazardous waste from the environment (Bose et al., 2021; Kumar et al., 2021). Utilizing the metabolic powers of microorganisms, bioremediation including mycoremediation converts environmental contaminants into less or non-hazardous forms, making it an appealing solution that requires little energy and chemical input (Al-Hawash et al., 2019; Mahmud et al., 2022). Fungal immobilization for dye decolorization and adsorption techniques are used to create an effective and economical therapy (González-González et al., 2022; Morin-Crini et al., 2022; Shahid et al., 2021). Usually, the absorbents are used to lower the dye concentration in wastewater and absorb toxic organic substances. The importance of fungi in environmental remediation is emphasized in this review, along with the necessity of further study and technical development to fully utilize their potential in tackling environmental issues worldwide (Fig. 2.1).

2.1.1 Bioremediation Using Fungi

Mycoremediation is a type of bioremediation where the area is decontaminated using fungi. The mycelium of fungi is responsible for breakdown, which is the main function of fungi in the ecosystem. Selecting the appropriate fungal species to target a particular pollution is known as mycoremediation. Fungi play a crucial role in the ecosystem by breaking down and changing both organic and inorganic substrates

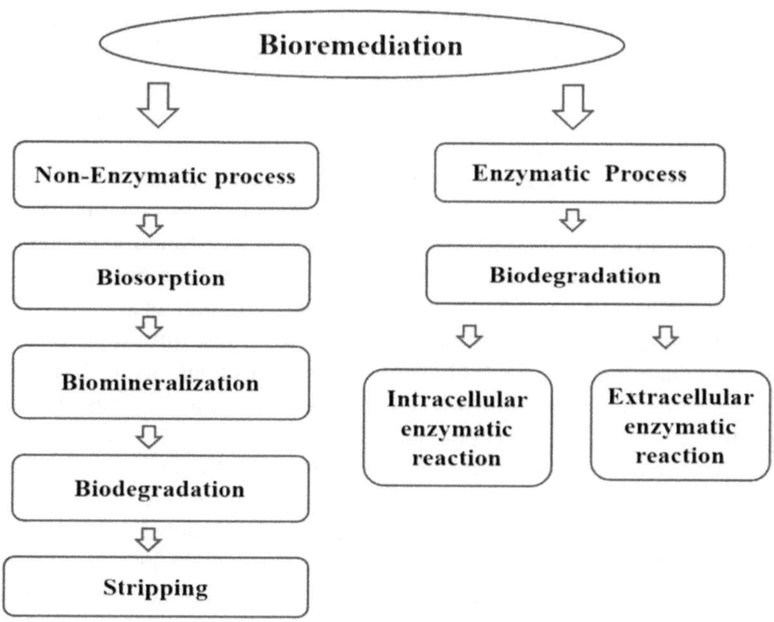

Fig. 2.1 Bioremediation mechanism (Tomar & Jajoo, 2021)

(Bastos & Magan, 2009). Nutrient removal using fungal species has been studied by various researchers (Beltrán-Flores et al., 2021). One of the best ways to clean up contaminated soil and water is through myco-remediation, which has been proven to be highly effective at decontaminating a variety of contaminants. It is also a financially and environmentally sound way to combat the growing problem of pollution in both aquatic and terrestrial environments. Fungi have several advantages, including strong growth, a vast network of hyphae, the production of multifunctional extracellular enzymes, a higher surface area to volume ratio, the ability to combat complex pollutants, the ability to adapt to changes in pH and temperature, and the presence of metal-binding protein (Baborová et al., 2006). Along with the fungal extracellular enzymes and their possible function in the myco-remediation process, the impact of various physical, chemical, and biological variables on the bioremediation efficacy of fungal species has been thoroughly evaluated. Fungi are unicellular organisms with a promising role in treating wastewater. Although they can also break down organic materials, certain filamentous fungus are naturally present in wastewater treatment systems as vegetative cells or spores. In terms of nutrient cycle, plant absorption, plant health, and animal diet, fungi play a major role. Climate change-induced changes in fungal growth will have a significant impact on ecosystem services (Bidartondo et al., 2018) and may altered the bioremediation potential. Fungal growth, reproduction, geographic distributions, and physiological activities are all directly impacted by climate change, while their habitat is indirectly impacted as well (Palazzotto & Weber, 2018). The process of bioremediation uses living things, primarily bacteria and microorganisms, to clean up contaminated areas. Pollution and other adverse environmental effects are examples of how our activities have affected the ecosystem's health. Fungi are known as natural decomposers that aid in the creation of soil because of their innate ability to break down harmful compounds and break down cellulose and lignin. In a fungal cell, mycoremediation can occur in three main locations: the surface, the extracellular environment, or the internal environment (Alzarhani et al., 2019). Because filamentous fungus species can adapt to harsh climates through mycelial growth, they also play a little role in waste degradation. Mycotoxin-producing filamentous fungus are also involved in solid waste management. The function of filamentous fungi in the bioremediation process, which can involve the bioseparation of suspended particles, is being clarified by recent research (Bahram et al., 2018). In addition to their unique medicinal and nutraceutical properties, fungus have the ability to clean up contaminated soils and absorb carbon. It entails using fungi's innate capacity for enzymatic breakdown to clear hazardous material. Plastics are long-lasting polymers that are not biodegradable. Pleurotus ostreatus has also reported that fungi have remedied them (Carrino-Kyker et al., 2016). Since mushrooms need these elements for their metabolic activities, their high rate of accumulation of heavy metals including lead, cadmium, magnesium, and nickel accounts for their benefits in bioremediation (Clausing et al., 2021). Through the processes of biodegradation, bioconversion, and biosorption, mushrooms revitalize contaminated soil. The synthesis of extracellular enzymes, such as ligninase, cellulases, and peroxidases, facilitates the breakdown of polycyclic aromatic hydrocarbons (PAHs)

(Deveautour et al., 2020). Waste lignin and cellulose can be used to grow mushrooms, a bioconversion end product. They have the ability to biotransform vegetable biomass into proteins, enzymes, and carbohydrates (beta-glucan). The process of using energy to assimilate heavy metals from an aqueous solution is known as biosorption. This procedure facilitates the removal of heavy metals. This mechanism involves fungal mass binding heavy metals on their surface. The low enzymatic degradation of chemicals may be the cause of toxicity and could be fatal for the mushrooms. Since these mushrooms are considered hazardous waste, they are no longer fit for human consumption. Since mycelial development takes place on agricultural waste, the entire biological process is neither costly nor harmful to the environment. In a highly irradiated environment, it is said to have shielding properties against free radicals (Naik, 2019). Fungi have a vital role in balancing the carbon circle because of their unique enzyme chain, which enables them to break down lignocellulose. Once broken, the hydrocarbon chains' nutrients are stored in cellulose and are available for plant use. There are fewer or harmless forms of harmful chemicals. Mycoremediation has not yet been marketed for widespread use, despite its immense promise. Because they can withstand harsh environments, fungi that thrive in extreme climates are particularly cost-effective for industrial applications (Durand et al., 2017). Their advantages over bacteria include their widespread distribution, higher growth percentage, hyphae network, ability to produce degrading enzymes, and capacity to accumulate metals (Table 2.1).

2.1.2 Advantages of Fungi in Wastewater Treatment

Large amounts of sludge, primarily bacterial biomass, are produced by conventional biological wastewater treatment systems. This sludge is of low value and costly to treat before disposal. According to Dissanayake et al. (2018), fungi are frequently grown in business as a source of numerous beneficial biochemicals. An economically feasible method for sustainable waste management may result from combining wastewater remediation with the recovery of valuable materials. According to these viewpoints, the fungal wastewater treatment method, which uses an inexpensive organic substrate as a feed to produce high-value fungal byproducts along with

Table 2.1 The wastewater enzymes degradation and producing fungi (Cameron et al., 2000)

Type of wastewater	Enzymes used for degradation	Enzymes producing fungi
Phenolic-laden wastewater	Lignin peroxidase, chloroperoxidase, manganese peroxidase, laccase	Phanerochaete chrysosporium, Caldariomyces fumago, Rhizoctonia praticola
Cyanide-rich wastewater	Cyanide hydratase	Gloeocercospora sorghi, Stemphylium loti
Food-processing wastewater	L-galactonolactone oxidase, Chitinase	Candida norvegensis, Serratia marcescens

concurrent wastewater remediation, can be a compelling substitute (Deveautour et al., 2020). A number of intrinsic benefits come with the fungal wastewater treatment process, such as (a) increased rates of breakdown of complex organic compounds found in wastewater because of the presence of particular fungal enzymes, (b) effective solid separation of the fungal biomass from the mixed liquor, and (c) the potential to recover valuable fungal byproducts.

2.2 Fungal Bioremediation

2.2.1 Basic Principles of Bioremediation

One innovative and promising technique for removing and reducing heavy metals from contaminated water and soil is bioremediation (Wang et al., 2017). Because bioremediation relies on the metabolic capabilities of microorganisms, it is acknowledged as the most fundamental and dependable method of removing contaminants, especially petroleum and its refractory components. According to Smith et al. (2018), microorganisms play a significant role in heavy metal bioremediation. It is possible to create organisms through genetic engineering that may be able to decrease several types of polycyclic hydrocarbons (PAHs).

2.2.2 Mycoremediation

Using fungi to extract toxins from various environmental elements, both living and dead, is known as mycoremediation. Mycoremediation is a low-cost technique that generates no hazardous waste. The choice and use of the appropriate fungus species for the target heavy metal or other contaminant determines the effectiveness of mycoremediation (Burton et al., 2019). This could be a workable and affordable way to address the expanding problem of water and soil contamination. A bioremediation method called mycoremediation uses fungi to remove harmful materials (S'anchez et al., 2018). Both macrofungi (mushrooms) and filamentous fungi (molds) might be present during the procedure. Mycoremediation can therefore be used as a biological agent for the degradation, transformation, or immobilization of pollutants.

2.3 Advantages of Fungal Bioremediation

Because of their diverse biological and ecological characteristics, fungi provide a number of advantages over bacteria in the bioremediation of pollutants. Because of these advantages, fungi are especially effective at treating a wide range of contaminants in a number of environmental settings.

2.3.1 Adaptability to Extreme Conditions

Fungi are incredibly adaptable and can thrive under extreme pH, temperature, and moisture conditions when bacterial activity is minimal. Fungi, such as Aspergillus niger and Penicillium chrysogenum, may grow in both highly acidic environments with pH as low as 2 and alkaline environments with pH as high as 11. This adaptability allows fungi to grow in environments that are unfriendly to most bacteria, such as acidic mine drainage sites or alkaline industrial waste areas.

2.3.2 Enzymatic Diversity and Efficiency

Numerous pollutants can be broken down by the diverse array of enzymes found in fungi. These include ligninolytic enzymes that may break down complicated chemical compounds, such as laccases, peroxidases, and oxidoreductases. Extracellular enzymes produced by white-rot fungi, such Phanerochaete chrysosporium, can break down lignin and a range of xenobiotic substances, such as chlorinated phenols and polycyclic aromatic hydrocarbons (PAHs) (Mony et al., 2020). Fungi's diverse enzymatic makeup allows them to break down difficult-to-digest pollutants that bacteria cannot.

2.3.3 Versatility in Bioremediation Applications

Fungi are flexible and can be used in a wide range of bioremediation applications, including soil, water, and air purification. They are suitable for both in situ and ex situ cleanup methods. In situ applications involve the direct cleanup of contaminated areas, using fungi to break down pollutants on-site. Ex situ applications involve transporting contaminated materials to a controlled environment for treatment. For example, fungi have been used successfully in bioreactors to treat wastewater contaminated with dyes and heavy metals, resulting in significant pollutant reductions (Morris et al., 2019).

2.3.4 Symbiotic Relationships

Plants and fungi commonly form symbiotic relationships that can enhance bioremediation procedures. For instance, mycorrhizal fungi work with plant roots to facilitate the intake and transmission of nutrients, including broken-down pollutants. By spreading the fungi's capacity for degradation throughout the plant's root system, this symbiosis not only enhances plant health and growth in contaminated soils but also raises the overall effectiveness of bioremediation (Mony et al., 2020).

2.4 Mycoremediation of Heavy Metals

A particular group of heavy metals with high densities and atomic weights, such as silver, mercury, manganese, copper, nickel, and antimony, make up the Earth's crust. Through contaminated food, soil, water, and air, they make their way up the food chain. The application of pesticides and fertilizers results in their incorporation into the foods we eat. Because heavy metals bioaccumulate, a large concentration of them may be harmful. Through industrial effluent that is dumped into bodies of water, they get into the water system and harm human health by causing cancer, bone defects, neurological diseases, liver and kidney damage, and digestive issues. Phosphate fertilizers, metals, petroleum products, and polyvinyl chloride (PVC) all contain cadmium. Batteries and wires, as well as the pollutants from burning coal, contain lead and mercury. Chronic illnesses such as Minamata disease, which was brought on by eating fish polluted with methylmercury in Japan, were caused by drinking tainted water (Parker et al., 2015). During excavation, cadmium was released into waterways, and rice cultivated in contaminated water led to biomagnification in the human population. This led to the spread of Itai-Itai disease throughout the region (Clausing et al., 2021). Heavy metals from water and industrial effluents could be successfully removed using living fungi (Blumenstein et al., 2015). Heavy metal toxicity is the cause of many other fatal illnesses. By removing heavy metals from the soil and storing them in their mycelia and fruiting bodies for future breakdown, fungi have contributed to the removal of heavy metals from the environment. Polluted habitats are restored by the use of macrofungi. Heavy metal mycoremediation uses macrofungal species such as *Pleurotus ostreatus*, *Agaricus bisporus*, *Boletus edulis*, and *Polyporus* sp. Because it can tolerate and retain heavy metals, the microfungus *Aspergillus niger* has been shown to be an excellent entity for chromium (Cr) remediation (Chen et al., 2018). The concentration of Pb was the highest at 75.81%, followed by Zn at 49.39%, Cu at 45.35%, and Ni at 25.20%. *Aspergillus flavus* was found to have heavy metal biosorption, with 22% Pb and 20% Cu aggregation. Several strains of Trichoderma, Penicillium, and Fusarium have been demonstrated in X-ray absorption studies to reduce and methylate arsenic. Drinking water eventually becomes contaminated by the dangerous carcinogen arsenic (As), which also makes it unfit for agricultural use. Aspergillus, Trichoderma, Rhizopus, and Chaetomium were among the organisms that were frequently seen; they are all potential candidates for arsenic mycoremediation. Glycerol, phosphate, and hexose transporters are thought to be the routes via which arsenic is absorbed (Tsai et al., 2009). The highest amounts of Cu, Cd, Mg, and Mn are found in the paper and pulp industry, which is a leader in heavy metal impurities. They have been seen to accumulate in larger numbers in Aspergillus and Mucor species. Abiotic stress conditions are favorable for Mucor species. Inhibiting ATPase activity can enhance the activity of Mucor circinelloides, which facilitates in-situ bioremediation. Fungi were the least affected of all the bacterial groups in a study on culturable soil microbial populations and the impact of arsenic and antimony on them (Chen et al., 2018) the impact of five heavy metals on Pleurotus tuber-regium growth: copper, zinc, lead, mercury, and cadmium. Unlike Hg, which hindered development, Cd accumulated well. Stipe morphology was affected by Pb (Akpaja et al., 2012). According to, species abundance declined as uranium content rose. Certain functional genes were activated as a result, and these genes can be integrated to gain a comprehensive understanding of ecological traits.

2.4.1 *Mycoremediation Using White-Rot Fungi*

The presence of lignin-modifying enzymes gives white-rot fungus (WRF) an unparalleled capacity to break down lignin, a heterogenous polyphenolic polymer. This makes it possible to specifically remove environmental contaminants such pesticides, herbicides, polychlorinated biphenyls (PCBs), and organochlorines. The name comes from the way the lignin breaks down and turns the wood white. Although several of these are ascomycetes, the majority are basidiomycetes. They need a substrate to thrive on pollutants like polycyclic aromatic hydrocarbons, especially in soil, because they are natural decomposers. By supplying substrate, their enzymes aid in development (Fennell et al., 2011; Stottmeister et al., 2003). Ligninolytic enzymes aid in the biotransformation of organic pollutants. *Fomes fomentarius*, *Ganoderma lucidum*, *Pleurotus ostreatus*, *Trametes* sp., *Lentinula edodes*, *Trichoderma viride*, *Phellinus pini*, and *Rhizopus* sp. are a few examples of white-rot fungus. The mycoremediation of palm kernels, crude oil, and petroleum is assessed for Pleurotus pulmonarius. WRF breaks down petroleum molecules via a number of methods, including their enzymatic talent. When heavy metal bioaccumulation took place, the nutritional value increased. Assays for biodegradation have also been carried out to evaluate their potential for treating wastewater. Coriolus versicolor may biologically decolorize the textile dye everzol turquoise blue, according to a 2000 study by Kapdan et al. Additionally helpful in the breakdown of industrial dyes is Trametestrogii, which was found in Tunisia (Mechichi et al., 2006). By breaking down PCB, a soil pollutant, Pleurotus ostreatus and Irpexlacteus can generate a range of transformation products, including hydroxylated PCBs and chlorobenzoates. Outperforming the other taxa, Pleurotus ostreatus invaded the region (Stella et al., 2017). For dye decolorization, irpex lacteus seems to provide an alternative to chemicals. On a nitrogen-rich substrate, the fungus generated laccase enzymes and lignin phosphate (LiP). The fungus proved great for mycoremediation because it developed swiftly and resisted soil microorganisms. The potential of Lentinus subnudus to clean up crude oil spills has been investigated in Nigeria.

2.4.2 *Enzymes Used by WRF in Bioremediation*

Scientists have conducted substantial research on lignin, the primary fungus enzyme that aids the mycoremediation process. The fragmented lignin produces a large number of degradation products, which are taken by hyphae and further metabolized outside of the intracellular fungal mesh. *Phanerochaete chrysosporium* uses extracellular enzymes such as lignin peroxidase and glyoxal oxidase to degrade lignin. Glyoxal oxidase appears to assist activate lignin peroxidase by oxidizing metabolites and reducing oxygen to water. Lignin peroxidase oxidizes non-phenolic aromatic nuclei in lignin (Kirk et al., 1992).

2.4.3 Ligninolytic Fungal Enzymes

WRF's enzymes have a role in its utility. A glycated heme protein called lignin peroxidase facilitates the oxidation of unsaturated substances with planar rings that resemble lignin in a manner that is dependent on hydrogen peroxide. Therefore, a variety of molecules, including non-phenolic aromatic compounds, can be oxidized with high redox potential. The lignin enzyme's refractory nature makes it challenging to break down. By employing the hydrogen peroxidase method, manganese peroxidase oxidizes manganese (+2) to manganese (+3). The oxidation of phenolic substrates is facilitated by the Mn (III) state of the enzyme. The primary enzyme involved in the breakdown process is laccase. Even without hydrogen peroxide, these multicopper oxidase enzymes can still work (Viswanath et al., 2014). Laccase probably oxidizes a wide spectrum of aromatic and non-aromatic compounds, notwithstanding their short shelf life. They are principally responsible for the oligomerization and polymerization of aromatic compounds. Laccase combined with ultrasound enhances dye removal precision in wastewater. Nanobiotechnological studies with laccases for biosensor cell implantation have been undertaken (Goncalves et al., 2015). Other enzymes involved in mycoremediation include cellulases (Trichoderma species), hemicellulases, pectinases, chitinases (Fusarium species), amylases (*Aspergillus niger* and *Penicillium* species), and proteases. WRF uses agricultural waste as a substrate to synthesize the aforementioned enzymes. The enzyme laccase is used by Trametes versicolor to break down tribromophenol (TBP) (Donoso et al., 2008). Copper is utilized to alleviate water pollution and mineralizes lignin. The usage of lignocellulose is made possible by a complex and well-coordinated relationship between termites and fungi. Extracellular phenol oxidases are produced by the symbiotic fungus Termitomyces albuminosus. MnP is encoded by two genes, Tam 1 and Tam 2. They have manganese (Mn II) binding sites, which confirm MnP encoding, and an essential amino acid needed for peroxidase action. Termites and a fungus have a symbiotic connection that facilitates the breakdown of lignin and the full biorecycling of plant litter (Belanger et al., 2021). Since the quantities of catalase and polyphenol oxidase decrease in soil contaminated by oil, they could be used to assess bioremediation. Different amounts of oil contamination were found in the soil (Bi et al., 2019).

2.4.4 Non-ligninolytic Fungal Enzymes

Fungi employ glutathione S-transferase and cyt P450-dependent monooxygenases in addition to hydrolytic enzymes to degrade pollutants. PAH metabolism is defined by Sutherland et al. (1995) as the oxidation of aromatic rings to arene oxide. Additionally, dioxygenase enzymes have been discovered. In addition, two fungal cytochrome P450 monooxygenases that were separated from Fusarium oxysporum were reproduced. Both catalysts were found to be effective in producing ω-hydroxyl fatty acids by Geml et al. (2021).

2.4.5 Mycoremediation of Polycyclic Aromatic Hydrocarbons (PAHs)

PAHs are heterogeneous hydrocarbons with numerous aromatic rings that make up the majority of organic contaminants. It results from the partial breakdown of organic materials released by petroleum spills, incinerators, and incomplete burning of coal and wood (Kadri et al., 2017). Valentin et al. (2007) conducted an experiment to predict the ability of Bjerkandera fungus species to enhance the degradation of hazardous substances such as pyrene, dibenzothiophene, and phenalene in a slurry reactor. Pleurotus ostreatus aids in PAH elimination. Coprinus comatus basidiocarp obtained from worthless paper with 47.9 ppm lead contamination showed 16.2 ppm lead uptake from the pulp (Dulay et al., 2012). Pleurotus tuber-regium removed petroleum hydrocarbons with an efficiency of 20%, 18.7%, and 18.8% at contamination rates of 1.0%, 2.5%, and 5.0%, respectively. Meanwhile, at the same contamination levels, over 3 months, the removal efficiency climbed to roughly 40%, 39%, and 38%, respectively (in Pleurotus tuber-regium), which was the greatest. Pleurotus pulmonarius, on the other hand, exhibited the lowest remediation capacity. Pleurotus tuber-regium has a substantially higher efficacy in eradicating heavy metals and hydrocarbon compounds than Pleurotus ostreatus and Pleurotus pulmonarius. Pharmaceutical chemicals (PhC) persist in water bodies and cause poisoning. Wastewater treatment systems are inefficient in their removal. The need for fungi-based biological treatments to eliminate PhC has been recognized as a result of researchers' efforts (Durand et al., 2017). WRF, Trametes versicolor, also degrades PhC, which includes naproxen, codeine, diazepam, and metoprolol. X-ray spectrophotometric investigations revealed that immobilization and precipitation could be used by fungal strains to decode heavy metal uptake, which occurs alongside the biosorption process. Ganoderma lucidum is useful in PAH cleanup. The pioneering work on the use of Trichoderma asperellum H15 strains to degrade polycyclic aromatic compounds in soil was established. Phenanthrene decomposition in severely contaminated soil was found to be around 79.9% after 2 weeks. Two forms of aromatic hydrocarbons, anthracene and benzathine, have been discovered to be mycoremediated by fungal species isolated in polluted coastal saline deposits. GC/

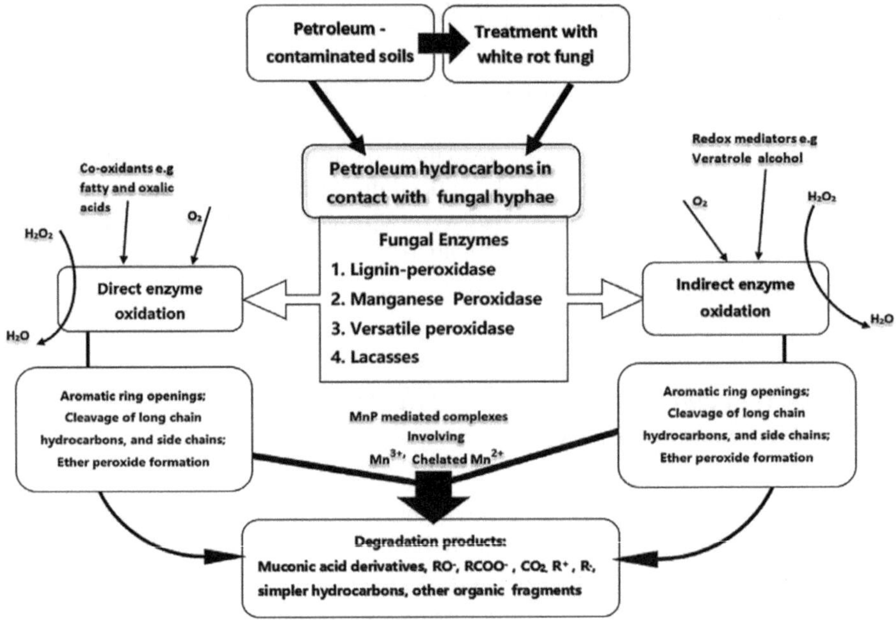

Fig. 2.2 Mechanism employed in mycoremediation of petroleum-contaminated soil (Adapted from Crawford et al., 2019)

MS tests demonstrated that Fusarium solani strains breakdown them to produce ortho-phthalic acid. Unbound laccase has been detected extracellularly. Pleurotus ostreatus (OST-1) showed promising results in removing organochlorines such as DDT, HCH, Aldrin, and Dieldrin (Ferlian et al., 2018). Trichoderma viride has been shown to remove cyclodienes like aldrin and dieldrin. Their hydrophobic nature limits their removal ability (Francioli et al., 2021). Two saprophytic microfungi, Trichoderma hamatum and Rhizopus arrhizus, are DDT-tolerant and produce better outcomes (Fig. 2.2). They displayed considerable metabolic activity for the depletion of carbon sources in the presence of an organochlorine (DDT) in soil. Possession of antioxidant enzymes to cope with the chemical stress caused by DDT exposure (Bergeron et al., 2019).

2.5 Conclusion and Future Prospective

Mycoremediation is an environmentally friendly method of cleaning up contaminated areas and detoxifying dangerous compounds. Reforms in science and technology are needed to get a better understanding of many occurrences. To summarize, fungal bioremediation, specifically mycoremediation, is an innovative and environmentally benign method of cleaning up the environment. Fungi have shown great promise and possibility for understanding their mechanisms and rehabilitating contaminated environments. Mycoremediation, which uses the unique metabolic abilities of particular fungi, is a low-cost, environmentally friendly, and adaptable way of eliminating a wide range of contaminants such as pesticides, heavy metals, hydrocarbons, and others. This strategy's effectiveness is demonstrated by fungi's capacity to convert complex pollutants into simpler, less harmful forms or elements that other species can break down more easily. The science of fungal bioremediation is still growing, with the discovery of new fungus species with amazing remediation abilities and the explanation of complex biochemical routes of pollutant destruction. Mycoremediation's potential for widespread application is enhanced by the fungi's resilience and adaptability in a wide range of environmental conditions. Mycoremediation has enormous promise, but there are still some hurdles to overcome, such as variability in effectiveness depending on environmental conditions, toxins being treated, and fungus species used. Additional research and development are required to optimize and scale this approach for large-scale, practical applications. Finally, mycoremediation is a new, long-term, and growing approach that holds great potential for environmental restoration, providing hope for polluted site cleanup while avoiding ecological damage. With continued research, development, and application, mycoremediation has the potential to significantly reduce environmental pollution and promote healthier, more balanced ecosystems. To summarize, much more light needs to be shed on the role of macrofungi in bioremediation, which is currently a massively underexplored area. With the global growth of unexpected environmental concerns, the popularization of mycoremediation approaches is becoming more important.

Acknowledgements We would like to express our gratitude to Department of Agronomy and Central library Bidhan Chandra Krishi Viswavidyalaya, Mohanpur, Nadia, West Bengal for the unwavering support in conducting the review work. Moreover, we would like to acknowledge the support of Dr. Binoy Sarkar, Senior Research Fellow, Future Industries Institute, University of South Australia for his constructive review.

References

Akpaja E.O., Nwogu N.A. & Odibo E. A. (2012). Effect of some heavy metals on the growth and development of Pleurotus tuberregium. Mycosphere. 3(1), 57–60.

Alzarhani, A. K., Clark, D. R., & Underwood, G. J. C. (2019). Are drivers of root-associated fungal community structure context specific? *The ISME Journal, 13*, 1330–1344.

Al-Hawash, A.B.; Zhang, X.; Ma, F. Removal and biodegradation of different petroleum hydrocarbons using the filamentous fungus Aspergillus sp. RFC-1. Microbiologyopen 2019, 8, e00619.

Baborová, P., Möder, M., Baldrian, P., Cajthamlová, K., & Cajthaml, T. (2006). Purification of a new manganese peroxidase of the white-rot fungus irpex lacteus, and degradation of polycyclic aromatic hydrocarbons by the enzyme. *Research in Microbiology, 157*, 248–253.

Bahram, M., Hildebrand, F., & Forslund, S. K. (2018). Structure and function of the global topsoil microbiome. *Nature, 560*, 233–237.

Bastos, A. C., & Magan, N. (2009). Trametes versicolor: Potential for atrazine bioremediation in calcareous clay soil, under low water availability conditions. *International Biodeterioration & Biodegradation, 63*, 389–394.

Belanger, N., Collin, A., Khlifa, R., & Lebel-Desrosiers, S. (2021). Balsam fir and American beech influence soil respiration rates in opposite directions in a sugar maple forest near northern range limit. *Frontiers in Forests and Global Change, 4*, 51.

Beltrán-Flores, E., Sarrà, M., & Blánquez, P. (2021). Pesticide bioremediation bytrametes versicolor: Application in a fixed-bed reactor, sorption contribution and bioregeneration. *Science of the Total Environment, 794*, 148386.

Bergeron, M. J., Feau, N., & Stewart, D. (2019). Genome-enhanced detection and identification of fungal pathogens responsible for pine and poplar rust diseases. *PLoS One, 14*, e0210952.

Bi, Y., Xie, L., & Wang, J. (2019). Impact of host plants, slope position and subsidence on arbuscular mycorrhizal fungal communities in the coal mining area of north-Central China. *Journal of Arid Environments, 163*, 68–76.

Bidartondo, M.I., Ellis, C., Kauserud, H., Kennedy, P.G., Lilleskov, E., Suz, L., & Andrew, C. (2018). Climate change: fungal responses and effects. In: Willis, Katherine J. (Ed.), State of the World's Fungi 2018. *Royal Botanical Gardens*, Kew, UK, pp. 62–69.

Blumenstein, K., Albrectsen, B. R., & Martin, J. A. (2015). Nutritional niche overlap potentiates the use of endophytes in biocontrol of a tree disease. *BioControl, 60*, 655–667.

Bose, S., Kumar, P. S., Vo, D.-V. N., Rajamohan, N., & Saravanan, R. (2021). Microbial degradation of recalcitrant pesticides: A review. *Environmental Chemistry Letters*. 19, 3209 –3228.

Burton, J. L., Peterson, S. M., & Smith, E. A. (2019). Degradation of polycyclic aromatic hydrocarbons by *Phanerochaete chrysosporium. Journal of Environmental Chemical Engineering, 7*(5), 4262–4271.

Cameron M.D., Timofeevski, S., & Aust S.D. (2000). Enzymology of Phanerochaetechrysosporium with respect to the degradation of recalcitrant compounds and xenobiotics. *Applie Microbiology and Biotechnology*. 54(6),751–758.

Carrino-Kyker, S. R., Kluber, L. A., & Petersen, S. M. (2016). Mycorrhizal fungal communities respond to experimental elevation of soil pH and P availability in temperate hardwood forests. *FEMS Microbiology Ecology, 92*, fiw024.

Chen, W., Xu, R., & Wu, Y. (2018). Plant diversity is coupled with beta not alpha diversity of soil fungal communities following N enrichment in a semi-arid grassland. *Soil Biology & Biochemistry, 116*, 388–398.

Clausing, S., Likulunga, L. E., & Janz, D. (2021). Impact of nitrogen and phosphorus addition on resident soil and root mycobiomes in beech forests. *Biology and Fertility of Soils, 57*, 1031–1052.

Crawford, K. M., Bauer, J. T., & Comita, L. S. (2019). When and where plant-soil feedback may promote plant coexistence: A metaanalysis. *Ecology Letters, 22*, 1274–1284.

Deveautour, C., Power, S. A., & Barnett, K. L. (2020). Temporal dynamics of mycorrhizal fungal communities and co-associations with grassland plant communities following experimental manipulation of rainfall. *Journal of Ecology, 108*, 515–527.

Dissanayake, A. J., Purahong, W., & Wubet, T. (2018). Direct comparison of culture-dependent and culture-independent molecular approaches reveal the diversity of fungal endophytic communities in stems of grapevine (*Vitis vinifera*). *Fungal Diversity, 90*, 85–107.

Donoso, C., Becerra, J., Martínez, M., Garrido, N. &d Silva, M. (2008). Degradative ability of 2,4,6-tribromophenol by saprophytic fungi Trametes versicolor and Agaricus augustus isolated from Chilean forestry. *World Journal of Microbiology and Biotechnology*. 24, 961–968.

Dulay, R.M.R., Parungao, A.G. IV, Kalaw, S.P. and Reyes. R.G. (2012). Aseptic cultivation of Coprinus comatus (O. F. Mull.) Gray on various pulp and paper wastes. Mycosphere, 3(3), 392–397.

Durand, A., Maillard, F., & Foulon, J. (2017). Environmental metabarcoding reveals contrasting belowground and aboveground fungal communities from poplar at a hg phyto management site. *Microbial Ecology, 74*, 795–809.

Fennell, D., Du, S., Liu, F., Liu, H., & Haggblom, M. (2011). Dehalogenation of polychlorinated Dibenzo-p-dioxins and dibenzofurans, polychlorinated biphenyls, and brominated flame retardants, and potential as a bioremediation strategy. In *Environmental biotechnology and safety* (pp. 135–149). Elsevier.

Ferlian, O., Cesarz, S., & Craven, D. (2018). Mycorrhiza in tree diversity– Ecosystem function relationships: Conceptual framework and experimental implementation. *Ecosphere, 9*, e02226.

Francioli, D., van Rijssel, S. Q., & van Ruijven, J. (2021). Plant functional group drives the community structure of saprophytic fungi in a grassland biodiversity experiment. *Plant and Soil, 461*, 91–105.

Geml, J., Morgado, L. N., & Semenova-Nelsen, T. A. (2021). Tundra type drives distinct trajectories of functional and taxonomic composition of arctic fungal communities in response to climate change–results from long-term experimental summer warming and increased snow depth. *Frontier of Microbiology, 12*, 490.

González-González, R. B., Flores-Contreras, E. A., & Iqbal, H. M. N. (2022). Bio-removal of emerging pollutants by advanced bioremediation techniques. *Environmental Research, 214*, 113936.

Goncalves, I., Silva, C. & Cavaco-Paulo, A. (2015). Ultrasound enhanced laccase applications. Green Chemistry. 17, 1362–1374.

Goodell, B., Winandy, J. E., & Morrell, J. J. (2020). Fungal degradation of wood: Emerging data, new insights and changing perceptions. *Coatings, 10*, 1210.

Kadri, T., Rouissi, T., Kaur Brar, S., Cledon, M., Sarma, S. & Verma, M. (2017). Biodegradation of polycyclic aromatic hydrocarbons (PAHs) by fungal enzymes: A review. *Journal of Environmental Science*. 51, 52–74.

Kirk, T.K., Lamar, R.T. & Glaser, J.A. (1992). Biotechnology and Environmental Science–Molecular Approaches. In: Mongkolsuk, S., Lovett, P.S. and Trempy, J.E. (Eds.), Proceedings of International Conference on Biotechnology and Environmental Science: Molecular Approaches; 1990 August 21–24, Bangkok, Thailand. New York: Plenum Press, pp. 131–138.

Kumar, M., Yadav, A. N., Saxena, R., Paul, D., & Tomar, R. S. (2021). Biodiversity of pesticides degrading microbial communities and their environmental impact. *Biocatalysis Agricultural Biotechnology*. 31, 101883.

Mahmud, T., Sabo, I. A., Lambu, Z. N., Danlami, D., & Shehu, A. A. (2022).Hydrocarbon degradation potentials of fungi: A review. *Journal of Environmental Bioremediation and Toxicology*. 5, 50 –56.

Mansouri, A., Cregut, M., Abbes, C., Durand, M.-J., Landoulsi, A., & Thouand, G. (2017). The environmental issues of DDT pollution and bioremediation: A multidisciplinary review. *Applied Biochemistry and Biotechnology, 181*, 309–339.

Mechichi, H.A., Mechichi, T., Dhouib, A., Sayadi, S., Angel Martínez, T. & Martínez, M.J. (2006). Laccase purification and characterization from Trametes trogii isolated in Tunisia: decolorization of textile dyes by the purified enzyme. *Enzyme and Microbial Technology*. 39 (1), 141–148.

Mony, C., Vandenkoornhuyse, P., & Bohannan, B. J. M. (2020). A landscape of opportunities for microbial ecology research. *Frontiers in Microbiology, 11*, 2964.

Morin-Crini, N., Lichtfouse, E., Fourmentin, M., Ribeiro, A. R. L., Noutsopoulos, C., & Mapelli, F. (2022). Removal of emerging contaminants from wastewater using advanced treatments. A review. *Environmental Chemistry Letters, 20*, 1333–1375.

Morris, E. K., Morris, D. J. P., & Vogt, S. (2019). Visualizing the dynamics of soil aggregation as affected by arbuscular mycorrhizal fungi. *The ISME Journal, 13*, 1639–1646.

Naik, B. S. (2019). Functional roles of fungal endophytes in host fitness during stress conditions. *Symbiosis, 79*, 99–115.

Netherway, T., Bengtsson, J., Krab, E. J., & Bahram, M. (2021). Biotic interactions with mycorrhizal systems as extended nutrient acquisition strategies shaping forest soil communities and functions. *Basic and Applied Ecology, 50*, 25–42.

Palazzotto, E., & Weber, T. (2018). Omics and multi-omics approaches to study the biosynthesis of secondary metabolites in microorganisms. *Current Opinion in Microbiology, 45*, 109–116.

Parker, I. M., Saunders, M., & Bontrager, M. (2015). Phylogenetic structure and host abundance drive disease pressure in communities. *Nature, 520*, 542–544.

Pujari, M., & Kapoor, D. (2021). Heavy metals in the ecosystem: Sources and their effects. In V. Kumar, A. Sharma, & A. Cerdà (Eds.), *Heavy metals in the environment*. Elsevier.

Roy, P., Mohanty, A. K., & Misra, M. (2022). Microplastics in ecosystems: Their implications and mitigation pathways. *Environmental Science Advances, 1*, 9–29.

S'anchez, J. R., Rivas, F., Torres, P. A., & Rodríguez-Couto, S. (2018). Comparative analysis of fungal and bacterial bioremediation of PAH-contaminated soil. *Applied Microbiology and Biotechnology, 102*(15), 6609–6618.

Shahid, M. K., Kashif, A., Fuwad, A., & Choi, Y. (2021). Current advances in treatment technologies for removal of emerging contaminants from water: A critical review. *Coordination Chemistry Reviews, 442*, 213993.

Smith, H. E., Tiedje, J. E., & Brown, M. T. (2018). Fungal phosphate solubilization and its role in heavy metal bioremediation. *Fungal Ecology, 36*, 103–112.

Stella, T, Covino S. & Cvancarova M. (2017). Bioremediation of longterm PCB-contaminated soil by white-rot fungi. *Journal of Hazardous Materials*. 324, 701–10.

Stottmeister, U., Wießner, A., Kuschk, P., Kappelmeyer, U., Kästner, M., Bederski, O., Müller, R., & Moormann, H. (2003). Effects of plants and microorganisms in constructed wetlands for wastewater treatment. *Biotechnology Advances, 22*, 93–117.

Sutherland, J.B., Rafii, F., Khan, A., & Cerniglia, C.E. (1995). Mechanisms of polycyclic aromatic hydrocarbon degradation. In: Young, L.Y. and Cerniglia, C.E. (Eds.) Microbial transformations and degradation of toxic organic chemicals. Wiley-Liss, New York, pp. 269–306.

Tomar, R. S., & Jajoo, A. (2021). Enzymatic pathway involved in the degradation of fluoranthene by microalgae Chlorella vulgaris. Ecotoxicology. 30 (2), 268–276.

Tsai, S.L., Singh, S. & Chen, W. (2009). Arsenic metabolism by microbes in nature and the impact on arsenic remediation. Current Opinion in Biotechnology. 20, 659–667.

Turusov, V., Rakitsky, V., & Tomatis, L. (2002). Dichlorodiphenyltrichloroethane (DDT): Ubiquity, persistence, and risks. *Environmental Health Perspectives, 110*, 125–128.

Valentin, L., Lu-Chau, T.A., Lopez, C., Feijoo, G., Moreira, M.T. & Lema, J.M. (2007). Biodegradation of dibenzothiophene, fluranthene, pyrene, and chrysene in a soil slurry reactor by the white-rot fungus Bjerkandera sp. BOS55. *Process Biochemistry*. 42, 641–648.

Viswanath, B., Rajesh, B., Janardhan, A., Kumar, A. P., and Narasimha, G. (2014). Fungal laccases and their applications in bioremediation. Enzyme Research 2014, 1–21.

Wagg, C., Schlaeppi, K., & Banerjee, S. (2019). Fungal-bacterial diversity and microbiome complexity predict ecosystem functioning. *Nature Communications, 10*, 1–10.

Wang, R. C., Chen, H., Chen, C., & Wang, H. (2017). Fungal bioremediation of cadmium contaminated soils by *Aspergillus niger*. *Environmental Science and Pollution Research, 24*(18), 15168–15175.

Zeghal, E., Vaksmaa, A., Vielfaure, H., Boekhout, T., & Niemann, H. (2021). The potential role of marine fungi in plastic degradation – a review. *Frontiers in Marine Science*, 8.

Chapter 3
Economical Taxonomy and Adaptive Mechanisms of Fungi in Wastewater Treatment System

Suman Kumar, Maneesh Kumar, Shashank Nand Tiwari, and Pratima Gupta

Abstract Fungi are essential in wastewater treatment systems and have different taxonomic classifications and adaptation strategies that allow them to thrive under these demanding conditions. The taxonomic diversity of fungi in wastewater treatment includes species such as Aspergillus, Penicillium and Trichoderma, which are known for their effective degradation of organic pollutants. In terms of wastewater, these fungi have the mechanism in them that allows them to survive in high concentration of nutrients and toxic compounds. Among these are the production of extracellular enzymes for the breakdown of high molecular weight organic substrates, the development of thick cell walls that can endure extreme toxic chemical hazards in the surrounding environment, and the ability to form biofilms for effective uptake and also protection from damaging environmental conditions. This chapter covers the fungi and their mechanisms of biological processes and heavy metal binding properties used in the removal of heavy metals and other contaminants from the effluent. The adaptations of the fungi have economic significance in that they allow for low-cost and eco-friendly measures for the treatment of wastewater which is more effective in cleaning contaminants within the water without using much power or producing excess sludges compared to traditional processes based on treatment with bacteria.

S. Kumar (✉)
Department of Microbiology, All India Institute of Medical Sciences, Deoghar, Jharkhand, India

State-Virus Research and Diagnostic Laboratory, All India Institute of Medical Sciences, Deoghar, Jharkhand, India

M. Kumar · S. N. Tiwari
State-Virus Research and Diagnostic Laboratory, All India Institute of Medical Sciences, Deoghar, Jharkhand, India

P. Gupta
Department of Microbiology, All India Institute of Medical Sciences, Deoghar, Jharkhand, India

C. M. Hussain et al. (eds.), *Fungi in Waste Water Treatment*, Springer Water,
https://doi.org/10.1007/978-3-032-12693-1_3

Keywords Fungal enzymes · Wastewater treatment · *Aspergillus* · Contaminants · Biodegradation pathways

3.1 Introduction

Fungi are an important group of organisms widely utilized in the removal of contaminants from wastewater. Currently, other fungal phyla contain Ascomycota, Basidiomycota, and Mucoromycota. Hence, fungi are highly ecologically significant in a sewage treatment plant because they play a major part in the natural biodegradation process of organic matter synthesis, which obviates the build-up of contaminants. Owing to the harsh conditions present in wastewater treatment plants for a prolonged period, some fungi have developed cytological and physiological adaptations for survival in such stressful environments (Vaksmaa et al., 2023). They can enter into chemical interactions through fusion reactions. Biocenoses can move through passive processes and can facilitate the exchange of nutrients during enzymatic degradation. These organisms' weather extreme wastewater conditions by constructing strong walls, rapidly absorbing nutrients, developing filaments for biofilm formation, and secreting enzymes to break down complex organic molecules (Nandan, 2017; Risse-Buhl et al., 2017). They act as natural helpers in the remediation of industrial wastewater containing persistent contaminants, including stubborn substances that bacteria cannot resist. Filamentous fungi such as *Aspergillus*, *Penicillium*, and *Trichoderma* have filamentous or hyphal bodies that form a network during food uptake or enzyme production. Because of these filamentous properties, they are also able to form biofilms to disinfect wastewater and remove suspended solids and organic material (More et al., 2010; Naher et al., 2021). Yeasts are unicellular fungi that multiply rapidly and are resistant to changes in environmental conditions. Some of the yeasts such as Saccharomyces cerevisiae and Candida albicans can be used in treatment processes for extremely contaminated organic wastewater (Silva et al., 2019; Walker & White, 2017). They can also tolerate or eliminate extremely toxic organic pollutants and heavy metals. Various other fungal species have been proven their fair share as biosorbents in removing heavy metal ions or other pollutants from wastewater. Yeast is a very rapidly reproducing and fair adapting single-cell form of fungus. Some of these are used in extreme conditions of organic wastewater pollution for instance *Saccharomyces cerevisiae* and *Candida albicans* (Hualpa-Cutipa et al., 2023; Mohiuddin et al., 2024; Schlüter & Schauer, 2017). Capable of tolerating or excluding high concentrations of toxic organic contaminants and heavy metals. These include other forms of fungi that show promise in the biosorption of metals in wastewater treatment. In alkaline fungi, the synergistic effect will treat a very active ecosystem environment that will further increase effectiveness in pollutant removal and acceptable water quality. The two fungi have shown a very strong versatility toward the different living conditions that are available to them. They adapt to diverse conditions and challenges, including pH and temperature fluctuations, as well as nutrient availability (Dinakarkumar et al., 2024).

After several systematic modifications or evolutionary developments, the biofilm promotes the exchange of genetic material with other microbial species that later increases the efficiency of biodegradation pathways. It also aids in the clean-up of suspended solids via easy filtration. Besides additive bio-remediation, fungi can help with other organisms for symbiotic relationships to further boost treatment. Because of their metabolic diversity, fungi can disintegrate refractory, resistant environmental toxicants, commonly not subject to degradation by bacteria (Dinakarkumar et al., 2024; Ganguly et al., 2024). These, together with the great metabolic capabilities of fungi, guarantee the economic and efficient use of fungi in wastewater treatment (Rambabu et al., 2020). Various other fungal species with specialized functions contribute to the diversified economic taxonomy of fungi in these systems. This chapter elaborates on and presents an overview of the importance of economic taxonomy and fungal adaptation in municipal wastewater treatment plants; it justifies these efforts. Therefore, the need to pay more attention and carry out more research on the specific role of various fungal kinds in bioremediation becomes apparent. This can help improve and better the treatment of industrial effluents due to better insights into the peculiarities in their adaptations and applications.

3.2 Significance of the Economic Taxonomy of Fungi

Fungal classification is important not only for assessing the significance of fungi but also for promoting conservation. Fungi are mostly found in acidic environments but there is a potential for their use in almost neutral industrial effluents. They are given a particular adaptive capacity that helps them thrive in wastewaters that are hostile to most treatment systems. Such a capacity can survive and operate effectively even in conditions where there are extreme gradients of pH levels, high salinity, and low oxygen areas (El-Gendi et al., 2021). The economic taxonomy of fungi and their mechanisms of acclimatization become important in the treatment of wastewaters. As, the case in these advanced ecological geopolitics, the pollution control mechanisms have to be cost-effective, environmentally friendly, and aggressive. The issue of economic utilization of biomass and its remedial works, which is very important in the research data provided, has also been tackled (Kesari et al., 2021; Sar et al., 2024). A general analysis of food processing, textiles, pharmacy, and the paper industry, wherein the waste nutrients are utilized by filamentous fungi and converted into useful products, found that such approaches are applicable. A further topic to be addressed is the role of fungi in the production of lignolytic enzymes as an important skill for the treatment of wastewater by metallic salt adsorption (Kumar, Bhoi, & Kumar, 2024; Negi & Das, 2023). Many of the advantages that fungal biorefinery holds for example the production of single-cell proteins rich in protein, the production of biolipids, enzymes, organic acids, etc. This enables environmental sustainability and even better aids economic development (Chatterjee & Mohan, 2022). There are significant challenges that must be addressed with great care when great-scale mushroom biorefinery processes are undertaken to treat wastewater. These range from identifying the suitable conditions for growing edible mushrooms for commercial purposes to the effective and continuous extraction of nutrients from different types of wastewaters and disposing of excess biomass. In addition, the viability of such systems in practical use at both industrial and economic levels must be assessed to safeguard operations considering the cost of running such systems, energy costs, and earnings from the wastes produced (Antunes et al., 2020; Jayaraman et al., 2024). There are also technical aspects underlying certain challenges, such as the control of pH and prevention of bioreactor contamination in fermenters of large volumes fuelled by divergently focused silos. It is also necessary to perform related assessments of the possibility of large-scale operations without violating any standards or causing detrimental effects on the environment (Ranjan et al., 2024). Within this context, researchers and industry stakeholders can identify these hurdles and formulate appropriate interventions to mitigate the challenges of the commercial rollout of mushroom-based technologies for wastewater treatment (Chandrashekhar & Rao, 2010; Navina et al., 2024).

Before being reused or discharged into natural waters, effluents or industrial effluents must undergo treatment to ensure that they meet the prescribed standards for clean water and also to ensure that otherwise harmful solids are removed. This in turn is necessary to achieve an acceptable standard (Kumar, Kumari, et al., 2024; Silva, 2023). The use of filamentous fungi in the treatment of effluents or raw sewage has been emphasized in comparison to treatment with bacteria, which is, however, much less common. Even though there are several strains of fungi that produce enzymes as well as surface proteins that play an important role in the degradation or absorption of pollutants present in water, such situations are very rare. This is the reason why fungi can synthesize a variety of reactive enzyme systems in addition to their filamentous structure. They are located in different habitats and substrates (Dinakarkumar et al., 2024; El-Gendi et al., 2021; Sar et al., 2024). And they are well known, the function of fungi is to rot biological materials in nature. With these properties, organic wastes are converted into useful products to manufacture high-value products. Furthermore, filamentous fungi are more advantageous when compared with other groups of microorganisms. Recent studies indicate that filamentous fungi are also very effective in the bioremediation of industrial effluents (Table 3.1). For instance, filamentous fungi can endure harsher environments than bacteria, can produce higher biomass, and are easier to recover than bacteria. Different types of higher-value biomass could have a significant impact on the economics of wastewater treatment. Fungi could provide this advantage over bacteria in wastewater treatment (Kumar, Bhoi, & Kumar, 2024; Kumar, Kumari, et al., 2024). The biomass produced by wastewater treatment with fungi has the potential to be much more valuable than the biomass produced by the bacterial-activated sludge process. Fungi can be used to produce useful biochemicals and also serve as a source of protein. Commercial aseptic cultivation of fungi using expensive substrates yields a variety of high-value biochemicals (Kumar, Kumari, et al., 2024; Kumari et al., 2024; Sar

Table 3.1 Involvement of various fungi in waste water treatment

Fungal species (filamentous)	Wastewater type	Key pollutants treated	Mechanism of action	References
Aspergillus flavus	Petroleum refinery and industrial wastewater	Petroleum hydrocarbons, heavy metals	Biodegradation of hydrocarbons; biosorption of metals	Ameen et al. (2024)
Aspergillus Niger	Textile and dye industry wastewater	Synthetic dyes, heavy metals (lead, cadmium, mercury)	Biosorption via cell wall binding; enzyme secretion (laccase, peroxidase)	Singh et al. (2022), Ameen et al. (2024)
Aspergillus oryzae	Textile and chemical wastewater	Azo dyes, phenolic compounds	Laccase production and biosorption	Santos and Corso (2014)
Cladosporium spp.	Oil refinery and organic wastewater	Petroleum hydrocarbons, organic solvents	Hydrocarbon degradation via enzyme secretion	Gaur et al., 2022 Rani et al. (2024)
Fusarium oxysporum	Petrochemical and dye industry wastewater	Hydrocarbons, azo dyes, heavy metals	Enzymatic breakdown of hydrocarbons; biosorption of metals	Chaurasia et al. (2023)
Lentinula edodes	Pulp and paper industry wastewater	Phenolic compounds, lignin, dyes	Enzymatic breakdown with laccase and versatile peroxidase	Kumla et al. (2020), Vara and Karnena (2020)
Mucor racemosus	Tannery and leather industry wastewater	Chromium, nickel, cadmium	Chitosan-rich cell wall biosorption of heavy metals	Shankar et al. (2014)
Penicillium chrysogenum	Electroplating and metal industry wastewater	Heavy metals (zinc, copper, lead, nickel)	Metal biosorption via cell walls rich in polysaccharides	Shakya et al. (2016)
Phanerochaete chrysosporium	Pesticide and agrochemical wastewater	Pesticides, herbicides, dyes	Enzymatic degradation with lignin peroxidase and manganese peroxidase	Prajapati et al. (2023)
Rhizopus arrhizus	Mining and smelting industry wastewater	Heavy metals (cadmium, lead, mercury)	Biosorption through fungal cell wall components	Kajla et al. (2021), Jamir et al. (2024)
Saccharomyces cerevisiae	Chemical and metal processing wastewater	Heavy metals (chromium, lead), synthetic dyes	Metal biosorption and dye adsorption through cell wall polysaccharides	Blaga et al. (2021), Malik et al. (2023)
Trametes versicolor	Pharmaceutical and textile wastewater	Pharmaceutical residues, synthetic dyes	Laccase and manganese peroxidase-mediated biodegradation	Aragaw et al. (2024)

et al., 2024). Challenges do not hinder fungi, but they thrive in such situations by using metabolic pathways such as biosorption, where heavy metals are bound by fungal cell walls, or digestion of complex organic aflatoxins by secreted extracellular enzymes (Kumar, Kumari, et al., 2024). The system's specificity is an advantage since it can be employed against a broader spectrum of contaminants without the need for extra chemical agents, thereby enhancing the cost-effectiveness of the treatment process. While treating waste waters laden with different types of toxins, it is possible to make use of fungi since they are friendly, cheap in application as opposed to treating them with aggressive measures and this helps in minimizing operational costs.

Fungi help in the bioremediation of industrial wastewater utilizing fungicidal effects, and they do play a predominant role in degradation or biosorption is pivotal. Aflatoxigenic fungi like *Aspergillus flavus* are applied to petroleum refinery wastewater for hydrocarbon degradation or biosorbent heavy metals and then can enhance the environmental issues related to this type of pollutants reliably. *Aspergillus niger* synthesized laccase (produced by fermentation) and hydroxy phenyl derivative, which are used mainly, in the textile industry to degrade obscene dyes such as methylene blue also it delivers certain enzymes like peroxidase could reduce levels of harmful heavy metals ions lead, cadmium or mercury via cell wall biosorption from wastewater (Ameen et al., 2024; Kumla et al., 2020). In the same manner, *Aspergillus oryzae* has an incredible bio-potential for the decolorization of azo dyes and phenolic compounds present in textile and chemical wastewater through the production of laccase followed by biosorption (Rajhans et al., 2020; Santos & Corso, 2014). *Cladosporium* spp. helps treat wastewater from oil refineries by producing enzymes that break down petroleum hydrocarbons and solvents. It is wide range of applicability makes it degrade hydrocarbons, and azo dyes while interacting and binding to heavy metals thus showing utility in the petrochemical industries as well as the dye industry (Gaur et al., 2022; Rajhans et al., 2020; Rani et al., 2024). For instance, Lentinula edodes (Shiitake) is important for the industry because its enzymes help treat wastewater from pulp and paper production where it is used to break down lignin and phenolic substances and colours due to laccase or versatile peroxidase activity (Kumla et al., 2020). Various types of adsorbents have been used to remove heavy metals found in tanneries, for example, nickel and cadmium, using Mucor racemosus (Shankar et al., 2014). In contrast, Penicillium chrysogenum has been effective in treating waste solutions from electroplating, which is due to its glycan containing structural component that is capable of removing zinc, copper, lead and nickel (Kumar, Kumari, et al., 2024; Prajapati et al., 2023). Bioremediation of heavy metal contamination using fungi is advantageous compared to other chemical treatments due to lowered costs and eco-friendliness of such methods. Nonetheless, further investigation is required in order to improve these biosorption processes for the purpose of economic scale industrial use. It is envisaged that these fungi-based rehabilitation processes can be integrated into wastewater treatment plants for industrial operations thus providing a sustainable measure towards curb metal pollution. This presents an opportunity for *Penicillium chrysogenum* to be deployed in several different types of pollution cases because it can remove several heavy metals at the same time. Areas of Research might investigate enhancement of metal recovery capabilities of fungal biomass and elicit qualitative genetic change or other forms of growth condition changes to increase polysaccharide output (Saikia et al., 2023).

Phanerochaete chrysosporium also has applications in the treatment of agrochemical wastewater containing pesticides and herbicides by degrading them with lignin peroxidase, and manganese peroxidase processes (Prajapati et al., 2023; Shakya et al., 2016). Additional instances comprise *Rhizopus arrhizus*'s characteristics in which toxic elements such as cadmium, lead, and mercury are binding. This is through compositional materials of the cell wall in host mining wastewaters (Jamir et al., 2024). In the case of *Saccharomyces cerevisiae*, it was used in a biosorption process where heavy metal ions from waste stream dyes were due to the use of its yeast's cell wall polysaccharides (Blaga et al., 2021). In textile and pharmaceutical wastewater, *Trametes versicolor* is finally used to degrade synthetic dyes or pharmaceutical residues by laccase and manganese peroxidase therefore other fungi that can produce enzymes are efficient agents for clean industrial effluents treatment. Fungi are the most important in the bioremediation of industrial wastewater due to biodegradation and biosorption (Aragaw et al., 2024; Malik et al., 2023).

For example, *Aspergillus flavus* has been implemented for hydrocarbon degradation and heavy metals removal from effluents of petroleum refineries successfully carrying out the environmental clean-up of hydrocarbon and other pollutants by either binding or degrading these pollutants (Rani et al., 2024; Saravanan et al., 2021). *Aspergillus niger* is also used in the bioremediation of textiles industries where synthetic dyes and heavy metals such as lead and cadmium, mercury is found. This microorganism has a high tendency to cell wall biosorption of the mentioned metals and secreting bioremediation enzymes such as laccase/peroxidases for cleaning the dye house and metal-based sewage. In contrast, even further in wastewater treatment, especially chemical treatment of textile effluent, *Aspergillus oryzae* is one of the key performers when it comes to the degradation of chemical compounds from azo dyes up to phenolic compounds thanks to laccase production in addition to biosorption. In addition, *Cladosporium* spp. can also degrade petroleum hydrocarbons or organic solvents helping to treat the aqueous waste generated by an oil refinery (Bala et al., 2022; Ranjan et al., 2024; Singh et al., 2022; Tak et al., 2022).

Owing to its capability to chelate heavy metal pollutants, *Fusarium oxysporum* is used in the petrochemical and dye industries for the degradation of crude oil and azo dyes, respectively (Chaurasia et al., 2023). *Lentinula edodes* (shiitake mushroom) is likely the most promising for the treatment of tannery wastewater because of their developed enzyme system with laccase and versatile peroxidases. These enzymes effectively degrade lignin, phenolic compounds, and dyes, which are characteristically present in tannery wastewater (Singh, 2017). The production of these enzymes provides an advantage for *Lentinula edodes* in the breakdown of these harmful substances, thus making it a safe and effective alternative for tannery effluent treatment. The application of *Lentinula edodes* in tannery effluent treatment reduces the environmental burden of leather production and provides a useful fungal biomass. Overcoming pollution concerns is made easier through the use of waste to wealth strategies where waste is converted into useful products (Kumar & Hashmi, 2021; Kumla et al., 2020). This is further enhanced by the shiitake mushroom's ability to colonize different types of growth substrates suggesting direct cultivation of tannery waste stock for developing an economically integrated treatment system (Kumla et al., 2020). In addition, *Mucor racemosus* contains a large amount of chitosan within its cell walls, which can be used for the biosorption of metals such as chromium, nickel, and cadmium found in tannery effluents (Shankar et al., 2014). On the same note, *Penicillium chrysogenum* deals with waste from electroplating by use of biosorption of metal salts (zinc, copper, lead, nickel) through its matrix of polysaccharides. The specific value of *Phanerochaete chrysosporium* rests in handling agrochemical effluents by the breakdown of pesticides and herbicides through laccase and manganese peroxidase in that order (Saikia et al., 2023; Shakya et al., 2016). In mining effluents, *Rhizopus arrhizus* uses its wall components to bind heavy metals (cadmium, lead, mercury), whereas in the case of *Saccharomyces cerevisiae*, heavy metal and organic dye removal from wastewaters involved in metal processing industries are achieved with the help of these yeast cell wall polysaccharides. Eventually, *Trametes versicolor* found application in the treatment of wastewater troubled by the pharmaceutical and textile industries, employing laccase and manganese peroxidase to actively attack pharmaceutical wastes and artificial dyes, respectively; thus, these fungi are very useful for bioremediation of industrial wastewater (Jamir et al., 2024; Latif et al., 2023; Malik et al., 2023; Vara & Karnena, 2020).

3.3 Biodegradation and Biosorption Abilities of Fungi

Fungal biodegradation and biosorption processes involved in wastewater treatment have a whole range of different, yet significant, operations for achieving effective pollutant breakdown through natural capability of the fungi. The wastewater is inoculated at the beginning with selected fungus species such as *Aspergillus niger* or *Trametes versicolor* (Dinakarkumar et al., 2024). These fungi absorb and degrade many types of contaminants. After inoculation, the connecting of the waste substances into the fungal wall occurs; commonly, it is heavy metals or organic contaminants. This process is called biosorption. The walls are laden with chitin and other binding agents; they will capture the contaminants from the water and serve as natural adsorbents (Gajewska et al., 2022; Saravanan et al., 2021). After biosorption, several fungi secrete enzymes and produce powerfully extracellular enzymes like peroxidases and laccases, which are essential to the degradation of the polychlorinated aromatic compounds by their chemical enzymatic mechanism (Fig. 3.1). These enzymes convert complex chemicals, like dyes or drug residues, into nontoxic and simple substances (Vishnoi & Dixit, 2019).

Trametes versicolor is a good example of fungi that produce laccase enzymes capable of breaking down synthetic pigments in textile wastewater. By breaking down the hazardous substances into smaller, less hazardous compounds, such a

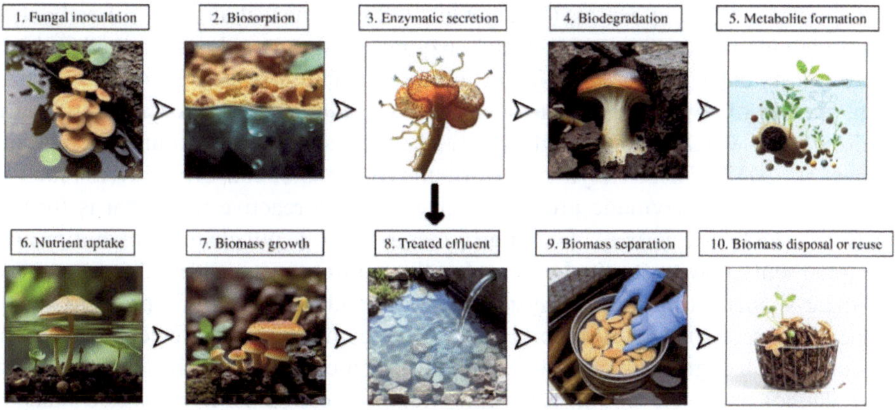

Fig. 3.1 Different path ways for the biodegradation and biosorption abilities of fungi

degradation process will liberate metabolites that help in the filtration of the water. The products formed from the degradation process can then be taken by fungi as nutrients to enhance their growth (Saravanan et al., 2021). As the fungal cells proliferate, their ability to absorb and metabolize certain pollutants increases, generating biomass. Treated wastewater refers to sufficiently purified water from further release or reuse, which is made safe because of its far lowered content of pollutants (Rajendran et al., 2024; Vara & Karnena, 2020). Separation of the biomass completes the process by removing the fungal biomass from the treated water, thus eliminating any remaining contaminant. This biomass can be either disposed of or reused. If there are other dangerous substances like heavy metals present in the biomass, they may need to be disposed of safely. Conversely, should biomass be primarily employed to absorb organic pollutants, it may then be reused in composting or in the generation of bioenergy (Ayele et al., 2021; Ramesh et al., 2024). The role of fungi in biological treatment of wastewater is important since fungi can protect this process from pollution by assimilating and mineralizing the pollutants at each and every stage. This technique helps avoid the undesirable accumulation of toxic waste and provides a green solution for processing waste from industries.

3.4 Fungal Enzymes and Their Role in Pollutant Breakdown

The ability of fungi to synthesize a wide array of enzymes is one of the key aspects because these enzymes can degrade and bio-remediate several pollutants found in the waste waters. Laccases, peroxidases, and cytochrome P450s are some of the enzymes that have been very effective in the degradation of complex organic matter and foreign organic compounds as xenobiotics. The range of activities of fungal enzymes would also be one of their merits in wastewater treatment as most of these enzymes possess the ability to act on common pollutants such as dyes, drugs, insecticides, and heavy metals (Khan et al., 2023; Kumari et al., 2021). Lastly, it is advantageous that fungi secrete these enzymes to the outside of the cells which increases its capacity for deployment in bioremediation strategies. As an illustration, species of *Aspergillus flavus* and *Cladosporium* spp. produce special alkane hydroxylases which act on straight-chain alkanes that are readily found in crude oil (Liu et al., 2022). These enzymes perform the catalysis of converting a less reactive bond through an enzymatic process to give rise to a reactive bond that is further broken down by the activity of other enzymes aiding oil spill clean-up operation and the treatment of petrochemical wastes. In other similar enzymes, amylase is present in spear-shaped strains of *Aspergillus oryzae* and *Penicillium* spp. these helps achieve the reduction of BOD biochemical oxygen demand of treated wastewater through starch degradation thus lowering the organic loading. Such is the case of an aryl alcohol oxidase found in fungi like *Pleurotus ostreatus*, *Bjerkandera adusta*, and *Trametes versicolor*, which plays a role in the degradation of lignin (Liu et al., 2022; Serrano et al., 2020). The mechanism not only supports the regeneration of different lignin-degrading enzymes but also assists in scavenging persistent organic pollutants. In addition, these enzymes are also used during bioremediation and attest to the efficiency of microbial systems in cleaning up the environment. It refers to their capacity to convert organic materials into simpler organic compounds (Rajendran et al., 2024). The research has mainly been oriented towards using such systems for delignification of lignin. However, these systems may be utilized to degrade structurally related organic pollutants as well. This adaptability is very encouraging concerning the application of bioprocessing enzymes for environmental remediation in various ecosystems. In addition, these pathways provide an insight into the development of microbial degradation mechanisms and their improvement using biotechnology. Other enzymes, especially, cellulases derived from Trichoderma reesei and Aspergillus fumigatus respectively, are used to act on cellulose and other solids present in the effluent to lower the total organic load. The cytochrome P450 monooxygenases of *Aspergillus flavus* and *Cladosporium* spp. aid in hydrocarbon primary oxidation where aliphatic and aromatic hydrocarbons are converted to less complex components for easier and further decomposition (Rani et al., 2024; Rezaei & Moghimi, 2024). Epoxide hydrolases in the aforementioned fungi convert toxic and unbiodegradable epoxides to diols which are more readily consumed by fungi thus helping to reduce toxicity (Saini & Sareen, 2017). In addition, laccases found in fungi such as *Agaricus bisporus*, *Pleurotus ostreatus*, and *Trametes versicolor* are enzymes with multi-copper oxidases capable of working with large organic molecules (Abo Nahas et al., 2021; Kues & Ruhl, 2011). Such enzymes help reduce the pollution of water bodies by facilitating the breakdown of synthetic colours, phenolic compounds, and PAHs from wastewater (Kues & Ruhl, 2011). Such enzymes act in conjunction with many other pathways of fungal metabolism to degrade complex mixtures of contaminants (Table 3.2). Mechanisms involving the work of cytochrome P450 monooxygenases, epoxide hydrolases, and laccases enable fungi to overcome the biological degradation of these complicated pollutants that cannot be effectively handled by other microorganisms (Rani et al., 2024).

Table 3.2 Fungal enzymes actions on pollutant

Fungal enzymes	Fungal species	Role in pollutant breakdown	References
Alkane hydroxylases	*Aspergillus flavus*, *Cladosporium* spp.	Specific decomposition of straight-chain alkanes (components of crude oil), transform alkanes into alcohols, enabling additional breakdown by other enzymes	Prenafeta-Boldu et al. (2006), Sarsan et al. (2024)
Amylase	*Aspergillus oryzae*, *Penicillium* spp.	Breaks down starches, lowering biochemical oxygen demand (BOD) in wastewater	Dhevagi et al. (2021), Sar et al. (2024)
Aryl alcohol oxidase	*Pleurotus ostreatus*, *Bjerkandera adusta*, *Trametes versicolor*, *Phanerochaete chrysosporium*	Enhances the oxidation of alcohols, which in turn participates in the process of lignin decomposition and supports the metabolism of foreign substances. Produces hydrogen peroxide that is applied by MnP and LiP, thus facilitating the oxidation of contaminants in an indirect manner	Hatakka (1994), Serrano et al. (2020), Urlacher and Koschorreck (2021), Abdelhamid et al. (2024)
Cellulase	*Trichoderma reesei*, *Aspergillus fumigatus*	Breaks down cellulose and other organic matter, which lowers the amount of organic matter in wastewater	Zhao et al. (2018)
Cytochrome P450 monooxygenases	*Aspergillus flavus*, *Cladosporium* spp.	Aliphatic and aromatic hydrocarbons are first oxidized to introduce oxygen, which makes them more reactive and susceptible to additional enzymatic breakdown	Mohamed et al. (2024), Rani et al. (2024)
Epoxide hydrolases	*Aspergillus flavus*, *Cladosporium* spp	Detoxification and breakdown of epoxides, convert epoxides to diols, which are easier for fungi to metabolize, reducing toxicity	Wallraf (2018)
Laccase	*Agaricus bisporus*, *Aspergillus flavus*, *Cladosporium* spp., *Pleurotus ostreatus*, and *Trametes versicolor*	It facilitates the oxidative degradation of the various complex organic molecules including synthetic dyes phenols and endocrine disruptors. Polycyclic aromatic hydrocarbons PAHs; a multi-copper oxidase that disassembles complex aromatic systems, thereby minimizing toxicity	Arregui et al. (2019), Abo Nahas et al. (2021), Gałązka et al. (2023)
Lignin peroxidase (LiP)	*Trametes versicolor*, *Phanerochaete chrysosporium*	Degradation of lignin and large recalcitrant compounds, certain aromatic pollutants; oxidizes large compounds through radical formation, breaking down pollutants resistant to conventional treatments	Ali et al. (2019), Singh et al. (2021)

(continued)

Table 3.2 (continued)

Fungal enzymes	Fungal species	Role in pollutant breakdown	References
Lipase	*Candida rugosa*, *Rhizopus arrhizus*	Breaks down fats and oils, helping to reduce grease and organic load in wastewater	Mahboob et al. (2022)
Manganese peroxidase (MnP)	*Phanerochaete chrysosporium*, *Pleurotus eryngii*	Assists in the degradation of recalcitrant organics by degrading lignin like materials, PAHs and coloring agents. Utilizes manganese ions (Mn^{2+}) as a catalyst for oxidation, extremely efficient towards phenolic substances and their derivatives	Jha (2019)
Peroxidase	*Aspergillus flavus*, *Aspergillus niger*, *Cladosporium* spp., *Phanerochaete chrysosporium*	Hydrogen peroxide is employed in the production of reactive species. In addition, the process of oxidative breakdown makes use of polycyclic aromatic hydrocarbons such as PAHs, extremely stable compounds, textured structural components, colouring agents, and several organic substances containing a phenol group	Ali et al. (2019), Shanmugapriya et al. (2019)
Protease	*Aspergillus flavus*, *Rhizopus* spp.	Degrades proteins, reducing nitrogen content and BOD in wastewater.	Ameen et al. (2024)
Tyrosinase	*Aspergillus oryzae*, *Trametes versicolor*	Enzymes, assist in breaking down toxic pollutants are involved in the oxidation processes of phenolic compounds.	Mónica and Jaime (2019), Shanmugapriya et al. (2019)
Versatile peroxidase (VP)	*Trametes versicolor*, *Phanerochaete chrysosporium*	The ability to act on a wide range of substrates including both phenolic and non-phenolic substrates, including lignin, bears the traits of both laccase and manganese peroxidase enzymes, thus providing versatility in degradation of various pollutant bearing materials encountered in industry	Yao et al. (2023), Raychaudhuri and Behera (2022)

Enzymes named Lignin peroxidase (LiP) and manganese peroxidase (MnP) are useful in breaking down the industrial waste materials that are hard to degrade (Abdelhamid et al., 2024). These enzymes can be obtained from the fungi and are able to break down organic compounds that are unnatural in environment as well as native hydrocarbons that are significant to the return of nutrients in a given ecosystem. They are also able to withstand different environmental extremes; different pH and temperatures enhancing their use in various bioremediation approaches. This also encourages new strategies for use in biotechnological applications such as the design of environmental biosensors and the production of green catalysts for industrial processes. The fungi Trametes versicolor and *Phanerochaete chrysosporium* produce LiP, an enzyme capable of digesting bulky lignin pollutants and aromatic bodies through the formation of radicals; making possible the decomposition of materials that are otherwise treated poorly (Abdelhamid et al., 2024; Thomas et al., 2020). Enzymes such as lipases derived from *Candida rugosa* and *Rhizopus arrhizus* can also help in the management of wastewater by biodegrading fats and oils to control the amount of grease and organic waste in the treatment plants (Nimkande & Bafana, 2022). Furthermore, the versatile peroxidase (VP) enzyme obtained from *Trametes versicolor* and *Phanerochaete chrysosporium* embodies both lignin peroxidase (LiP) and manganese peroxidase (MnP) which increases its substrate specificity to both phenolic and non-phenolics thus advantageous in removing impurities in robust industrial effluents. Proteolytic enzymes made by *Aspergillus flavus* and *Rhizopus* species are beneficial in protein breakdown in effluents thereby reducing the nitrogen concentration and BOD in treatment plants (Ameen et al., 2024; Parmar, 2020). Finally, tyrosinase from *Aspergillus oryzae* and *Trametes versicolor* aids in degrading phenolic compounds, while peroxidase-producing fungi use hydrogen peroxide to produce (Sen & Nigam, 2022)

3.5 Cost-Effectiveness of Fungal-Based Wastewater Treatment

At a global scale, my core competencies and training incorporate the latest available information to the year 2030, reaffirming my continued evolution in 2023 as a baseline and also with a more conservative prediction of the next developmental stages for my practice. Due to poverty encountered in developing countries, there are restrictions in gaining advanced technologies and therefore, mushroom-based systems are preferred (Mosnier et al., 2023). In the country India, industrial uses of fungi like Aspergillus niger have been exploited not only for the treatment of waste from textile dying, but also achieving low costs and effective pollutant removal (Prabhu et al., 2024). Unlike chemical coagulation and advanced oxidation processes, fungal treatments are less energy-intensive and in turn, more cost-effective and sustainable in regions with scarce energy resources.

Moreover, small and medium-sized enterprises (SMEs) in developing countries stand to benefit particularly from these systems owing to their lower initial capital and operation costs. The utilization of cost-effective treatments is accompanied by the investigation of fungus as a novel approach in Europe (Raychaudhuri & Behera, 2022). Scandinavian nations incorporated organisms like Trametes versicolor into the systems of wastewater treatment plants to dissolve drug residues. This helped in reducing the use of costly chemical solutions. These microorganisms produce a variety of extracellular enzymes that aid in the degrading of persistent organic pollutants thus saving on the sludge disposal and secondary treatment costs. It has been confirmed that the application of fungi can prolong the operational period of wastewater treatment facilities thus resulting in economic gains from the lowered upkeep and renewal expenses (Arvanitoyannis & Tserkezou, 2008; Sarsan et al., 2024). China has also been employing the cultivation of mushrooms for cleaning up wastewater in some of its regions and this is more so in waste waters that are generated by the copper mining activities. One of the uses of fungi which includes *Penicillium* species is for biosorption where metals are fixed into place and in this case, they can be removed at a far lower cost than through chemical means (Pundir et al., 2024). This method is attractive for business situations as wastewater treatment processes can be used to harness precious metals from wastewater which optimizes cost-based factors. Countries have begun to employ treatment systems based on mushrooms that help in environmental protection while lowering the costs of treatment deepening the embedding of such systems into the global strategy for management of wastewater treatment. This technique provides an answer for businesses that need new technology for dealing with waste—effective, cheap, and eco-friendly (Prabhu et al., 2024; Thomas et al., 2020).

3.6 Adaptive Mechanisms of Fungi in Harsh Wastewater Conditions

Fungi exhibit fortifiable adaptability to environmental stressors such as those found in waste waters, through several means, thus occupying a hyper-extreme niche. This physiology is complemented by morphologically adaptive strategies that aid survival against extreme pH, high toxicity levels, and nutrient imbalance. One of the adaptations relates to changes in the compositions of cell walls of adapted fungi to protect their cells and regulate unwanted materials influx into the cell (Parrilli et al., 2022). Some species have thicker cell walls developed or modified their levels of chitin synthesis to reduce the effect of chemical stress. Apart from this, fungi also possess the ability to manufacture enzymes which can inactivate or degrade toxins found in waste water turning them into harmless compounds or even using them for energy generation. This enzymatic plasticity not only improves their sustainability but also shows a glimpse of the promise of fungi in bioremediation (Shanmugapriya et al., 2019). Yet another important adaptive strategy is the development of efflux pumps that prevent the accumulation of harmful chemicals in fungi. In addition, these pumps play an important role in the survival of environmental toxins, especially metals, which are commonly found in industrial wastewater (Dwivedi, 2023; Patil et al., 2024). In addition, fungi can form biofilms, which are protective surfaces under extreme conditions, but also provide better adhesion in wastewater treatment systems. The development of biofilms further helps in the deposition of extracellular enzymes which dramatically increases the rate of pollutant degradation. A few fungal species also have variability in their growth characteristics and growth rates as a result of environmental factors. Chlamydospores or sclerotia may form out of them. Fungi metabolize their products and alter that depending on the ecological condition they are in, which may include changing the nutrition available (Dwivedi, 2023; Kumar, Bhoi, & Kumar, 2024). This metabolic versatility will enable them to survive even during the constantly changing conditions that characterize most industrial activities including how wastewater is treated. It is evidenced by their genetic diversity coupled with their ability to withstand adverse conditions that fungi can adapt well to the processes used in wastewater treatment. Fungal species show a high rate of evolution and are often genetically adaptable to the extent that they can find defence mechanisms against many destructive substances and environmental challenges. For instance, some fungi possess expel pumps that help in getting rid of harmful compounds thereby surviving in the dirtiest environments. Moreover, some fungi practice horizontal uptake of genes from other organisms present in the wastewater ecosystems which are useful to them. This makes it easier for them to degrade the wastes in the water since their resilience and adaptability are increased. Besides, the ability to build a biofilm has a protective benefit that aids the organism in surface attachment in a wastewater treatment facility and also boosts the biosorption of contaminants (Ameen et al., 2024; d'Enfert et al., 2021; Thomas et al., 2020). Utilizing fungi in the treatment of wastewater has quite a number of advantages. First of all, fungi are very versatile and can thrive in different types of wastewaters, be it alkaline or acidic, hot or cold, toxic or non-toxic. These organisms secrete a number of enzymes, for example, laccases and peroxidases, that can biotransform or detoxify these adverse conditions; in some instances, even feeding on the toxic materials (Kumar, Kumari, et al., 2024). Thus, bioremediation with respect to fungi can be said to be a promising field. Due to their enormous enzymatic diversity, they can assist with the clean-up of persistent organic pollutants, dyed products, and heavy metals thus relevant in bioremediation. Their metabolic machinery can adjust to the changes in the availability of nutrients and grow both in conditions of excess nutrients and in the absence of nutrients (Ameen et al., 2024; Silva, 2023). The operational costs of fungal treatment systems are often lower than those incurred in the use of chemical treatments since they are less resource-dependent and mainly self-propelling. This approach is beneficial to the environment because it reduces the need for chemical agents and also controls secondary pollution. Certain fungi serve this function by accumulating resources from wastewater in the form of precious metals as excessive heavy metals and thus permitting their recycling (Ameen et al., 2024; Sar et al., 2024). Appreciating how these phenomena occur and their advantages allows us to go beyond just looking at fungi from a biodiversity perspective and also offers ways of enhancing processes used in treating wastewater. The ecological flexibility of fungi, the economic side and low impact on the environment are the factors which can be regarded as a response towards the creation of more sustainable strategies for wastewater management (Obaideen et al., 2022).

3.7 Integration of Fungi in Hybrid Wastewater Treatment Systems

Fungi have an important function in hybrid wastewater treatment systems due to their specific characteristics which support processes involving bacteria. Thanks to remarkable enzymatic machinery, these eukaryotic organisms are capable of degrading a myriad of complex organic pollutants, even the most recalcitrant substances that usually elude bacterial degradation. In cooperative systems, filamentous fungi are frequently introduced within existing treatment systems such as activated sludge or constructed wetlands, in order to increase the treatment efficiency and the variety of pollutants that can be removed (Sar et al., 2024). The filamentous growth of numerous fungal species provides another benefit in the form of biofilms with a large surface area to volume ratio, which can increase the efficiency of nutrient and pollutant absorption. In addition, it is operationally feasible to have the fungi undergo the different phases of the treatment because fungi can withstand environments of different pH and oxygen levels. Some species also exhibited the capacity to sequester and bio-accumulate heavy metals, making their use in the reduction of toxic metals in wastewater plausible. Such hybrid models utilize fungi-bacteria for remediation which, besides improving water quality, speeds up the rate of degradation of the pollutants (Ameen et al., 2024; Silva, 2023).

3.7.1 Fungi-Bacterial Consortia

Fungal-bacterial consortia have lost global attention because they have been exploited solely for treating textile and dye industry effluents. Fungi species such as *Trametes versicolor* and *Aspergillus niger* have remarkable ability to degrade organic substances which are difficult to treat by conventional treatments processes. Bacteria then use these intermediate combination products created by the fungus as a source of growth, cleaving them into subunits which are less toxic (Karthika et al., 2024; Pundir et al., 2024). Such partnership enhances the rate and extent of pollutants' elimination compared to use of fungi or bacteria individually. The modern advancement is derived from the convergence of fungi and bacteria functionalities in the treatment of wastewater and it aims at revolutionizing the process of removing the waste products from the manufacturing or municipal plants. In this sense, the issue of global water resource pollution, which is becoming a more and more important problem, can be addressed by utilizing the advantages of both microbes for more efficient and less expensive water cleaning systems. The recent technique involving the use of Fungi and Bacteria in waste water Treatment, will enhance the dispersion of the wastes and enrich the compound breakdown irrespective of the waste's location (Pundir et al., 2024). Options are available that are cheap and friendly towards the environment with the use of active substances. By harnessing the biological capabilities of these microorganisms, this method can assist in the preservation of the water ecosystem and hence, this technology could be significant for future generations. This fungal-bacterial combination gives a novel solution to wastewater that stands to enhance water treatment operations and offer a more sustainable method for addressing the world's ecological degradation. Such methods can lead to large feedback on costs as well as reduce effects to the environment where they are implemented in both industries and municipalities. If utilized extensively, this technology presents endless possibilities to benefit water and eventually help improve human health and protect global ecology (Navina et al., 2024; Thirumalaivasan et al., 2024).

3.7.2 Fungi-Activated Sludge Systems

Fungi-activated sludge systems are effective in treating municipal and industrial waste water. For example, in such systems *Phanerochaete chrysosporium* and other fungal strains are used in conventional activated sludge processes in order to promote decomposing of highly resistant organic compounds such as persistent pollutants of Lignin-like structure, which are not readily degraded by bacteria alone. This is because fungi come in as enzymes that act on these problematic molecules, which in turn raises the rates of pollutant removal and optimizes the utilization of sludge up for management of complex wastes (Akerman-Sanchez & Rojas-Jimenez, 2021). In such systems fungi and bacteria work in a synergistic manner not only for the purpose of elimination of pollutants, but also in aiding improvement in settling characteristics of sludge and also reducing the amount of excess sludge produced (Nancharaiah & Reddy, 2018). In addition, it has also been found that such communities of fungi and bacteria incorporated within the active sludge are capable of responding to changes in the influent composition as well as environmental variations. Therefore, improving the treatment process. This integrated treatment method improves wastewater treatment efficiency as a whole but also helps in lowering the operational costs and energy use for the treatment plants. The enthusiasm for this new way of treating wastewater has led to the development of innovative reactors that may be able to enhance the interactions between fungi and bacteria and the production of enzymes. The application of membrane bioreactors or packed-bed reactors and others are being tried in order to utilize the full potential of the fungal-bacterial cooperative processes (Akkoyunlu et al., 2021). Moreover, specific substrate degrading enzymes can be engineered into chosen fungal strains, which improves the efficiency of pollutant degradation within sewage treatment facilities (Dwivedi, 2023; Pundir et al., 2024).

3.7.3 Fungi-Algae Systems

In integral systems of fungi and algae, these organisms are simply applied in treatment of effluents rich in nutrients like agricultural excess by use of fungi and algae together. Algae take in excess nutrients such as nitrogen and phosphorus whereas fungi assist in organic pollutants breakdown and heavy metals adsorption. This type of association serves the purpose of removing organic and nutrient loads from waste, helps concerned about the excess loading of nutrients in the form of effluents, and aims for a composite strategy in water treatment (Gururani et al., 2022). The treated effluent contains a wide variety of products, some of which are present at extremely toxic levels. Furthermore, a considerable volume of inorganic as well as organic nutrients run into the surrounding hence the COD and BOD are high. The loadings of phosphorus (P) and nitrogen (N) into water systems in excess cause eutrophication which in turn creates problems such as more solid waste produced and bad smelly air (Chandran et al., 2023; Mohsenpour et al., 2021; Nancharaiah & Reddy, 2018). It encourages the growth of harmful bacteria that other aquatic animals could depend on and which in turn makes water that is free from the bacterium to the surrounding regions very unsafe in health standards thereby increasing diseases in the areas adjacent to the waste release. Heavy metals comprise the most pervasive category of wastes found in the wastewater. Their direct or indirect component inclusion such as inhalation, ingestion, or skin contact even after toxic level threshold is low, poses health wellbeing issues and potential cancer risks (Gupta, 2024; Katpatal et al., 2024). Also, improved augmentation of the wastewater treatment processes is promoted by the symbiotic character of the fungi-algae system. This method enhances the water quality improvement, and can be used to produce useful biomass, which may be used to produce bio-energy or fed to hydrocarbon ferments.

3.7.4 Fungi-Membrane Bioreactors (F-MBR)

Fungi-membrane bioreactors (F-MBRs) are very efficient for treating high-strength industrial, pharmaceutical, and dye effluents. In these systems, fungi like *Aspergillus flavus*, are integrated with membrane filtration units in such a way that both the components work hand-in-hand. The fungus breaks down the hazardous organic materials introduced, while the membrane retains the fungi preventing its wash out and hence a greater biodegradation activity is achieved (Selim et al., 2021). The approach mitigates the level of pollutants present while ensuring that there is still active fungal population to continue responding to the pollutant. Incorporating membrane bioreactor systems to supplement the degradation of fungal organisms, F-MBRs has several advantages over conventional waste management systems. Such systems have demonstrated better handling of recalcitrant pollutants, ease of operation and decreased solid waste effluent (Bala et al., 2022). As well, the mycelium held in the membrane may serve a variety of purposes from producing bio fuels or being a storehouse for useful metabolites (Thirumalaivasan et al., 2024). Even more, the efficiency of course F-MBR systems can be improved by optimizing some parameters, for example, hydraulic retention time, pH, temperature and so on.

Also, the choice of fungi species and membrane types is crucial for effective pollutant removal and smooth operation of the system. Such aspects may include assessing the viability and operational performance of F-MBR systems with a view to their commercial implementation in the future or assessing the performance of the treatment processes over time (Chandran et al., 2023; Jijingi et al., 2024; Selim et al., 2021).

3.7.5 Fungi-Biosorbent Hybrid Systems

Unusual fungal biosorbents have been utilized for the effective elimination of heavy metal pollutants present in wastewater. In this case, the fungi Mucor racemosus and Penicillium chrysogenum were used along with biochar the biosorbent material. Fungi serve to degrade organic contaminants, while biosorbent materials serve to cowl with the heavy metals thus enabling the highest efficiency to both organic and inorganic pollutants removal in a single treatment phase (Ramya et al., 2021). The effectiveness of combining various types of fungi with the natural absorbent materials in setups gives a bright prospect for complete treatment of the wastewater. This approach not only addresses pollutants concurrently but is also promising in shortening the treatment period and costs associated with the whole process (Dusengemungu et al., 2020). Moreover, the inclusion of such systems within the existing treatment plants can significantly enhance their performance and increase the handling capacity of complex industrial effluents.

3.7.6 Fungi-Oxidant Hybrid Systems

Biological agents and chemical oxidants are often combined to treat oil refineries or petrochemical plants' effluents. An example will be the use of fungi such as Cladosporium spp. with hydrogen peroxide or ozone. Hydrocarbon degrading fungi are responsible for the primary degradation of pollutants while oxidants degrade the resistant hydrocarbon chains ensuring that the process is accelerated and completed to the fullest in the breakdown of complex hydrocarbons (Ghosh et al., 2023). The introduction of fungi into the oxidant systems created hybrid systems that proved to be more efficient and complete in the treatment process considering that the compounds were non-biodegradable. Such improvements in the treatments can also be done by deploying particular fungal strains that are effective in respect to the pollutants and the environmental settings (Chandran et al., 2023). Several support materials can be used to immobilize fungus and increase efficiency in treatment by enhancing its functional dependability (Kumla et al., 2020; Najim et al., 2024).

3.7.7 Fungi-Anoxic/Anaerobic Systems

In fungi-anoxic/anaerobic systems for example, fungi complement anaerobic bacteria in the treatment of organic and hydrocarbon laden waste waters. When organisms of this level treat organic wastes, they convert them into relatively simple chemical compounds biosynthetic activity of bacteria and fungi degrade more complex compounds like hydrocarbons, which bacterial degraded (Thirumalaivasan et al., 2024). Hence this strategy helps in the lowering of biochemical oxygen demand (BOD) and chemical oxygen demand (COD), thus improving the treatment efficiency for severely polluted wastewater. These hybrid systems combine the efficacy of fungi and its other components for effective waste water treatment (Zahmatkesh et al., 2023). The mycobacterial association in these hybrid systems is also applicable in a different setting: that of effluent and its treatment. Such interactions among the microorganisms can be used in soil remediation as well where fungal species and bacteria degrade pollutants present in soils. This model of development is also applicable in factories discharging different effluents into rivers after their respective manufacturing processes (Vaksmaa et al., 2023).

3.8 Conclusion

Fungal treatment in combined systems of wastewater treatment allows to eliminate difficult and recalcitrant contaminants from effluents of industrial and municipal origin sustainably. Fungus has unique enzymatic and biosorptive abilities that are capable of degrading and eliminating contaminants which cannot be removed by the conventional treatment process. Stages of the biodegradation process are considered impeded by the ability of the fungi to degrade complex organic substrates while simple regeneration of constituents occurs with bacteria. In waste management, the application of fungi, especially Ascomycota, Basidiomycota, and Mucoromycota, is due to their tenacity and resilience characteristics. Wastewater is broken down by microorganisms which utilize organic trash and toxic metals. In detoxifying plants, it is possible to modify the cell morphology and functions to enable survival in conditions of low pH and high salinity or low oxygen concentration. *Aspergillus*, *Penicillium*, *Trichoderma*, and *Saccharomyces cerevisiae* accelerate the formation of biofilms, increase the enzymatic activity, and accelerate waste degradation. Enzymatic activity of fungi makes bacterial destruction of pollutants more efficient by providing additional degradation capabilities hence improving the bioremediation alternatives. Fungi are biological agents for stabilization ponds often classified by economic purposes. Fungi are able to treat wastewaters and at the same time grow single-cell protein, enzymes, and organic acids, hence making the processes of averting pollution both active and economically sustainable. It is necessary to carry out research aimed at the amelioration of both the techniques used for the treatment of wastewater using fungi and the reasons hindering the scaling up of such processes, and explore the potential of certain fungi. These technological advancements will not only boost the efficacy of bioremediation but will also promote eco-friendly practices as well ease the problem associated with the industrial treatment of wastewater. The creativity of these hybrid systems lies in the diverse enzymatic activities, environmental tolerance, and numerous metabolic capabilities of fungi. These fungi are less toxic for the environment than the chemical treatments and they can biodegrade complex pollutants. In the conditions of the increasing industrial pollution, a hybrid system of wastewater treatment with filamentous fungi is very sustainable. They can considerably reduce the environmental burden, improve the water, and allow for the remediation techniques that are profitable both for the industry and for the ecology. Fungal activity sludge Dynamics enhances the degradation of lignin-like recalcitrant to wastewater treatment. Fungal algal systems take advantage of excess nutrients and organic pollutants from agricultural waste waters in order to save aged water bodies from nuisance eutrophication. In contrast to pure systems, hybrid fungal-based waste water treatment systems are more effective in dealing with industrial and municipal complex pollutants. These strategies manage industrial waste water treatment in a cost efficient, effective and sustainable manner, thus allowing for industrial development and environmental conservation to co-exist.

Acknowledgement We tribute our heartily thanks to our Prof (Dr.) Saurabh Varshney, Executive Director and CEO of AIIMS Deoghar for his valuable suggestions during the preparation of the chapter.

References

Abdelhamid, M. A., Khalifa, H. O., Yoon, H. J., et al. (2024). Microbial immobilized enzyme biocatalysts for multipollutant mitigation: Harnessing nature's toolkit for environmental sustainability. *International Journal of Molecular Sciences, 25*(16), 8616.

Abo Nahas, H. H., Mansour, S. A., Abo Nouh, F. A., et al. (2021). Fungal Laccases to Where and Where? In *Industrially important fungi for sustainable development: Volume 2: Bioprospecting for biomolecules* (pp. 205–262). Springer.

Akerman-Sanchez, G., & Rojas-Jimenez, K. (2021). Fungi for the bioremediation of pharmaceutical-derived pollutants: A bioengineering approach to water treatment. *Environmental Advances, 4*, 100071.

Akkoyunlu, B., Daly, S., & Casey, E. (2021). Membrane bioreactors for the production of value-added products: Recent developments, challenges and perspectives. *Bioresource Technology, 341*, 125793.

Ali, M., Husain, Q., & Ishqi, H. M. (2019). Fungal peroxidases mediated bioremediation of industrial pollutants. In *Fungal bioremediation* (pp. 22–61). CRC Press.

Ameen, F., Alsarraf, M. J., & Stephenson, S. L. (2024). Bioremediation petroleum wastewater and oil-polluted soils by the non-toxigenic indigenous fungi. *World Journal of Microbiology and Biotechnology, 40*(11), 336.

Antunes, F., Marçal, S., Taofiq, O., et al. (2020). Valorization of mushroom by-products as a source of value-added compounds and potential applications. *Molecules, 25*(11), 2672.

Aragaw, T. A., Bogale, F. M., & Tesfaye, E. L. (2024). Oxidative ligninolytic enzymes and their role in textile dye biodegradation: A comprehensive review. *Water Practice & Technology, 19*(9), 3598–3630.

Arregui, L., Ayala, M., Gómez-Gil, X., et al. (2019). Laccases: Structure, function, and potential application in water bioremediation. *Microbial Cell Factories, 18*, 1–33.

Arvanitoyannis, I. S., & Tserkezou, P. (2008). Cereal waste management: Treatment methods and potential uses of treated waste. In *Waste management for the food industries* (pp. 629–702). Elsevier.

Ayele, A., Haile, S., Alemu, D., et al. (2021). Comparative utilization of dead and live fungal biomass for the removal of heavy metal: A concise review. *The Scientific World Journal, 2021*(1), 5588111.

Bala, S., Garg, D., Thirumalesh, B. V., et al. (2022). Recent strategies for bioremediation of emerging pollutants: A review for a green and sustainable environment. *Toxics, 10*(8), 484.

Blaga, A. C., Zaharia, C., & Suteu, D. (2021). Polysaccharides as support for microbial biomass-based adsorbents with applications in removal of heavy metals and dyes. *Polymers, 13*(17), 2893.

Chandran, P., Suresh, S., Balasubramain, B., et al. (2023). Biological treatment solutions using bioreactors for environmental contaminants from industrial waste water. *Journal of umm Al-Qura University for Applied Sciences, 11*, 3.

Chandrashekhar, H., & Rao, J. V. (2010). An overview of fermenter and the design considerations to enhance its productivity. *Pharmacology, 1*, 261–301.

Chatterjee, S., & Mohan, S. V. (2022). Fungal biorefinery for sustainable resource recovery from waste. *Bioresource Technology, 345*, 126443.

Chaurasia, P. K., Nagraj, S., et al. (2023). Fungal assisted bio-treatment of environmental pollutants with comprehensive emphasis on noxious heavy metals: Recent updates. *Biotechnology and Bioengineering, 120*(1), 57–81.

d'Enfert, C., Kaune, A. K., Alaban, L. R., et al. (2021). The impact of the fungus-host-microbiota interplay upon Candida albicans infections: Current knowledge and new perspectives. *FEMS Microbiology Reviews, 45*(3), fuaa060.

Dhevagi, P., Ramya, A., Priyatharshini, S., et al. (2021). Industrially important fungal enzymes: productions and applications. In *Recent trends in mycological research: Volume 2: Environmental and industrial perspective* (pp. 263–309). Springer.

Dinakarkumar, Y., Gnanasekaran, R., Reddy, G. K., et al. (2024). Fungal bioremediation: An overview of the mechanisms, applications and future perspectives. *Environmental Chemistry and Ecotoxicology, 6*, 293–302.

Dusengemungu, L., Kasali, G., Gwanama, C., et al. (2020). Recent advances in biosorption of copper and cobalt by filamentous fungi. *Frontiers in Microbiology, 11*, 582016.

Dwivedi, S. K. (2023). Fungi mediated detoxification of heavy metals: Insights on mechanisms, influencing factors and recent developments. *Journal of Water Process Engineering, 53*, 103800.

El-Gendi, H., Saleh, A. K., Badierah, R., et al. (2021). A comprehensive insight into fungal enzymes: Structure, classification, and their role in mankind's challenges. *Journal of Fungi, 8*(1), 23.

Gajewska, J., Floryszak-Wieczorek, J., Sobieszczuk-Nowicka, E., et al. (2022). Fungal and oomycete pathogens and heavy metals: An inglorious couple in the environment. *IMA Fungus, 13*(1), 6.

Gałązka, A.;, Jankiewicz, U., & Szczepkowski, A. (2023). Biochemical characteristics of laccases and their practical application in the removal of xenobiotics from water. *Applied Sciences, 13*(7), 4394.

Ganguly, D., Kumar, P., Kumari, A., et al. (2024). Importance of microbial consortia and green chemistry in the removal of xenobiotics from the environment. In *Role of green chemistry in ecosystem restoration to achieve environmental sustainability* (pp. 11–21). Elsevier.

Gaur, V. K., Tripathi, V., & Manickam, N. (2022). Bacterial-and fungal-mediated biodegradation of petroleum hydrocarbons in soil. In *Development in wastewater treatment research and processes* (pp. 407–427). Elsevier.

Ghosh, S., Rusyn, I., Dmytruk, O. V., et al. (2023). Filamentous fungi for sustainable remediation of pharmaceutical compounds, heavy metal and oil hydrocarbons. *Frontiers in Bioengineering and Biotechnology, 11*, 1106973.

Gupta, P. (2024). *Environmental health and occupational safety*. CRC Press.

Gururani, P., Bhatnagar, P., Kumar, V., et al. (2022). Algal consortiums: A novel and integrated approach for wastewater treatment. *Water, 14*(22), 3784.

Hatakka, A. (1994). Lignin-modifying enzymes from selected white-rot fungi: Production and role from in lignin degradation. *FEMS Microbiology Reviews, 13*(2-3), 125–135.

Hualpa-Cutipa, E., Acosta, R. A. S., Luis-Alaya, B., et al. (2023). Extremophilic yeasts and their potential in bioremediation of polluted environments. In *Advances in yeast biotechnology for biofuels and sustainability* (pp. 521–542). Elsevier.

Jamir, I., Ezung, L. Y., Merry, L., et al. (2024). Heavy metals clean up: The application of fungi for biosorption. *Geomicrobiology Journal, 41*(3), 201–212.

Jayaraman, S., Yadav, B., Dalal, R. C., et al. (2024). Mushroom farming: A review focusing on soil health, nutritional security and environmental sustainability. *Farming System, 2*(3), 100098.

Jha, H. (2019). Fungal diversity and enzymes involved in lignin degradation. In *Mycodegradation of lignocelluloses* (pp. 35–49). Springer.

Jijingi, H. E., Yazdia, S. K., Abakr, Y. A., et al. (2024). Evaluation of membrane bioreactor (MBR) technology for industrial wastewater treatment and its application in developing countries: A review. *Case Studies in Chemical and Environmental Engineering, 10*, 100886.

Kajla, S., Nagi, G. K., & Kumari, R. (2021). Microorganisms employed in the removal of contaminants from wastewater of iron and steel industries. *Rendiconti Lincei. Scienze Fisiche e Naturali, 32*, 257–272.

Karthika, A., Seenivasagan, R., Kasimani, R., et al. (2024). Microbial technologies for sustainable textile effluent treatment: A review. *Journal of Environmental Chemical Engineering, 12*, 113275.

Katpatal, A., Tandulkar, D., Kularkar, A., et al. (2024). Accumulation, processing, and destruction of waste: Environmental toxicities, levels, and assessment method. In *Advances in energy from waste* (pp. 81–113). Woodhead Publishing.

Kesari, K. K., Soni, R., Jamal, Q. M. S., et al. (2021). Wastewater treatment and reuse: A review of its applications and health implications. *Water, Air, & Soil Pollution, 232*, 1–28.

Khan, M. F., Hof, C., Niemcová, P., et al. (2023). Recent advances in fungal xenobiotic metabolism: Enzymes and applications. *World Journal of Microbiology and Biotechnology, 39*(11), 296.

Kues, U., & Ruhl, M. (2011). Multiple multi-copper oxidase gene families in basidiomycetes-what for? *Current Genomics, 12*(2), 72–94.

Kumar, M., Bhoi, B., & Kumar, H. (2024). Fungi's involvement in metal NPS synthesis and environmentally sustainable practices. In *Biogenic wastes-enabled nanomaterial synthesis: Applications in environmental sustainability* (pp. 149–170). Springer Nature.

Kumar, M., Kumari, P., Kumari, A., et al. (2024). Occupancy of flavonoids in plants, pathogen, and opportunistic fungi interactions. In *Opportunistic fungi, nematode and plant interactions: Interplay and mechanisms* (pp. 125–145). Springer Nature.

Kumar, S., & Hashmi, M. Z. (Eds.). (2021). *Biological approaches to controlling pollutants*. Woodhead Publishing.

Kumari, A., Kumar, M., & Bhoi, B. (2024). Role of bacterial degradation in lignocellulosic biomass for biofuel production. In *Valorization of biomass wastes for environmental sustainability: Green practices for the rural circular economy* (pp. 303–315). Springer Nature.

Kumari, R., Singh, A., & Yadav, A. N. (2021). Fungal enzymes: degradation and detoxification of organic and inorganic pollutants. In *Recent trends in mycological research: Volume 2: Environmental and industrial perspective* (pp. 99–125). Springer.

Kumla, J., Suwannarach, N., Sujarit, K., et al. (2020). Cultivation of mushrooms and their lignocellulolytic enzyme production through the utilization of agro-industrial waste. *Molecules, 25*(12), 2811.

Latif, W., Ciniglia, C., Iovinella, M., et al. (2023). Role of white rot fungi in industrial wastewater treatment: A review. *Applied Sciences, 13*(14), 8318.

Liu, H., Xu, L., Bao, X., et al. (2022). Microbial fermentation for biodegradation and biotransformation of waste plastics into high value–added chemicals. In *Biomass, biofuels, biochemicals* (pp. 395–412). Elsevier.

Mahboob, S., Tahir, K., & Ali, S. (2022). A systematic overview on the upstreaming, downstreaming and industrial applications of microbial lipases. *International Journal of Biology and Biotechnology, 19*, 171–182.

Malik, S., Bora, J., Nag, S., et al. (2023). Fungal-based remediation in the treatment of anthropogenic activities and pharmaceutical-pollutant-contaminated wastewater. *Water, 15*(12), 2262.

Mohamed, H. I., Aal, M. H. A., & El-Mahdy, O. M. (2024). Fungal metabolites and their role in remediation of environmental pollution. In *Fungal secondary metabolites* (pp. 283–315). Elsevier.

Mohiuddin, O., Harvey, A., Ledesma, M. T. O., et al. (2024). Bioremediation of waste by yeast strains. *Electronic Journal of Biotechnology., 69*, 30–42.

Mohsenpour, S. F., Hennige, S., Willoughby, N., et al. (2021). Integrating micro-algae into wastewater treatment: A review. *Science of the Total Environment, 752*, 142168.

Mónica, M. G. A., & Jaime, M. Q. (2019). Phenoloxidases of fungi and bioremediation. In *Fungal bioremediation* (pp. 62–90). CRC Press.

More, T. T., Yan, S., Tyagi, R. D., et al. (2010). Potential use of filamentous fungi for wastewater sludge treatment. *Bioresource Technology, 101*(20), 7691–7700.

Mosnier, A., Schmidt-Traub, G., Obersteiner, M., et al. (2023). How can diverse national food and land-use priorities be reconciled with global sustainability targets? Lessons from the FABLE initiative. *Sustainability Science, 18*(1), 335–345.

Naher, L., Fatin, S. N., Sheikh, M. A. H., et al. (2021). Cellulase enzyme production from filamentous fungi Trichoderma reesei and aspergillus awamori in submerged fermentation with rice straw. *Journal of Fungi, 7*(10), 868.

Najim, A. A., Radeef, A. Y., Al Doori, I., et al. (2024). Immobilization: The promising technique to protect and increase the efficiency of microorganisms to remove contaminants. *Journal of Chemical Technology & Biotechnology, 99*, 1707–1733.

Nancharaiah, Y. V., & Reddy, G. K. K. (2018). Aerobic granular sludge technology: Mechanisms of granulation and biotechnological applications. *Bioresource Technology, 247*, 1128–1143.

Nandan, S. B. (2017). *Dynamics of biocenosis and its spatio-temporal variability in Maranchery Kole wetland, Kerala* (Doctoral dissertation, Cochin University of Science and Technology).

Navina, B. K., Velmurugan, N. K., Kumar, P. S., et al. (2024). Fungal bioremediation approaches for the removal of toxic pollutants: Mechanistic understanding for biorefinery applications. *Chemosphere, 350*, 141123.

Negi, B. B., & Das, C. (2023). Mycoremediation of wastewater, challenges, and current status: A review. *Bioresource Technology Reports, 22*, 101409.

Nimkande, V. D., & Bafana, A. (2022). A review on the utility of microbial lipases in wastewater treatment. *Journal of Water Process Engineering, 46*, 102591.

Obaideen, K., Shehata, N., Sayed, E. T., et al. (2022). The role of wastewater treatment in achieving sustainable development goals (SDGs) and sustainability guideline. *Energy Nexus, 7*, 100112.

Parmar, M. (2020). Development of stress tolerant strain of versatile peroxidase producing fungus for increasing paddy straw digestibility (Doctoral dissertation, Punjab Agricultural University, Ludhiana).

Parrilli, E., Tutino, M. L., & Marino, G. (2022). Biofilm as an adaptation strategy to extreme conditions. *Rendiconti Lincei. Scienze Fisiche e Naturali, 33*(3), 527–536.

Patil, A., Chakraborty, S., Yadav, Y., et al. Bioremediation strategies and mechanisms of bacteria for resistance against heavy metals: A review. *Bioremediation Journal*. https://doi.org/10.1080/10889868.2024.2375204

Prabhu, N., Sabour, A. A. A., Rengarajan, S., et al. (2024). Analysis of the remediation competence of aspergillus flavus biomass in wastewater of the dyeing industry: An in-vitro study. *Environmental Research, 252*, 118705.

Prajapati, A. V., AbdelRahim, S. K., Dave, S. R., et al. (2023). Mycosorption: A sustainable approach towards an emerging concern for removal of heavy metal (s) contaminant from aquatic systems. *Journal of Molecular Liquids, 394*, 123712.

Prenafeta-Boldu, F. X., Summerbell, R., & Sybren de Hoog, G. (2006). Fungi growing on aromatic hydrocarbons: Biotechnology's unexpected encounter with biohazard? *FEMS Microbiology Reviews, 30*(1), 109–130.

Pundir, A., Thakur, M. S., Prakash, S., et al. (2024). Fungi as versatile biocatalytic tool for treatment of textile wastewater effluents. *Environmental Sciences Europe, 36*(1), 1–15.

Rajendran, S., Kalairaj, A., & Senthilvelan, T. (2024). A comprehensive review on enzymatic decolorization of various azo dyes using laccase for the abatement of industrial pollution. *Biomass Conversion and Biorefinery, 15*, 13079–13101.

Rajhans, G., Sen, S. K., Barik, A., et al. (2020). Ligninolytic enzyme system of white-rot fungi: A natural approach to bioremediation and detoxification of azo dyes in textile wastewater. *Environmental Engineering and Management Journal, 19*, 1983–2002.

Rambabu, K., Banat, F., Pham, Q. M., et al. (2020). Biological remediation of acid mine drainage: Review of past trends and current outlook. *Environmental Science and Ecotechnology, 2*, 100024.

Ramesh, R., Saran, A., Damodaran, G., et al. (2024). Compostable wastes generated from biomass. In *New technologies for energy transition based on sustainable development goals: Factors contributing to global warming* (pp. 461–493). Springer Nature.

Ramya, D., Kiruba, N. J. M., & Thatheyus, A. J. (2021). Biosorption of heavy metals using fungal biosorbents–a review. In *Fungi bio-prospects in sustainable agriculture, environment and nano-technology* (pp. 331–352). Elsevier.

Rani, M. H. S., Nandana, R. K., Khatun, A., et al. (2024). Three strategy rules of filamentous fungi in hydrocarbon remediation: An overview. *Biodegradation, 35*, 833–861.

Ranjan, P., Kumar, M., Bhardwaj, H., et al. (2024). Application of enzymes in biomass waste management. In *Valorization of biomass wastes for environmental sustainability: Green practices for the rural circular economy* (pp. 189–205). Springer Nature.

Raychaudhuri, A., & Behera, M. (2022). Microbial degradation of lignin: Conversion, application, and challenges. In *Development in wastewater treatment research and processes* (pp. 195–219). Elsevier.

Rezaei, Z., & Moghimi, H. (2024). Fungal-bacterial consortia: A promising strategy for the removal of petroleum hydrocarbons. *Ecotoxicology and Environmental Safety, 280*, 116543.

Risse-Buhl, U., Anlanger, C., & Kalla, K. (2017). The role of hydrodynamics in shaping the composition and architecture of epilithic biofilms in fluvial ecosystems. *Water Research, 127*, 211–222.

Saikia, S., Yadav, M., Hoque, R. A., et al. (2023). Bioremediation mediated by manganese peroxidase–an overview. *Biocatalysis and Biotransformation, 41*(3), 161–173.

Saini, P., & Sareen, D. (2017). An overview on the enhancement of enantioselectivity and stability of microbial epoxide hydrolases. *Molecular Biotechnology, 59*, 98–116.

Santos, G. C., & Corso, C. R. (2014). Comparative analysis of azo dye biodegradation by aspergillus oryzae and Phanerochaete chrysosporium. *Water, Air, & Soil Pollution, 225*, 1–11.

Sar, T., Marchlewicz, A., Harirchi, S., et al. (2024). Resource recovery and treatment of wastewaters using filamentous fungi. *Science of the Total Environment, 951*, 175752.

Saravanan, A., Kumar, P. S., Jeevanantham, S., et al. (2021). Effective water/wastewater treatment methodologies for toxic pollutants removal: Processes and applications towards sustainable development. *Chemosphere, 280*, 130595.

Sarsan, S., Kodaparthi, A., & Birru, S. (2024). Microbial enzymes in plastic degradation. In *Microbial essentialism* (pp. 207–242). Academic.

Schlüter, R., & Schauer, F. (2017). Biotransformation and detoxification of environmental pollutants with aromatic structures by yeasts. In *Yeast diversity in human welfare* (pp. 323–369). Springer.

Selim, M. T., Salem, S. S., Mohamed, A. A., et al. (2021). Biological treatment of real textile effluent using aspergillus flavus and fusarium oxysporium and their consortium along with the evaluation of their phytotoxicity. *Journal of Fungi, 7*(3), 193.

Sen, A., & Nigam, A. (2022). Bioengineering for decolorization of synthetic dyes in textile effluents using microbial enzymes. *Journal of Scientific Research, 66*(3), 66–78.

Serrano, A., Carro, J., & Martínez, A. T. (2020). Reaction mechanisms and applications of aryl-alcohol oxidase. In *The enzymes* (Vol. 47, pp. 167–192). Academic.

Shakya, M., Sharma, P., Meryem, S. S., et al. (2016). Heavy metal removal from industrial wastewater using fungi: Uptake mechanism and biochemical aspects. *Journal of Environmental Engineering, 142*(9), C6015001.

Shankar, D., Sivakumar, D., & Yuvashree, R. (2014). Chromium (VI) removal from tannery industry wastewater using fungi species. *Pollution Research, 33*(3), 505–510.

Shanmugapriya, S., Manivannan, G., Selvakumar, G., et al. (2019). Extracellular fungal peroxidases and laccases for waste treatment: recent improvement. In *Recent advancement in white biotechnology through fungi: Volume 3: Perspective for sustainable environments* (pp. 153–187). Springer.

Silva, A., Delerue-Matos, C., Figueiredo, S. A., et al. (2019). The use of algae and fungi for removal of pharmaceuticals by bioremediation and biosorption processes: A review. *Water, 11*(8), 1555.

Silva, J. A. (2023). Wastewater treatment and reuse for sustainable water resources management: A systematic literature review. *Sustainability, 15*(14), 10940.

Singh, A., Pal, D. B., Mohammad, A., et al. (2022). Biological remediation technologies for dyes and heavy metals in wastewater treatment: New insight. *Bioresource Technology, 343*, 126154.

Singh, A. K., Bilal, M., Iqbal, H. M., et al. (2021). Lignin peroxidase in focus for catalytic elimination of contaminants—A critical review on recent progress and perspectives. *International Journal of Biological Macromolecules, 177*, 58–82.

Singh, L. (2017). Biodegradation of synthetic dyes: A mycoremediation approach for degradation/decolourization of textile dyes and effluents. *Journal of Applied Biology & Biotechnology, 3*(5), 430–435.

Tak, Y., Kaur, M., Tilgam, J., et al. (2022). Microbes assisted bioremediation: A green technology to remediate pollutants. In *Bioremediation of environmental pollutants: Emerging trends and strategies* (pp. 25–52). Springer.

Thirumalaivasan, N., Gnanasekaran, L., Kumar, S., et al. (2024). Utilization of fungal and bacterial bioremediation techniques for the treatment of toxic waste and biowaste. *Frontiers in Materials, 11*, 1416445.

Thomas, B., Mathew, D., & Devaky, K. S. (2020). Nature's green catalyst for environmental remediation, clean energy production, and sustainable development. In *Chemistry and chemical engineering for sustainable development* (pp. 107–137). Apple Academic Press.

Urlacher, V. B., & Koschorreck, K. (2021). Pecularities and applications of aryl-alcohol oxidases from fungi. *Applied Microbiology and Biotechnology, 105*, 4111–4126.

Vaksmaa, A., Guerrero-Cruz, S., Ghosh, P., et al. (2023). Role of fungi in bioremediation of emerging pollutants. *Frontiers in Marine Science, 10*, 1070905.

Vara, S., & Karnena, M. K. (2020). Fungal enzymatic degradation of industrial effluents–a review. *Current Research in Environmental & Applied Mycology, 10*(1), 417–442.

Vishnoi, N., & Dixit, S. (2019). Bioremediation: new prospects for environmental cleaning by fungal enzymes. In *Recent advancement in white biotechnology through fungi: Volume 3: Perspective for sustainable environments* (pp. 17–52). Springer.

Walker, G. M., & White, N. A. (2017). Introduction to fungal physiology. In *Fungi: Biology and applications* (pp. 1–35). Wiley.

Wallraf, A. M. (2018). Protein engineering of laccase for improved resistance towards cosolvents. Doctoral dissertation, Dissertation, RWTH Aachen University.

Yao, W., Cai, D., Huang, F., et al. (2023). Promoting lignin exploitability in compost: A cooperative microbial depolymerization mechanism. *Process Safety and Environmental Protection, 174*, 856–868.

Zahmatkesh, S., Gholian-Jouybari, F., Klemeš, J. J., et al. (2023). Sustainable and optimized values for municipal wastewater: The removal of biological oxygen demand and chemical oxygen demand by various levels of geranular activated carbon-and genetic algorithm-based simulation. *Journal of Cleaner Production, 417*, 137932.

Zhao, C., Deng, L., & Fang, H. (2018). Mixed culture of recombinant Trichoderma reesei and aspergillus Niger for cellulase production to increase the cellulose degrading capability. *Biomass and Bioenergy, 112*, 93–98.

Chapter 4
Removal of Heavy Metal and Degradation of Organic Compounds in Fungi Waste Water

Hani Nasser Abdelhamid

Abstract Water treatment via pollutant removal could be achieved using fungi cells. This book chapter summarizes the removal of heavy metals and degradation of organic compounds in fungi wastewater. Through distinct enzymatic pathways targeting different polymers, it presents contemporary insights into various fungal species contributing to organic degradation, including drugs, plastics, dyes, and pharmaceuticals. The fungi species such as Eurotiomycetes contain the most significant plastic degraders in the Fungi kingdom. Moreover, the existence of particular pro-oxidant ions may enhance the degradation process. The enzyme-mediated oxidation or hydrolysis generates functional groups that enhance the hydrophilicity of polymers, hence degrading the high molecular weight polymer into lower molecular weight fragments. Among several fungi species, white rot fungi have been demonstrated to be the most efficient microbes in decomposing artificial pigments and dyes. The future trends of the series on organic degradation, such as plastic using fungi, will be included. This book chapter highlighted the key factors affecting fungi performance.

Keywords Pathogen · Heavy metal · Removal · Degradation of organic compounds in fungi in waste water treatment

4.1 Introduction

Fungi are omnipresent and found in terrestrial, freshwater, and marine ecosystems, making challenge for environment and health issuses (Abdelhamid, 2020; Abdelhamid et al., 2020; Abdelhamid & Mahmoud, 2023; Kumaran et al., 2020; Lockhart et al., 2023). Fungi display many lifestyles and can adapt to various redox

---

H. N. Abdelhamid (✉)
Department of Chemistry, College of Science, Imam Mohammad Ibn Saud Islamic University (IMSIU), Riyadh, Saudi Arabia
e-mail: hnabdelhamid@imamu.edu.sa

C. M. Hussain et al. (eds.), *Fungi in Waste Water Treatment*, Springer Water,
https://doi.org/10.1007/978-3-032-12693-1_4

situations. Most fungi favor oxic settings (presence of oxygen (O_2)); certain fungal species reside in oxygen-minimum zones. Additionally, anaerobic fungi have been recognized and primarily examined inside gut microbiomes. They can be classified based on morphology as unicellular (e.g., yeasts and flagellated cells like zoospores); filamentous (e.g., molds and mushrooms); or dimorphic (existing in two forms: yeast-like single cells or hyphae). They exist as free-living organisms in mutualistic symbiotic relationships (such as mycorrhizas, lichens, and gut microbiomes) or parasitic pathogens affecting many plants and animals, including humans. The phylogeny and taxonomy of the fungal kingdom remain contentious. Specific fungal phyla are infrequently sampled, and our comprehension of the profoundly branching groupings remains inadequate. The number of fungal species may vary between 2 to 4 million (Hawksworth & Lücking, 2017), although only 100,000 have been documented (Wu et al., 2019). The reader can use an online database, www.marinefungi.org, to search for marine fungal species (Jones et al., 2019). Fungi are also essential for several industrial sectors, including medical and biotechnological applications. For example, penicillium fungi can be used for antibiotic production, i.e., penicillin. *Saccharomyces cerevisiae* can be used in baking and bioethanol production. Fungi are essential sources for enzyme production, such as amylase and protease. They can also be applied to degrade organic pollutants, e.g., hydrocarbons, via the 'Mycoremediation' process.

Water pollution represents a challenge. It can be caused due to several pollutants, including metals species, drugs, dyes, etc. The contaminants can be inorganic, organic, or hybrid molecules (Abdelhamid & Mathew, 2022; Etman et al., 2018; Nasser Abdelhamid & Mathew, 2021). Metals, including heavy metals, play an essential role in living organisms. The term "heavy metal" is defined in literature as metallic elements with a density exceeding water density (1 g/cm^3), specifically more significant than 5 g/cm^3. An excess and a deficiency of critical metal ions are detrimental to plants.

Additionally, certain heavy metals are harmful at all concentrations and impede cell functionality for humans, plants, and microorganisms. Heavy metals are enduring in the environment and are incapable of biodegradation. Consequently, heavy metal contamination poses a significant environmental, agricultural, and human health hazard. Thus, removing water pollutants is essential for water treatment (Abdelhamid & Mathew, 2021a, 2021b; Chen et al., 2022; Nasser Abdelhamid et al., 2022).

Several fungi species were reported for effective degradation of organic pollutants such as plastics, including *Aspergillus* (*nidulans, flavus, glaucus, Aspergillus oryzae, nomius*), *Penicillium griseofulvum*, *Bjerkandera adusta*, *Phanerochaete chrysosporium*, *Cladosporium cladosporioides*, *Pleurotus (abalones, ostreatus, and eryngii)*, and *Agaricus bisporus* which also helps in degradation of plastics by growing on them. The capacity of fungi for organic pollutant degradation depends on the enzymatic activity inside the fungi. These non-specifically enzymes function, allowing fungi to decompose lignin and several refractory chemicals and xenobiotics, including dyes. Fungal and oomycete pathogens exhibit a comparatively high tolerance to heavy metals relative to other microbial species, such as bacteria (Gajewska et al., 2022).

This book chapter outlines the removal of heavy metals and the degradation of organic pollutants in or using fungal effluent. It presents contemporary insights into several fungal species that facilitate the organic breakdown of drugs, plastics, dyes, and pharmaceuticals through specialized enzyme pathways. The fungus class Eurotiomycetes includes the most significant plastic degraders in the Fungi kingdom. Moreover, the existence of particular pro-oxidant ions, such as heavy metals, may expedite the degradation process. The enzyme-mediated oxidation or hydrolysis generates functional groups that enhance the hydrophilicity of polymers, hence degrading the high molecular weight polymer into lower molecular weight fragments. White rot fungi are recognized as effective microorganisms for degrading synthetic pigments and dyes among various fungal species. Future trends in the series on organic degradation will include incorporating fungi for plastic disintegration. Key factors affecting fungal performance will be highlighted.

4.2 Fungi Microorganisms

Fungi are a diverse collection of eukaryotic organisms categorized within the kingdom of Fungi, distinct from plants, animals, and protists. They execute vital ecological, industrial, and environmental roles by digesting organic matter, producing bioactive compounds, and interacting with various ecosystems. Fungi exhibit a unique cellular structure and life cycles, making them versatile in natural and artificial environments, including wastewater systems. They are classified based on their reproductive structures, spore generation, and genetic characteristics (Fig. 4.1). The standard categories consist of (1) Ascomycota (Sac Fungi), e.g., *Aspergillus, Penicillium*, *Fusarium*, and *Saccharomyces*; (2) Basidiomycota (Club Fungi), e.g., mushrooms, rusts, and smuts. Reproduce through basidia and are crucial for lignin degradation; (3) Zygomycota, e.g., molds like Rhizopus. They are responsible for food spoilage and industrial fermentation processes.; (4) Mucoromycota, e.g., Mucor (for the soil nutrient, Fig. 4.1); (5) Chytridiomycota (can be found in aquatic environments, they feature flagellated spores); 6) Deuteromycota (Fungi Imperfecti).

Fungi exhibit unique structural and metabolic characteristics that enhance their survival and adaptability (Fig. 4.1). The cell wall consists of chitin, glucans, and proteins, providing rigidity and protection. Fungi with filamentous morphology exhibit hyphae, which constitute a network known as mycelium. The spores in fungi can serve as reproductive units, enabling dissemination and survival in challenging conditions. Fungi can release enzymes such as cellulases, ligninases, and proteases. These released enzymes can break down organic compounds or complicated molecules such as lignin.

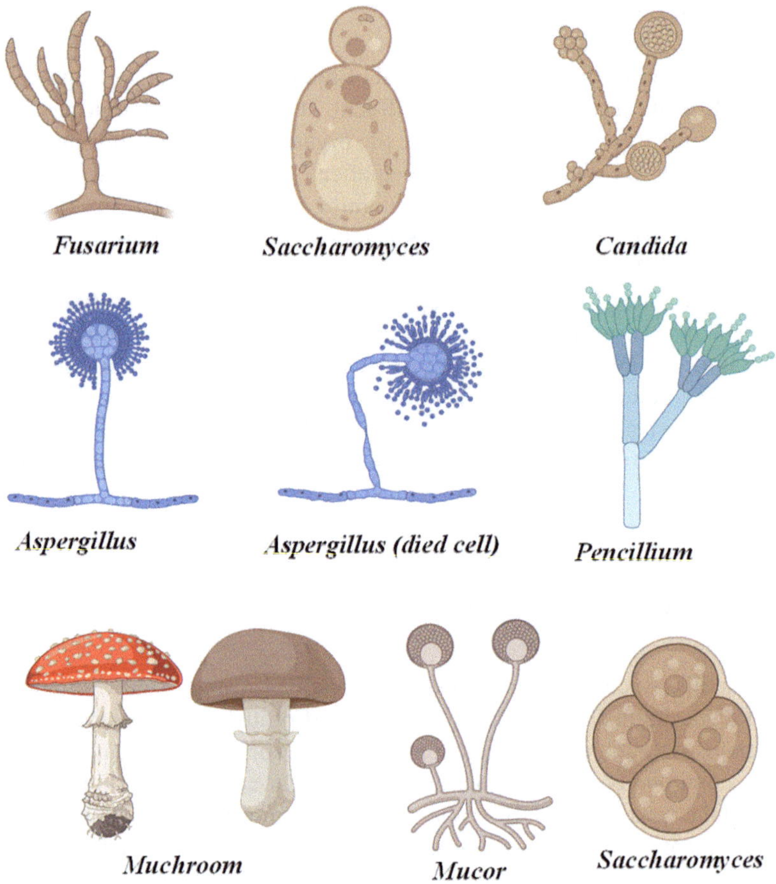

Fig. 4.1 Common fungi reported in the literature

4.3 Presence of Fungi in Wastewater

Fungi are ubiquitous in wastewater treatment systems. They are crucial for the decomposition of organic matter and the elimination of contaminants. Their presence is attributed to their ability to withstand harsh conditions and breakdown-resistant compounds. Fungi degrade complex organic pollutants, including phenols and dyes, through enzymatic processes. Specific fungi can be used as sequesters for heavy metals with the potential to perform heavy metal transformation. Fungi, e.g., *Aspergillus* and *Fusarium*, demonstrate potential for the degradation of microplastics and other persistent pollutants (Table 4.1). This chapter summarizes numerous features and recent developments regarding removing heavy metals (Ghosh et al., 2023), volatile organic compounds (VOCs) (Marycz et al., 2022), and degradation of organic compounds (e.g., pharmaceutical drugs (Akerman-Sanchez &

Table 4.1 summarizes the removal of various pollutants in fungal wastewater treatment, including advantages and disadvantages

Type	Examples	Removal/degradation mechanism	Advantages	Disadvantages
Heavy metals	Pb^{2+}, Hg^{2+}, Cd^{2+}, Cr^{3+}	Biosorption, bio precipitation bioaccumulation, enzymatic reduction	– High efficiency at low concentrations – Low-cost biomass – Environmentally friendly	– Limited capacity for high concentrations – Disposal of metal-laden biomass – Potential toxicity to fungi
Dyes	Azo dyes, methylene blue	Laccase-catalyzed degradation, adsorption	– Efficient degradation with enzymes – Reduction of color and toxicity – Cost-effective	– Enzyme production can be expensive – Limited by pH and temperature ranges
Plastics	Microplastics, polyethylene, polystyrene	Biodegradation, enzymatic oxidation	– Potential to break down into macroplastic (>5 mm), microplastics (1–5 mm), and nanoplastic (1 μm) – Environmentally sustainable	– Slow degradation rates – Limited to specific polymers – Need for pretreatment
Lignocellulose	Cellulose, hemicellulose, lignin	Enzymatic hydrolysis (cellulases, ligninases)	– Generates value-added products (bioethanol, bio-based chemicals) – Reduces waste	– Requires pretreatment – Enzyme production costs – Slow degradation for lignin
Pharmaceuticals	Ibuprofen, paracetamol	Enzymatic oxidation (peroxidases, laccases)	– Targets persistent pollutants – Effective at low concentrations	– Low biodegradability of some compounds – Requires optimization for complex mixtures
Drugs	Steroids	Enzymatic degradation, adsorption	– Reduces toxicity and ecological risks – Efficient at low concentrations	– May form toxic intermediates – Challenges in scaling up
Antibiotics	Tetracyclines,	Enzymatic degradation, sorption	– Reduces antimicrobial resistance – High removal efficiency	– Risk of resistant metabolites – enzyme inactivation by matrix effects
VOCs	Benzene, toluene, formaldehyde	Enzymatic oxidation (oxidases)	– Converts VOCs to CO_2 and water – Sustainable and clean process	– Limited to low concentrations – Sensitive to environmental conditions

Rojas-Jimenez, 2021), plastics (Vaksmaa et al., 2023), and hydrocarbons (Dinakarkumar et al., 2024)) in fungi wastewater (Table 4.1).

4.4 Removal of Heavy Metals and Degradation of Organic Compounds in Fungi Wastewater

Multicellular eukaryotic fungi serve as effective agents of bioremediation (Table 4.1). Mycoremediation has been coined for the bioremediation of hazardous organic compounds. It can be achieved using various fungal groups, e.g., white rot fungi, marine fungi, extremophilic fungi, filamentous fungi such as *Aspergillus* sp. and *Trichoderma* sp.; and symbiotic fungi like *Arbuscular mycorrhizal* fungi. Mycroremediation using fungi is a highly sustainable and environmentally benign method to remediate contaminated locations. Fungi, such as *extremophilic*, can be applied to harsh environmental conditions.

Fungi have demonstrated efficacy as organisms for the bioremediation of wastewater containing organic pollutants. Fungal biomass can be used as an absorber or a source of numerous extracellular enzymes necessary for biodegradation. Biosorption predominantly occurs in non-ligninolytic fungi, including *A. niger*, *A. oryzae*, *A. fumigatus*, and *Saccharomyces cerevisiae*, whereas the majority of research on the biodegradation of textile dyes has been conducted on white rot fungi.

4.4.1 Removal of Heavy Metals

Several fungi species were reported for removing heavy metal ions (Table 4.2). White-rot fungi species were reported for heavy metal contamination cleanup (Chen et al., 2022). They can be used for removing heavy metals such as lead (Pb^{2+}), cadmium (Cd^{2+}), copper (Cu^{2+}), zinc (Zn^{2+}), nickel (Ni^{2+}), mercury (Hg^{2+}), and chromium (Cr^{3+} or Cr^{6+}). Fungi species can be applied to live or dead cells to remove heavy metal ions. They can be used for the adsorption of toxic and nontoxic metal ions. They offer several biomolecules that can capture heavy metal ions. Some fungi species can cause the reduction of heavy metal ions with high oxidation state to low oxidation state. The reduction process could reduce the toxic effects of heavy metals with high oxidation states.

Phanerochaete chrysosporium enabled the remediation of cadmium (Cd^{2+}) and nickel (Ni^{2+}) via adsorption with efficiencies of 96.23% and 89.48%, with corresponding capacities of 71.43 mg/g and 46.50 mg/g for Cd^{2+} ions and Ni^{2+} ions, respectively at optimal conditions (at pH of 6, temperature of 36 °C and after 9 h of mixing)(Noormohamadi et al., 2019). Both live and dried biomass of *Phanerochaete chrysosporium* were reported to remove Cu ions with efficiencies of 40%–60% (Kahraman et al., 2005). Polyethyleneimine (PEI)-modified *Phanerochaete*

Table 4.2 Examples of fungi used for organic removal and their degradation mechanism

Fungal species	Target organic pollutants	Enzymatic activity	Mechanism of degradation
Phanerochaete chrysosporium	Lignin, phenols, dyes, antibiotics	Lignin peroxidase, manganese peroxidase, Laccase	Oxidative cleavage of lignin and aromatic rings through radical formation
Trametes versicolor	Dyes, pharmaceuticals, antibiotics		Electron transfer reactions oxidize phenolic and non-phenolic compounds, forming smaller by-products
Pleurotus ostreatus	Lignin, pharmaceuticals, dyes		Oxidation of aromatic rings and cleavage of complex polymeric structures
Aspergillus niger	VOCs, hydrocarbons, dyes	Laccase, cellulase, pectinase	Hydrolysis of complex organic pollutants and oxidative breakdown of aromatic compounds
Trichoderma reesei	Lignocellulose, pharmaceuticals	Cellulase, hemicellulase	Hydrolysis of cellulose and hemicellulose into sugars
Fusarium oxysporum	Pesticides, dyes	Peroxidases, esterase	Oxidative cleavage of pesticide structures and dye decolorization
Penicillium chrysogenum	Heavy metals (indirectly), pharmaceuticals	Glutathione S-transferase, laccase	Conjugation and oxidation of organic pollutants
Aspergillus flavus	Microplastics, hydrocarbons	Laccase, peroxidase	Oxidative breakdown of polymer chains, forming smaller degradable fragments
Schizophyllum commune	Pharmaceuticals, VOCs	Laccase, peroxidase	Oxidation of aromatic rings and cleavage of persistent organic pollutants
Cladosporium cladosporioides	VOCs, hydrocarbons	Oxidase, laccase	Oxidation of volatile organic compounds into CO_2 and H_2O

chrysosporium enhances the efficacy of hexavalent chromium (Cr^{6+}) elimination from wastewater(Chen et al., 2011). Under ideal conditions (pH of 3), approximately 344.48 mg/g of Cr^{6+} ions were adsorbed. The authors observed a reduction of Cr^{6+} ions into Cr^{3+} ions, with over 32.5% being transformed to trivalent chromium (Cr^{3+}). They also noticed a drop in the adsorption below or above the pH value of 3. They found that the adsorption process followed a pseudo-first-order model(Chen et al., 2011). *Phanerochaete chrysosporium* can also eliminate Pb^{2+} ions via extracellular adsorption and intracellular accumulation with adsorption capacities of 123.8 mg/g and 162.5 mg/g, respectively(Huang et al., 2017). It showed a maximum removal effectiveness of 91.3% when Pb^{2+} ions content in a system is 50 mg/L (Huang et al., 2017).

The heavy metal removal was reported using lived cells of *Pleurotus sajorcaju*, *Agaricus bitorquis (A. bitorquis)*, and *Ganoderma lucidum* for ions of Pb^{2+}, Cu^{2+}, Cr^{3+}, and Cr^{6+} in aqueous solutions and actual hazardous effluents (Hanif & Bhatti,

2015). *A. bitorquis* exhibited superior heavy metal ion remediation capacity relative to the other two strains. The maximal capacities of *A. bitorquis* for the remediation of Cr^{3+}, Pb^{2+}, Cr^{6+}, and Cu^{2+} were 226.6, 208.5, 207.3, and 205.1 mg/g, respectively. The adsorbed metals can be desorbed using sulfuric acid (0.1 M). *A. bitorquis* can effectively remove ions from textile industry effluent in the sequence of $Cr^{3+} > Pb^{2+} > Cr^{6+} > Cu^{2+}$ ions (Hanif & Bhatti, 2015).

Schizophyllum commune (*S. commune*), *Pleurotus ostreatus (P. ostreatus*), and *Leucoagaricus nautilus (L. naucinus)* were reported uranium adsorption. They exhibit uranium removal capabilities of 463.2 ± 38.1 μmol/g, 441.8 ± 79.4 μmol/g, and 123.3 ± 23.6 μmol/g, respectively (Wollenberg et al., 2021). *Schizophyllum commune* was identified as the sole fungus that emitted tryptophan and indole derivatives under the specified experimental conditions in the presence of uranium (VI) (Wollenberg et al., 2021). The authors visualized and localized the uranium associated with fungi after 48 hour incubation period using high-angle annular dark-field (HAADF)-scanning transmission electron microscopy (STEM), abbreviated as HAADF-STEM, for imaging and elemental analysis using energy-dispersive X-ray spectroscopy (EDXS) (Fig. 4.2). *S. commune* and *P. ostreatus* developed needle-like morphology comprising uranium, phosphorus, and nitrogen (Fig. 4.2). Nonetheless,

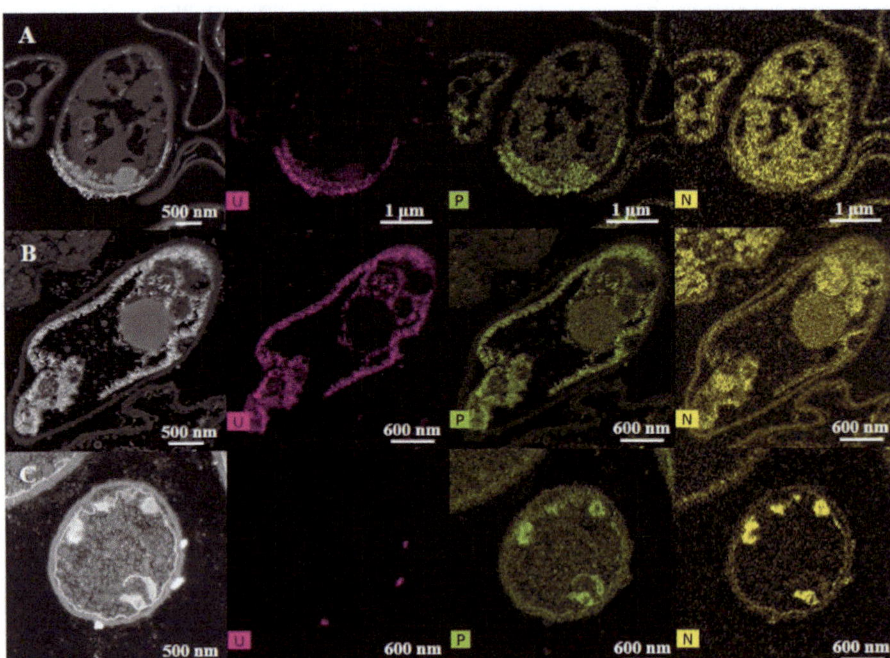

Fig. 4.2 Thin sections analysis using HAADF-STEM images and EDX analysis for (**a**) *S. commune*, (**b**) *P. ostreatus*, and (**c**) *L. naucinus* after 48 h incubation with mineral media with uranium (MMU); U, P, and N refer to uranium, phosphorous, and nitrogen, respectively. Figure reprinted with permission from Wollenberg et al. (2021). © 2021 Elsevier BV

the localization of the needles varied between the two fungus. In *P. ostreatus*, the needles are likely membrane-associated, as they are located within the cell membrane and surrounding cell organelles (Fig. 4.2). In contrast, the uranium was shown to be localized differentially in *L. naucinus*. The uranium was located at certain sites on the exterior of the cell membrane, which contained phosphorus, and nitrogen, and calcium. One can also observe a cluster of uranium within the observed needle-like (Fig. 4.2).

4.4.2 Degradation of Organic Pollutants Using Fungi

Water is widely polluted by organic molecules applied in the industrial sectors. Several organic pollutants were reported in water, including drugs, pharmaceuticals, dyes, and polymers (organic polymers and plastic species). They represent large classes of contaminants in water treatments (Table 4.2). Organic pollutants can be removed via adsorption. However, the efficiency of the adsorption process is low. Thus, the degradation of organic molecules into small fragments is widely used. The degradation process also enables the production of other molecules from organic pollutants. Organic molecule degradation can be beneficial for the circular economy.

Marine fungi, e.g., terrestrial and lacustrine fungal species, include a significant group of microorganisms that exhibit a high capability of metabolizing refractory chemicals, pollutants, and polymers (e.g., polyurethane (PUE), and polyethylene (PE)), including lignin, pesticides, hydrocarbons (Zeghal et al., 2021). *Maritime fungi* may significantly contribute to the breakdown of complex organic materials in maritime environments. Specific fungal strains from the water, such as *Zalerion maritimum*, demonstrated the potential to degrade plastics such as polyethylene (Zeghal et al., 2021).

4.4.3 Removal of Organic Dyes Using Fungi-Based Degradation

Fungi can be applied for the effective degradation of organic dyes (Saxena & Bharagava, 2020). White rot fungi are the common fungi species that can be used for organic dye degradation among several fungi species. The term "white rot" originates from the pale, bleached look of wood affected by these fungi caused by lignin degradation. White rot fungi are filamentous fungi that degrade wood and are found universally in nature. They are classified within the *phylum Basidiomycota* and include several notable taxa such as *Pleurotus*, *Phanerochaete*, *Trametes*, *Ganoderma*, and Lentinus. *Phanerochaete chrysosporium* was considered the first white-rot fungus studied for lignin biodegradation.

Nectriella pironii, isolated from an urban post-industrial textile green area, showed a high capacity to proliferate and maintain metabolic activity in the presence of dye effluent (Góralczyk-Bińkowska et al., 2021). It can eliminate organic dyes such as Reactive Red 120, Acid Orange 7, and the toxic aromatic amine o-tolidine, even at elevated doses (Góralczyk-Bińkowska et al., 2021). It showed the biotransformation of o-tolidine via hydroxylation. The mechanism of the fungi activity depends on the enzymatic activity of laccase and cytochrome P450 enzymes in the fungi that can cause pollutant breakdown. The inclusion of heavy metals in the growing media affected azo dye decolorization; however, the removal efficiency persisted at about 90%. The inclusion of all examined chemicals and landfill leachate in the medium promoted fungal biomass proliferation, with minimal growth restriction noted at 50 mg/L of o-tolidine. The fungus showed significant tolerance to hazardous substances, and its effective pollutant removal abilities render it a suitable choice for wastewater treatment systems that aim to eliminate harmful pollutants from the textile industry (Góralczyk-Bińkowska et al., 2021).

A review was published on the application of white rot fungus for dye decolorization, including its processes and the factors influencing dye decolorization (Herath et al., 2024). White rot fungus exhibits a notable capacity to decompose organic molecules such as dyes. The dye pollutant breakdown of white rot fungus is mainly due to an extracellular enzyme system comprising lignin peroxidase (LiP), manganese-dependent peroxidases (MnP), and laccases. Wang et al. reported the impact of temperature on the decolorization of Congo red using *Ceriporia lacerata* was analyzed within a temperature range of 4–45 °C(Wang et al., 2017). The decolorization rate can be increased with the temperature rise from 4 to 30 °C, achieving 88.50% at 30 °C after 48 h. The optimal dye decolorization occurred at a temperature of 30 °C. The decolorization of dye escalates with temperature due to the increased solubility of enzymes at elevated temperatures, enhancing the enzymatic activity of fungi in the dye degradation. Decolorization efficiency significantly diminishes at temperatures exceeding the optimal range, typically due to a decline in cell viability and enzyme denaturation, i.e., destroy of the three-dimensional structural.

4.4.4 Lignocellulose Degradation

Lignocellulose constitutes the principal component of biomass, accounting for around 50% of the plant matter generated by photosynthesis (e.g., photomass), and serves as the most prevalent renewable organic resource in soil. It comprises three polymer types: cellulose, hemicellulose, and lignin, which are intricately interwoven and chemically connected by non-covalent interactions and covalent cross-linkages. Lignocellulosic residues from wood, grass, agricultural and forestry waste, and municipal solid waste are particularly plentiful in nature and possess significant potential for bioconversion. The accumulation of lignocellulosic materials in substantial quantities, particularly where agricultural residues pose disposal challenges, leads to environmental degradation and the loss of potentially valuable resources for paper production, biomass fuel generation, composting, and animal and human feed, among other uses.

The capacity of fungi to decompose lignocellulosic substances is attributed to their effective enzymatic system (Sánchez, 2009). Fungi possess two categories of extracellular enzymatic systems: the hydrolytic system, which generates hydrolases for polysaccharide breakdown, and a distinctive oxidative extracellular ligninolytic system degrading lignin and cleaving phenyl rings. Using fungi in lignocellulose degradation offers low-cost bioremediation projects that might be attractive given their lignocellulose hydrolysis enzyme machinery. Fungi species such as white rot fungi can be classified into selective and non-selective white rot degraders. Non-selective white rot degraders can decompose all wood components, including cellulose, hemicellulose, and lignin, but selective white rot degraders specifically target lignin while preserving cellulose (Dashtban et al., 2010).

4.4.5 Plastic Degradation

Plastic become a fundamental material for daily living. There are several types of plastic, including polyethylene terephthalate (PET), polyethylene (PE), high-density PE (HDPE), low-density pe (LDPE) polyvinyl chloride (PVC), polypropylene (PP), polyurethane (PU), polystyrene (PS), polylactic acid (PLA), and nylon (Fig. 4.3). Global plastic output currently surpasses several hundred million tons per year. Plastics possess numerous materials, making them widely used, including inexpensive manufacturing expenses, chemical inertness, comparatively lightweight, and durability. The principal drawback of plastics is their exceedingly protracted natural breakdown. Some plastics exhibit high chemical stability and require several thousand years for degradation, causing environmental accumulation. Global plastic trash can be counted as several billion tons in landfills. Consequently, it is imperative to establish an efficient plastic biodegradation technique to expedite the natural decomposition rate of plastics.

Microbes capable of decomposing plastic were isolated from diverse sources, including the rhizosphere soil of mangroves (Fig. 4.3). Bacteria, fungi, and algae are the biological agents that naturally decompose plastic. Over 400 microorganisms have been recognized for degrading plastic. This is the inaugural paper of the series about plastic-degrading fungus. The evolutionary relationships of plastic-degrading fungi were examined utilizing genes such as ITS, LSU, SSU, TEF, RPB1, and RPB2, comprising 395 strains (Ekanayaka et al., 2022). Several classes within the fungal phyla can degrade plastic (Fig. 4.3) (Ekanayaka et al., 2022). The Eurotiomycetes encompass the highest quantity of plastic degraders within the Fungi kingdom.

The enzymatic activity of fungi plays a crucial role in plastic degradation. Several enzymes were reported, including cutinase, lipase, proteases, and lignocellulolytic enzymes (Fig. 4.3) (Srikanth et al., 2022). Additionally, the presence of specific

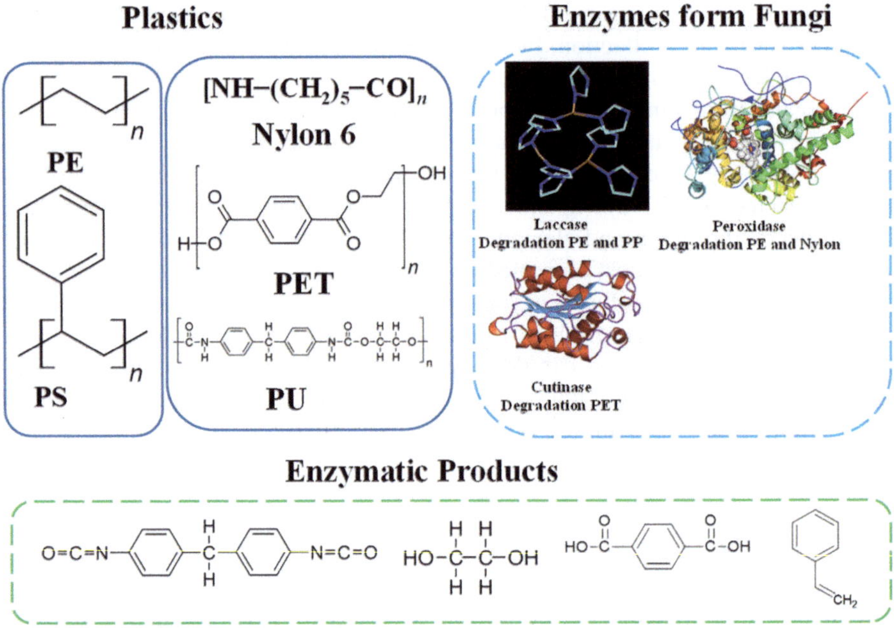

Fig. 4.3 Common plastic structure, enzymes derived from fungi, and the plastic enzymatic products

pro-oxidant ions might facilitate the degradation process. The enzyme-mediated oxidation or hydrolysis generates functional groups that enhance the hydrophilicity of polymers, hence degrading the high molecular weight polymer into lower molecular weight fragments. This results in the deterioration of plastics within a few days. Plastic decomposition is more efficient when photodegradation and thermo-oxidative mechanisms are concurrently active in biodegradation, accelerating and facilitating the process.

Marine fungi are typically under-researched, as seen by the comparatively few studies focused on fungi associated with marine plastic debris (MPD) (Jacquin et al., 2019). The exposure studies of polyethylene (PE) and polystyrene (PS) in the Baltic Sea and the Warnow River examined fungi colonization of these plastic surfaces (Kettner et al., 2019). De Tender et al. executed a 44-week experiment in which PE plastic sheets and dolly ropes were anchored near the sediment at a harbor and an offshore site in the North Sea (De Tender et al., 2017). Analyses of beta diversity in fungal community composition revealed statistically significant impacts of sample type (natural substrate versus plastic polymers), environment, and exposure duration. Fungi organisms were recognized as potential PE degraders in terrestrial ecosystems e.g., *Cladosporium cladosporioides* and *Fusarium redolens*.

The biodegradation of LDPE using fungi was reported using *Aspergillus niger* and *A. terreus* (Sáenz et al., 2019). The degradation method showed a high degradation rate without using co-substrates or photothermal treatment. The degradation of

LDPE under minimal nutritional conditions utilizing both LDPE and sugar as carbon sources. *A. niger* and *A. terreus* achieved weight loss of up to 30%. The presence of cracks and biomass accumulation on the LDPE surface indicated the capability of both fungal species to thrive in low nutritional conditions(Sáenz et al., 2019). However, the authors also reported enhancement in fungi colonization after adding an elevated level of sucrose content to expedite fungal development and enhance colonization on the LDPE film. They observed that Czapek media is an efficient option to enhance plastic decomposition. The in vitro breakdown of LDPE utilizing six strains of *Aspergillus* and two strains of *Fusarium* demonstrated that plastic biodegradation resulted from their enzymes. Polyethene was destroyed using *Pseudomonas*, *Streptomyces*, and *Aspergillus* by 37.09%, 46.16%, and 20.96%, respectively, throughout 6 months in liquid (shaker) culture (Usha et al., 2011). Conversely, *Pseudomonas*, *Streptomyces*, and *Aspergillus* show plastic degradation with the efficiency of 28.42%, 35.78%, and 16.84%, respectively (Usha et al., 2011).

Aspergillus sp. and *Lysinibacillus* sp., exhibiting exceptional capabilities to break down LDPE, were identified from landfill soils (Esmaeili et al., 2013). The biodegradation process was conducted over 126 days in soil utilizing both UV-irradiated and non-UV-irradiated pure LDPE films devoid of pro-oxidant chemicals, with and without the presence of mixed cultures of certain bacteria. The biodegradation in the un-inoculated treatments was slow, about 7.6% and 8.6% of the mineralization measured for the non-UV- and UV-irradiated LDPE, respectively, after 126 days. Conversely, biodegradation was significantly more effective with the selected microorganisms, achieving rates of 29.5% for UV-irradiated films and 15.8% for non-UV-irradiated films. The reduction in the carbonyl index was more pronounced for the UV-irradiated LDPE after biodegradation in soil inoculated with the investigated microorganisms. The analyses demonstrated that the chosen microorganisms could alter and colonize both forms of polyethylene. The fungi degradation is based on the use of polyethylene as the carbon source (Esmaeili et al., 2013).

White rot fungus exhibits maximal efficacy for plastic degradation. Lignin-degrading microbes have been extensively researched for plastic breakdown. Manganese peroxidase (MnP) is the main enzyme in polyethylene degradation. Partially pure manganese peroxidase (MnP) utilized for the treatment of polyethylene membranes in conjunction with Tween 80, manganese (II), and a manganese (III) chelator led to substantial degradation (Iiyoshi et al., 1998). The mechanism of plastic degradation using lignin-degrading fungi is based on the release of Laccases, which are primarily released by catalyzing the oxidation of diverse polyaromatic chemicals.

4.5 Factors Affecting Fungi Performance

Several factors affect the applications of fungi to remove pollutants (Fig. 4.4). These factors can be classified based on fungi, pollutants, experiment setup, and data evaluation or interpretation. The fungi properties should be considered, including

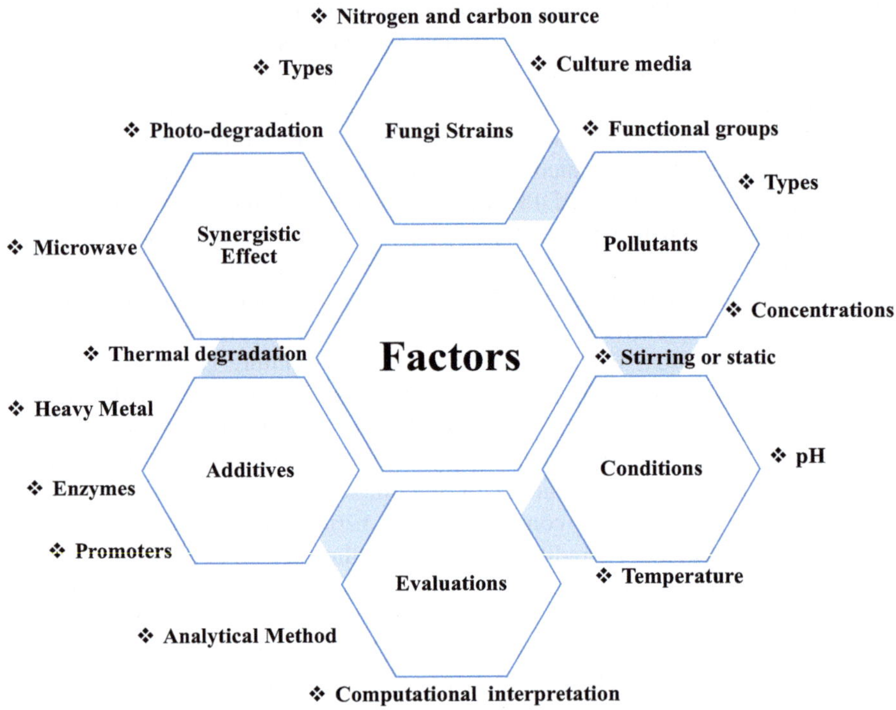

Fig. 4.4 Summary of factors affecting fungi performance for pollutants removal

growth rate, cell media, culture media components or additives, and extracellular enzyme production. Pollutant properties should include types (e.g., heavy metals, dyes, plastic, or drugs), initial concentration, and available functional groups. The experiment setups also play an essential role and should be considered, including pH, temperature, static or non-static (e.g., stirring or shaking), presence of heavy metals or promoters.

The chemical structure of organic pollutants affects the degradation of organic contaminants using fungi. Despite the substantial differences between plastics and lignocellulose, plastics do not possess hydrolyzable bonds, e.g., single bonds C-C, C-N, or C-O bonds, and exhibit more remarkable recalcitrance (i.e., degradation resistance) (Daly et al., 2021). Both comprise hydrophobic polymer mixtures featuring amorphous and crystalline regions, necessitating hydrolases and oxidoreductases for degradation.

The conditions of fungi cell culture affect the prepared cells' quality and performance during pollutant removal. Fungi secrete lignin-modifying enzymes during secondary metabolism. It is widely acknowledged that stressors, such as food deficiency, induce secondary metabolism in white rot fungi, producing lignin-modifying enzymes (Grelska & Noszczyńska, 2020). Consequently, nutrients like carbon and nitrogen must be restricted for effective dye decolorization (Fig. 4.4). Cell culture with low nitrogen and low glucose, e.g., Kirk's medium, is required. Moreover, a study demonstrated that adding nitrogen to the cell culture of *Bjerkandera adusta* did not significantly affect the percentage of dye degradation (Robinson et al., 2001). It is highly required to optimize the cell culture for fungi. Live biomass of *Phanerochaete chrysosporium* exhibited higher adsorption for Cu ions than dried biomass (Kahraman et al., 2005).

The cell culture for fungal cells is significantly influenced by pH, with fungus often thriving at low pH levels between 4 and 5. The pH values of a solution affect both the surface dye's chromophores and their binding sites in the fungal biomass. At lower pH, the fungal biomass, mostly chitin, displays a net positive charge, and these charged sites become available for binding anionic groups such as reactive dyes. pH value is also critical for the enzymatic activity of fungi cells or enzymes. The optimum pH for maximal laccase production by fungi was pH 5.0, and the enzyme yield declined dramatically above pH 7.0. *Daldinia concentrica* and *Xylaria polymorpha* showed 95% and 97% removal efficiencies for Cibacron brilliant red 3B-A dye under optimal pH 4.5 and 6.5 conditions, respectively (Bankole et al., 2018). Moreover, the fungal consortium showed a 99% decolorization rate at pH 4.5. Notably, the dye removal efficiency was reduced when the pH deviated significantly from the ideal values (pH of 4.5–6.5) (Bankole et al., 2018). It is crucial to remember that the pH of textile wastewater falls within the range of 6–9.5. Adjusting pH values for cell culture and experiments for pollutant removal are essential.

The ambient and solutions' temperatures are the factors that influence the enzymatic properties of the fungi cells. The biodegradation efficacy of white rot fungus is maximized at their optimal temperature, essential for growth, reproduction, and enzymatic activity. The optimum development temperature for most of the fungus is reported to be around 25–37 °C. Exceeding the optimal temperature results in diminished degradation activities due to reduced growth and reproduction rates and the inactivation of enzymes responsible for degradation. The decolorization rate of Reactive Black 5 using P. eryngii F032 was increased from 46.37% to 96.77% when the temperature rises from 15 to 40 °C (Hadibarata et al., 2013).

Fungi's performance for plastic degradation can be affected by several environmental factors such as humidity, temperature, pH, salinity, oxygen availability, sunlight, water, stress, and cultural conditions that profoundly influence polymer degradation, microbial populations, and enzymatic activity. Maximum CO_2 production and lignolytic activity were achieved when fungi were grown at the lowest pH. Polyester's chemical and physical properties substantially influence its biodegradability. *A. oryzae* caused a weight loss of 47.2% from LDPE after 3 months of incubation (Konduri et al., 2011).

Combining other treatments with fungi enhances plastic degradation via the synergistic effect. The biodegradation of PE is enhanced by pretreatment with ultraviolet radiation (photo-oxidation), additives, and antioxidants due to the chemical structure of PE resembling that of linear alkane monomers. The decrease in molecular weight and the oxidation of polymers depend on the interaction effects of biotic and abiotic factors, including thermal treatments and photo-oxidation. Thus, a photothermal or chemical pretreatment improves the polymer's surface hydrophilicity by incorporating additional groups, such as carbonyl groups, which are biodegradable by microorganisms. Organic pollutant decomposition can be more efficient when other methods, such as photodegradation and thermo-oxidative mechanisms, are concurrently active in biodegradation, accelerating and facilitating the process.

The location of plastic pollution affects fungi performance. The plastic-associated eukaryotic communities, including fungus, exhibited significant variation based on season and location. The makeup of the fungal population was markedly affected by location. The fungal diversity linked to plastics degradation in the surface waters of the Western South Atlantic and the Antarctic Peninsula were analyzed using three distinct molecular marker genes e.g., ITS2 and 18S rRNA for fungal identification (Lacerda et al., 2020). Sixty-four distinct fungal orders were linked to plastics throughout the examined marker genes. The primers specific to the 18S rRNA gene (V4 and V9 regions) successfully identified a greater quantity of Chytridiomycota. The ITS2 markers were identified several fungi species. The genus Aspergillus exhibited the highest abundance across all samples.

4.6 Conclusions

Fungal cells exhibit several advantages for removing heavy metals and degradation of organic pollutants, including high tolerance to heavy metals relative to other microbial species. They are ubiquitous in nature and are exceedingly diverse. Due to an increased cell-to-surface ratio, fungi have significant physical and enzymatic properties in the environment. Furthermore, the extracellular nature of the fungal enzymes facilitates the tolerance of elevated quantities of organic pollutants, e.g., dyes. Fungi cells exhibit a high rate of pollutant removal. As a result, numerous genera of fungi, i.e., living or dead forms, have been exploited for dye decolorization.

Genetic engineering could enhance the ability of fungi to biodegrade organic pollutants such as plastic. The fungi's performance can be improved using contemporary and advanced techniques for enzyme engineering, encompassing random or site-directed mutagenesis and advanced computer modeling. These technologies may offer genetically engineered robust enzyme systems that may provide an effective solution to reduce plastic waste. Developing fungus-based remediation systems can enhance the comprehension of the mechanisms involved in fungal-mediated pollution removal.

Selective degradation for certain organic pollutants can be achieved using fungi. For example, white root fungi exhibit a high potential for dye degradation. Meanwhile, marine fungi, e.g., Zalerion maritimum, demonstrate a high potential to degrade plastic products such as polyethylene. The experimental parameters should be optimized, including fungi species, selection of the pollutants, investigation of the optimal experimental setup, and analytical method.

References

Abdelhamid, H. N. (2020). Nanocytotoxicity using matrix-assisted laser desorption ionization mass spectrometry. *Future Microbiology, 15*, 385–387.

Abdelhamid, H. N., & Mahmoud, G. A. (2023). Antifungal and nanozyme activities of metal–organic framework-derived CuO@C. *Applied Organometallic Chemistry, 37*, e7011.

Abdelhamid, H. N., Mahmoud, G. A.-E., & Sharmouk, W. (2020). A cerium-based MOFzyme with multi-enzyme-like activity for the disruption and inhibition of fungal recolonization. *Journal of Materials Chemistry B, 8*, 7548–7556.

Abdelhamid, H. N., & Mathew, A. (2022). Cellulose-metal organic frameworks (CelloMOFs) hybrid materials and their multifaceted applications: A review. *Coordination Chemistry Reviews, 451*, 214263.

Abdelhamid, H. N., & Mathew, A. P. (2021a). Cellulose-based materials for water remediation: Adsorption, catalysis, and antifouling. *Frontiers in Chemical Engineering, 3*, 790314.

Abdelhamid, H. N., & Mathew, A. P. (2021b). In-situ growth of zeolitic imidazolate frameworks into a cellulosic filter paper for the reduction of 4-nitrophenol. *Carbohydrate Polymers, 274*, 118657.

Akerman-Sanchez, G., & Rojas-Jimenez, K. (2021). Fungi for the bioremediation of pharmaceutical-derived pollutants: A bioengineering approach to water treatment. *Environmental Advances, 4*, 100071.

Bankole, P. O., Adekunle, A. A., & Govindwar, S. P. (2018). Biodegradation of a monochlorotriazine dye, cibacron brilliant red 3B-A in solid state fermentation by wood-rot fungal consortium, Daldinia concentrica and Xylaria polymorpha. *International Journal of Biological Macromolecules, 120*, 19–27.

Chen, G.-Q., Zhang, W.-J., Zeng, G.-M., Huang, J.-H., Wang, L., & Shen, G.-L. (2011). Surface-modified Phanerochaete chrysosporium as a biosorbent for Cr(VI)-contaminated wastewater. *Journal of Hazardous Materials, 186*, 2138–2143.

Chen, L., Zhang, X., Zhang, M., Zhu, Y., & Zhuo, R. (2022). Removal of heavy-metal pollutants by white rot fungi: Mechanisms, achievements, and perspectives. *Journal of Cleaner Production, 354*, 131681.

Daly, P., Cai, F., Kubicek, C. P., Jiang, S., Grujic, M., Rahimi, M. J., Sheteiwy, M. S., Giles, R., Riaz, A., de Vries, R. P., Akcapinar, G. B., Wei, L., & Druzhinina, I. S. (2021). From lignocellulose to plastics: Knowledge transfer on the degradation approaches by fungi. *Biotechnology Advances, 50*, 107770.

Dashtban, M., Schraft, H., Syed, T. A., & Qin, W. (2010). Fungal biodegradation and enzymatic modification of lignin. *International Journal of Biochemistry and Molecular Biology, 1*, 36–50.

De Tender, C., Devriese, L. I., Haegeman, A., Maes, S., Vangeyte, J., Cattrijsse, A., Dawyndt, P., & Ruttink, T. (2017). Temporal dynamics of bacterial and fungal colonization on plastic debris in the North Sea. *Environmental Science & Technology, 51*, 7350–7360.

Dinakarkumar, Y., Ramakrishnan, G., Gujjula, K. R., Vasu, V., Balamurugan, P., & Murali, G. (2024). Fungal bioremediation: An overview of the mechanisms, applications and future perspectives. *Environmental Chemistry and Ecotoxicology, 6*, 293–302.

Ekanayaka, A. H., Tibpromma, S., Dai, D., Xu, R., Suwannarach, N., Stephenson, S. L., Dao, C., & Karunarathna, S. C. (2022). A review of the fungi that degrade plastic. *Journal of Fungi, 8*, 772.

Esmaeili, A., Pourbabaee, A. A., Alikhani, H. A., Shabani, F., & Esmaeili, E. (2013). Biodegradation of low-density polyethylene (LDPE) by mixed culture of *Lysinibacillus xylanilyticus* and *Aspergillus niger* in soil. *PLoS One, 8*, e71720.

Etman, A. S., Abdelhamid, H. N., Yuan, Y., Wang, L., Zou, X., & Sun, J. (2018). Facile water-based strategy for synthesizing MoO3– X nanosheets: Efficient visible light photocatalysts for dye degradation. *ACS Omega, 3*, 2193–2201.

Gajewska, J., Floryszak-Wieczorek, J., Sobieszczuk-Nowicka, E., Mattoo, A., & Arasimowicz-Jelonek, M. (2022). Fungal and oomycete pathogens and heavy metals: An inglorious couple in the environment. *IMA Fungus, 13*, 6.

Ghosh, S., Rusyn, I., Dmytruk, O. V., Dmytruk, K. V., Onyeaka, H., Gryzenhout, M., & Gafforov, Y. (2023). Filamentous fungi for sustainable remediation of pharmaceutical compounds, heavy metal and oil hydrocarbons. *Frontiers in Bioengineering and Biotechnology, 11*, 1106973.

Góralczyk-Bińkowska, A., Długoński, A., Bernat, P., Długoński, J., & Jasińska, A. (2021). Environmental and molecular approach to dye industry waste degradation by the ascomycete fungus *Nectriella pironii*. *Scientific Reports, 11*, 23829.

Grelska, A., & Noszczyńska, M. (2020). White rot fungi can be a promising tool for removal of bisphenol a, bisphenol S, and nonylphenol from wastewater. *Environmental Science and Pollution Research, 27*, 39958–39976.

Hadibarata, T., Adnan, L. A., Yusoff, A. R. M., Yuniarto, A., Rubiyatno, Zubir, M. M. F. A., Khudhair, A. B., Teh, Z. C., & Naser, M. A. (2013). Microbial decolorization of an azo dye reactive black 5 using white-rot fungus *Pleurotus eryngii* F032. *Water, Air, & Soil Pollution, 224*, 1595.

Hanif, M. A., & Bhatti, H. N. (2015). Remediation of heavy metals using easily cultivable, fast growing, and highly accumulating white rot fungi from hazardous aqueous streams. *Desalination and Water Treatment, 53*, 238–248.

Hawksworth, D. L., & Lücking, R. (2017). Fungal diversity revisited: 2.2 to 3.8 million species. *Microbiology Spectrum, 5*, 52.

Herath, I. S., Udayanga, D., Jayasanka, D. J., & Hewawasam, C. (2024). Textile dye decolorization by white rot fungi—A review. *Bioresource Technology Reports, 25*, 101687.

Huang, C., Lai, C., Xu, P., Zeng, G., Huang, D., Zhang, J., Zhang, C., Cheng, M., Wan, J., & Wang, R. (2017). Lead-induced oxidative stress and antioxidant response provide insight into the tolerance of Phanerochaete chrysosporium to lead exposure. *Chemosphere, 187*, 70–77.

Iiyoshi, Y., Tsutsumi, Y., & Nishida, T. (1998). Polyethylene degradation by lignin-degrading fungi and manganese peroxidase. *Journal of Wood Science, 44*, 222–229.

Jacquin, J., Cheng, J., Odobel, C., Pandin, C., Conan, P., Pujo-Pay, M., Barbe, V., Meistertzheim, A.-L., & Ghiglione, J.-F. (2019). Microbial ecotoxicology of marine plastic debris: A review on colonization and biodegradation by the "plastisphere". *Frontiers in Microbiology, 10*, 865.

Jones, E. B. G., Pang, K.-L., Abdel-Wahab, M. A., Scholz, B., Hyde, K. D., Boekhout, T., Ebel, R., Rateb, M. E., Henderson, L., Sakayaroj, J., Suetrong, S., Dayarathne, M. C., Kumar, V., Raghukumar, S., Sridhar, K. R., Bahkali, A. H. A., Gleason, F. H., & Norphanphoun, C. (2019). An online resource for marine fungi. *Fungal Diversity, 96*, 347–433.

Kahraman, S., Asma Hamamci, D., Erdemoglu, S., & Yesilada, O. (2005). Biosorption of copper(II) by live and dried biomass of the white rot fungi *Phanerochaete chrysosporium* and *Funalia trogii*. *Engineering in Life Sciences, 5*, 72–77.

Kettner, M. T., Oberbeckmann, S., Labrenz, M., & Grossart, H.-P. (2019). The eukaryotic life on microplastics in brackish ecosystems. *Frontiers in Microbiology, 10*, 538.

Konduri, M. K. R., Koteswarareddy, G., Rohini Kumar, D. B., Venkata Reddy, B., & Lakshmi Narasu, M. (2011). Effect of pro-oxidants on biodegradation of polyethylene (LDPE) by indigenous fungal isolate, aspergillus oryzae. *Journal of Applied Polymer Science, 120*, 3536–3545.

Kumaran, S., Abdelhamid, H. N., Hasan, N., & Wu, H.-F. (2020). Cytotoxicity of palladium nanoparticles against aspergillus Niger. *Nanoscience and Nanotechnology - Asia, 10*, 80–85.

Lacerda, A. L. D. F., Proietti, M. C., Secchi, E. R., & Taylor, J. D. (2020). Diverse groups of fungi are associated with plastics in the surface waters of the Western South Atlantic and the Antarctic peninsula. *Molecular Ecology, 29*, 1903–1918.

Lockhart, S. R., Chowdhary, A., & Gold, J. A. W. (2023). The rapid emergence of antifungal-resistant human-pathogenic fungi. *Nature Reviews. Microbiology, 21*, 818–832.

Marycz, M., Brillowska-Dąbrowska, A., Muñoz, R., & Gębicki, J. (2022). A state of the art review on the use of fungi in biofiltration to remove volatile hydrophobic pollutants. *Reviews in Environmental Science and Bio/Technology, 21*, 225–246.

Nasser Abdelhamid, H., Georgouvelas, D., Edlund, U., & Mathew, A. P. (2022). CelloZIFPaper: Cellulose-ZIF hybrid paper for heavy metal removal and electrochemical sensing. *Chemical Engineering Journal, 446*, 136614.

Nasser Abdelhamid, H., & Mathew, A. P. (2021). Cellulose-zeolitic imidazolate frameworks (CelloZIFs) for multifunctional environmental remediation: Adsorption and catalytic degradation. *Chemical Engineering Journal, 426*, 131733.
Noormohamadi, H. R., Fat'hi, M. R., Ghaedi, M., & Ghezelbash, G. R. (2019). Potentiality of white-rot fungi in biosorption of nickel and cadmium: Modeling optimization and kinetics study. *Chemosphere, 216*, 124–130.
Robinson, T., McMullan, G., Marchant, R., & Nigam, P. (2001). Remediation of dyes in textile effluent: A critical review on current treatment technologies with a proposed alternative. *Bioresource Technology, 77*, 247–255.
Sáenz, M., Borodulina, T., Diaz, L., & Banchon, C. (2019). Minimal conditions to degrade low density polyethylene by *Aspergillus terreus* and Niger. *Journal of Ecological Engineering, 20*, 44–51.
Sánchez, C. (2009). Lignocellulosic residues: Biodegradation and bioconversion by fungi. *Biotechnology Advances, 27*, 185–194.
Saxena, G., & Bharagava, R. N. (Eds.). (2020). *Bioremediation of industrial waste for environmental safety*. Springer.
Srikanth, M., Sandeep, T. S. R. S., Sucharitha, K., & Godi, S. (2022). Biodegradation of plastic polymers by fungi: A brief review. *Bioresources and Bioprocessing, 9*, 42.
Usha, R., Sangeetha, T., & Palaniswamy, M. (2011). Screening of polyethylene degrading microorganisms from garbage soil. *Libyan Agriculture Research Center Journal International, 2*, 200–204.
Vaksmaa, A., Guerrero-Cruz, S., Ghosh, P., Zeghal, E., Hernando-Morales, V., & Niemann, H. (2023). Role of fungi in bioremediation of emerging pollutants. *Frontiers in Marine Science, 10*, 1070905.
Wang, N., Chu, Y., Wu, F., Zhao, Z., & Xu, X. (2017). Decolorization and degradation of Congo red by a newly isolated white rot fungus, Ceriporia lacerata, from decayed mulberry branches. *International Biodeterioration & Biodegradation, 117*, 236–244.
Wollenberg, A., Kretzschmar, J., Drobot, B., Hübner, R., Freitag, L., Lehmann, F., Günther, A., Stumpf, T., & Raff, J. (2021). Uranium(VI) bioassociation by different fungi—A comparative study into molecular processes. *Journal of Hazardous Materials, 411*, 125068.
Wu, B., Hussain, M., Zhang, W., Stadler, M., Liu, X., & Xiang, M. (2019). Current insights into fungal species diversity and perspective on naming the environmental DNA sequences of fungi. *Mycology, 10*, 127–140.
Zeghal, E., Vaksmaa, A., Vielfaure, H., Boekhout, T., & Niemann, H. (2021). The potential role of marine fungi in plastic degradation—A review. *Frontiers in Marine Science, 8*, 738877.

Chapter 5
Biodegradation and Transformation of Per- and Polyfluoroalkyl Substances in Wastewater

Qingyang Liu

Abstract Per- and poly-fluoroalkyl substances (PFASs), a class of highly stable environmentally persistent organic compounds with multiple carbon-fluorine bonds exerts a series of negative consequences for the economy, industry, public health, and the ecosystem. It is expected that over 12,000 PFASs exist in the environment. These impacts are a global issue affecting various areas around the world, including developed and developing nations, emphasizing the need to manage emissions from different sectors across the globe. There are many current and upcoming regulations regarding the restriction of PFAS from various continents including the United States and European Union. Because PFASs are found to be highly resistant to chemical degradation strongly, the eliminations of PFASs in environmental matrixes pose a challenge to researchers and becomes subject to strict government regulations. This chapter includes the occurrences, toxicological, and ecological effects of PFASs in wastewater and the existing physicochemical remediation strategies of PFASs in wastewater. This chapter summarizes the recent advances in biological remediation strategies of PFASs with bacterial and fungal in the processes of transformation and degradation of PFASs in wastewater.

Keywords PFASs · Perfluorinated · Polyfluoroalkyl substances · Environmental pollution · Wastewater treatment · Toxicological effects · Biological remediation · Bacterial transformation · Fungal bioremediation · Perfluoroalkyl acids · Perfluorooctanesulfonic acid · Perfluorooctanoic acid · Advanced oxidation · Sorption · Activated carbon adsorption · Nanomaterials · Thermal destruction · Sonochemistry · Biodegradation · Biodegradation pathways · Regulatory frameworks · Eco-friendly remediation

Q. Liu (✉)
College of Ecology and Environment, Nanjing Forestry University, Nanjing, China
e-mail: qyliu@njfu.edu.cn

C. M. Hussain et al. (eds.), *Fungi in Waste Water Treatment*, Springer Water,
https://doi.org/10.1007/978-3-032-12693-1_5

5.1 Introduction

According to the definition from the Organization for Economic Co-operation and Development (OECD), perfluorinated and poly-fluoroalkyl substances (PFASs) are a type of chemical compound with more than one fluorinated methyl or methylene carbon atom and without any other atom (e.g., H, Cl, Br, and I) being attached to (Muir & Howard, 2006; OECD, 2021). It is noted that perfluoromethyl and perfluoromethylene are both saturated aliphatic groups. It is noticed that the carbon of the R–CF_2–O– or R–CF_2–Si– group (R is not H, Cl, Br, and I atom) is a perfluorinated methylene carbon (Butt et al., 2014). A perfluorinated methylene group can also be expressed as ">CF2", where ">" represents two single chemical bonds. A completely fluorinated carbon that is linked to the remaining molecule by a double bond is called a perfluorinated methylidene carbon atom (=CF_2). This distinguishing feature is essential for defining PFAS, because a chemical compound with a perfluorinated methane carbon moiety (>CF–) alone is in accord with the revised definition of PFASs (Cousins et al., 2020).

Perfluoroalkyl and polyfluoroalkyl substances can be divided into three categories, which are perfluoroalkyl substances, polyfluoroalkyl substances, and fluorinated polymers, respectively (Ahrens & Bundschuh, 2014; EPA, 2023a). The fully fluorinated perfluoroalkyl substances contain alkyl chain of perfluoroalkyl-substituted compounds, which includes (aliphatic) perfluorocarbons (PFCs), perfluoroalkyl acids (PFAA), perfluoroalkane sulfonyl fluorides, perfluoroalkane sulfonamides, perfluoroalkyl iodides, and perfluoroalkyl aldehydes (Fig. 5.1). Note that the alkyl chain of poly-fluoroalkyl substances is partly fluorinated with one or fluorine atoms attached to it, but not all H atoms are substituted for F atoms. In contrast, the chemical substances including perfluoroalkane sulfonamide derivates, fluorotelomer-based compounds, semifluorinated n-alkanes, and alkenes are subordinate to polyfluoroalkyl substances (Buck et al., 2011). Polymeric PFAS includes branched fluorinated polymers, fluorinated polymers, and polymer

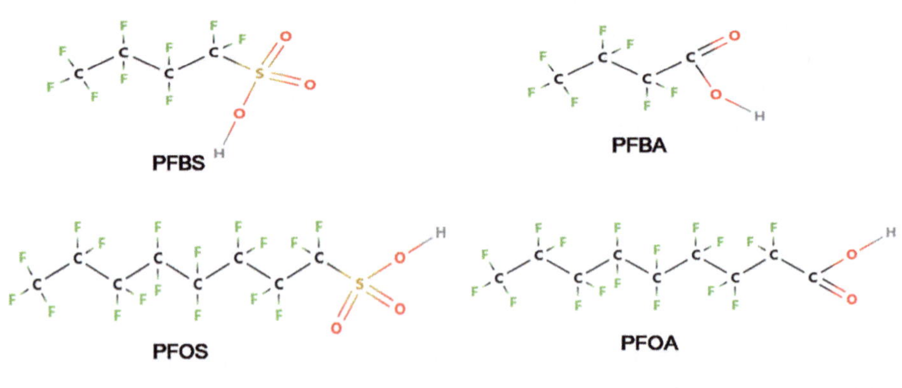

Fig. 5.1 Frequent detected PFAS species in wastewater. *PFBA* perfluorobutyric acid, *PFBS* nonafluoro-1-butanesulfonic acid, *PFOS* perfluorooctanesulfonic acid, *PFOA* perfluorooctanoic acid

perfluoropolyether. The polymeric PFAS is a kind of macromolecule with repeating monomers in linear and branched structures. The polymeric PFAS can be expressed using the formula CnF_{2n+1} (Grgas et al., 2023).

Since 2023, several regulations from the United States and European Union have been proposed, and intend to ban an entire class of PFAS based on the revised definitions of OECD because PFAS are persistent in the environment and could pose unprecedented negative consequences to the environment and human health (ECHA, 2023; EPA, 2023b; OECD, 2023). If the proposed regulations are effective, more than 12,000 PFAS can no longer be used in a variety of industries (e.g., electronics, household products, cosmetics, pharmaceuticals, pesticides, etc.) (Norén et al., 2021; Spyrakis & Dragani, 2023).

PFAS is not a chemical category that is composed of chemical compounds with similar chemical structures, but the term includes various compounds with diverse physiochemical, biological, and environmental properties (Wang et al., 2021). Though some PFAS are toxicological to the ecosystem, the toxicological characteristics of the vast majority of PFAS have not yet been determined (Polishchuk, 2017). For instance, fluorinated polymers do not behave the typical toxicological and environmental characteristics similar to those of other PFAS (Lampic & Parnis, 2020). Thus, it is possible to fully implement the ban on the use of PFAS in industrial applications due to sustainable development and economic reasons (Brusseau & Van Glubt, 2021). This is especially true for PFAS (e.g., fluorinated polymers and perfluoropolyether oils), which have not been linked to toxicological properties in ecosystems and human beings (Buck et al., 2011; Cousins et al., 2020).

The emergence of PFAS, particularly those with toxicological properties and currently in production and utilization requires a call to conventional monitor endeavors in different environment matrixes (e.g., air, soil, lake, and wastewater) (Thapa et al., 2024). The vast monitoring efforts to monitor the occurrences of PFAS with toxicological properties have been documented in developed and developing continents including the United States, the European Union, China, India, and Australia (Feng et al., 2021; Gomis et al., 2017; Norén et al., 2021). Current efforts are proceeding to detect the levels of PFAS in human blood because PFAS could be persistent in the human body and result in diseases (e.g., breast cancer, liver cancer, increased levels of cholesterol, apoptosis, etc.) (Petriello et al., 2022). Different remediation strategies including chemical and biological methods have been reported to reduce the concentration of PFAS in various environmental matrixes (Grgas et al., 2023; Veciana et al., 2022). The commonly used physicochemical remediation strategies of PFAS include sorption, thermal destruction, and advanced oxidation (Veciana et al., 2022). Due to the stability and chemical inertness of PFAS, some chemical remediation strategies require greater input costs and complex operations (Buck et al., 2011; Thapa et al., 2024). However, the studies on biological remediation strategies of PFAS are limited and the findings of different studies were conflicting.

This chapter provides an overview of the current advances in the occurrences of PFAS, and chemical and biological remediation strategies of PFAS in wastewater.

5.2 PFAS in Wastewater

5.2.1 *Sources*

Though PFAS has been widely used for over 70 years, the prevalence and bioaccumulation tendency of PFASs in the environment were not first reported until 2001 (Brunn et al., 2023). The detection of PFOS was observed in wildlife tissues (birds, fish, and mammals) globally (Cousins et al., 2020; Cousins et al., 2022). Later, some researchers found that many types of PFAS are widely distributed in the environment (Hori et al., 2006). The occurrences of PFAS were not only found in wildlife, but also in humans on different continents, including Canada, the United States, the European Union, China, Japan, and Australia (Cousins et al., 2022; Gomis et al., 2017). At present, the sources of PFAS in the urban environment include the products containing PFAS or created in the process with the use of PFAS at industrial sites, fire training or fire response locations near civil and military aviation facilities in the usage history of aqueous film-forming foams and film-forming fluoroproteins, industrial and wastewater treatment plants for managing wastewater contaminated with PFAS, as well as landfills serving as the disposal of wastes coated with PFAS (e.g., textiles and cooking materials) (Liou et al., 2010; Liu et al., 2023; Ochoa-Herrera et al., 2016).

5.2.2 *Levels*

PFASs can then be discharged into the environment from the sources through a geochemical cycling process, i.e., rainwater runoff, accidental leaks of wastewater, and spills into groundwater (Scott et al., 2009). It is reported that the levels of PFAS varied from <1 to 4300 $\mu g\ L^{-1}$ in groundwater with the most abundant compound of perfluorooctanesulfonic acid (PFOS), followed by perfluorohexane sulfonic acid and perfluorooctanoic acid (PFOA) (Miner et al., 2021). It is estimated that drinking water in 33 states of the United States and more than 16 million Americans face drinking water with PFAS pollution (Kurwadkar et al., 2022). In 2016, the United States Environmental Protection Agency (USEPA) issued an unenforceable lifetime health warning (HA) for perfluorooctanoic acid (PFOA) and perfluorooctane sulfonate (PFOS) in drinking water (Cousins et al., 2022; Kurwadkar et al., 2022). Because public water supply systems (PWSs) do not require the treatment of water exceeding EPA HAs and do not require routine measurement of PFAS, there is no detailed detection information on PFAS frequency in drinking water in the United States (Cousins et al., 2022). Other regions are also facing similar issues in drinking water with PFAS pollution (Bunting et al., 2021; Wang et al., 2022) (Table 5.1).

Table 5.1 The levels of detected PFAS in water across different countries

Area	PFAS	Level	Contaminated environment	References
China	PFOA and PFOS	Lower than 1.0 mg L^{-1}	Songhua River	Wang et al. (2022)
Italy	PFOS	253–6468 ng L^{-1}	Bermuda River	Kurwadkar et al. (2022)
Canada	PFOA	49 ng L^{-1}	Well and river	Scott et al. (2009)
Mount Everest Khumbu glacier	PFOS	10–26 ng L^{-1}	Snow and meltwater	Miner et al. (2021)
Europe	PFAS	2–14 ng L^{-1}	Groundwater	Bunting et al. (2021)
Global survey	PFAS	Higher than the health advisory levels of US EPA lifetime drinking water	Rainwater	Cousins et al. (2022)

5.2.3 Toxicological Properties

Exposure to PFAS at high concentrations can lead to diverse negative effects on human health, which are associated with increased incidences of thyroid diseases, breast cancer, liver cancer, elevated cholesterol levels, lung cancer, and testicular cancer (Blake et al., 2018; Crawford et al., 2017; Domingo & Nadal, 2017; Olsen & Ley, 2015). A study from the Florida Department of Health was carried out in populations under the exposure to PFAS at levels as low as parts per billion (ppb) during two periods, i.e., from 1996 to 2005 and from 2006 to 2015, respectively. The findings demonstrated that the incidence rate of bladder cancer and leukemia was greater than expected during the periods from 1996 to 2005, and the incidence rate of non-Hodgkin lymphoma, pancreas cancer, and bladder cancer was higher compared to the expected during the years from 2006 to 2015 (Tarapore & Ouyang, 2021). Laboratory studies have documented evidence of immune system modulation in the response to exposures to PFAS (Ochoa-Herrera et al., 2016). A study found that a decrease in the levels of diphtheria antibodies in children at 5 years old was associated with an increase in the levels of PFOS in maternal blood. The exposures to PFAS could lead to a higher likelihood of contracting diphtheria and tetanus at the age of 7 (Domingo & Nadal, 2017). It is found that C-reactive protein as a marker of human systemic inflammation is associated with the amounts of PFOS in the blood (Tarapore & Ouyang, 2021).

5.3 Physicochemical Remediation Strategies

Some traditional remediation methods, including those for hydrocarbons and chlorinated solvents, are acceptable for removing other organic pollutants in wastewater and have been proven ineffective in the removal of PFAS in wastewater (Liou et al.,

2010). Due to PFAS's resistance to chemical and biological degradation, completely removing PFAS from wastewater is challenging (Shahsavari et al., 2021). In the past decade, there has been a significant increase in studies on the degradation of PFAS using physical, chemical, and biological methods (Thapa et al., 2024).

5.3.1 Sorption

One of the effective techniques for removing PFAS from wastewater, especially long-chain PFAS, is activated carbon adsorption (AC) (Dickman & Aga, 2022). The influencing factors that dominate the adsorption between activated carbon and long-chain PFAS include target PFAS characteristics, water quality, carbon materials, and porosity, as well as contact time (Cantoni et al., 2021). The porosity of activated carbon is crucial for its adsorption effect. Generally, mesoporous-rich and microporous-rich carbons are in the priority selection of microporous carbons (Wu et al., 2020). When the contact time between PFAS and carbon is fixed, mesopores are useful in granular activated carbon (GAC) filters because they allow pollutants to rapidly diffuse to their final adsorption sites (micropores) (Cantoni et al., 2021).

Ultra-fine powder activated carbon (UPAC) is shown to be a better choice of absorbents compared to granular activated carbon (GAC) because UPAC has a larger specific surface area and faster adsorption rate relative to those of granular activated carbon (GAC) (Murray et al., 2019). When the overall input rate of PFAS is 10% of PFAS breach in polluted groundwater, the absorption capacity of UPAC is almost twice that of GAC (Murray et al., 2019). Further mass load testing revealed that the adsorption of PFAS on UPAC reached 2990 g g^{-1}, which is 480 times larger than GAC (Murray et al., 2019). The impacting factors in the absorption capacity are found to be the size of the adsorbent, pore capacity, and chain length of PFAS (Su et al., 2022).

Three types of nano-metals (zinc, copper, and nanocrystalline iron) were bonded onto multi-walled carbon nanotubes (MWCNTs) for removing PFOA compared to pure MWCNTs (Ahrens & Bundschuh, 2014; Liu et al., 2018; Mohd Azmi et al., 2021; Yang et al., 2020). The results indicated that the surface of MWCNTs has been successfully replaced by nano-metals (Liu et al., 2018). According to the adsorption isotherm, the adsorption capacity of PFOA on composite materials of metals and MWCNTs is higher than that of the reference MWCNTs (Ahrens & Bundschuh, 2014). The adsorption rate indicated that the adsorption of PFOA on metal-doped MWCNTs (MDCNTs) was caused by intraparticle diffusion. The effect of ion type (Ca^{2+}) and ion strength (0–100 mM) influenced the adsorption capacities of MDCNTs negatively, while the pH value of the wastewater significantly affected the binding degree between PFOA and MDCNTs (Yang et al., 2020). These findings provide a new approach to eliminating trace amounts of PFOA in aqueous media (Ahrens & Bundschuh, 2014; Liu et al., 2018; Mohd Azmi et al., 2021; Yang et al., 2020).

Several Fe-based metal-organic frameworks (MOFs) such as Fe-BTC, MIL-100-Fe, and MIL-101-F, are proven to remove PFOA in theoretical and experimental simulations of wastewater (Yang et al., 2020; Yassin et al., 2022). These studies mainly focus on the effectiveness of adsorption, the mechanism of interface contact, and the role of pores. Fe-BTC (418 mg g^{-1}) has a high binding potential with PFOA relative to that of MIL-100-Fe (349 mg g^{-1}), and MIL-101-Fe (370 mg g^{-1}) (Yang et al., 2020). This result could be ascribed to that Fe-BTC has more Lewis acid sites. The XPS Spectroscopy of MOF-PFOA demonstrated that the binding energies of Fe2p are decreasing and those of F1s are enhancing, indicating the process of hydrophobic interactions and Lewis acid-base complexation (LAB) in the adsorption (Yang et al., 2020). Iron nuclei and benzene exist in organic molecules and act as corresponding electron acceptors and donors (Yang et al., 2020).

5.3.2 *Thermal Destruction*

Sonochemistry is a scientific technique that uses acoustic fields to induce chemical reactions in aqueous media (Kurwadkar et al., 2022). Due to the introduction of sound waves, bubbles in the solution will collapse, resulting in a phenomenon known as cavitation (Pokhrel et al., 2016). This collapse leads to an increase in temperature, followed by thermal decomposition of PFAS at the interface between the bubbles and surrounding water (Moriwaki et al., 2005).

In the atmosphere of argon, it was found that the half-life of PFOS is 43 min, and the half-life of PFOA is 22 min using sonochemical degradation (Moriwaki et al., 2005). The presence of PFOS and PFOA undergo sonochemical degradation into their components through the fission of perfluorocarbon chains. This technology will help reverse the environmental hazards caused by PFOS and PFOA, as it can reduce the toxicity of chemicals and extend their perfluorocarbon chain (Moriwaki et al., 2005).

A prior study investigated the sonochemical deterioration of PFOS and PFOA in underground groundwater in landfills (Cheng et al., 2008). Due to the presence of organic components (e.g., dissolved organic matter), the acoustic dissociation rates of PFOA and PFOS decreased, leading to competitive adsorption at the interface between bubbles and water. However, in the remediation process of landfill water, the use of ozone oxidation and ultrasound significantly enhanced the mineralization of PFOA and PFOS. Vecitis et al. (2008) reported the acoustic degradation of PFOA and PFOS in water. Their method has been proven to effectively convert PFAS with concentrations ranging from 10 nM and 10 M into carbon dioxide, carbon monoxide, fluoride, and sulfate. The presence of sulfate or bicarbonate ions can impede sonication, while, when exposed to ultrasonic radiation, perchlorates or nitrates in the solution enhance the mineralization of PFAS.

PFOA could be removed from wastewater using an electrical discharge plasma-based method to achieve the goal of high removal efficiency and high effectiveness (Pokhrel et al., 2016). Under a high energy rate (76.5 W input power) process, the removal percentages of PFOA in 1.4 L of aqueous liquid were observed to be 90% and 25%, respectively, over a period of 30 min (Liu et al., 2023). Compared with other conventional technologies, the electrical discharge plasma-based method has achieved greater rates of PFOA elimination and defluorination. Using the electrical discharge plasma-based method, the removal of PFOA could be treated along with other co-pollutants. The co-pollutants were found to have a minor effect on the treatment process, and PFOA can be rapidly decomposed. Based on the chemical analysis of the residues, only about 10% of PFOA are converted to shorter chain perfluorooctanoic acid, while the rest were in the process of mineralization (Stratton et al., 2017).

5.3.3 Advanced Oxidation

PFOA exhibits strong bioaccumulation and can be efficiently removed in water by heating or UV oxidation of persulfate ($S_2O_8^{2-}$) (Brunn et al., 2023). However, to use this method, the system must be irradiated at high temperatures (70–90 °C), and a large amount of energy must be consumed (Lee et al., 2012). Lee et al. (2012) attempted to investigate the oxidative effect of persulfate on low-concentration PFOA with low temperatures (20–40 °C). The experiment includes the effects of the dosage of persulfate (20–200 mM) and pH (pH = 2.5, 3.6, 7.1, and 11.0) on the reaction rate within this temperature range. Under the initial pH of 2.5, the PFOA product containing 200 mM persulfate was almost decomposed after being exposed to 40 °C and 30 °C for 72 h and 215 h, respectively. After 648 h, even at 20 °C, the decomposition rate reached as high as 80.5%. At lower pH values or higher doses of persulfate, faster decomposition of PFOA was observed. Through a series of kinetic studies on the oxidation of PFOA (20, 26, 30, and 40 °C) by persulfate, a model was established to accurately predict the reduction rate of PFOA at different reaction times in the low-temperature range of 20–40 °C (Lee et al., 2012).

5.4 Biological Remediation Strategies

As is well known, microorganisms, especially bacteria, can metabolize various compounds (Shahsavari et al., 2021). Several studies in recent years have shown that certain microbial species or mixed cultures are capable of digesting specific PFAS, such as precursors of perfluoroalkyl acids (PFAAs), PFOS, and perfluorooctanoic acid (PFOA) (Grgas et al., 2023; Liou et al., 2010; Shahsavari et al., 2021). Although the "biodegradation triangle" is composed of PFAS varieties, the current genetic composition of microorganisms and environmental conditions play a crucial role in transformation (Ahrens & Bundschuh, 2014; Cousins et al., 2020). Successful biodegradation cases provide hope for the treatment of biological PFAS in wastewater (Kurwadkar et al., 2022).

5.4.1 *Biodegradation and Transformation of PFAS Using Bacteria*

Although it has been proven that both aerobic and anaerobic bacteria can decompose PFAS, aerobic biotransformation is more common (Wang et al., 2009; Wang et al., 2011). Early research tended to center on this pathway because PFOA and perfluorooctane sulfonic acid (PFOS) were considered to be easy to undergo biotransformation relative to other perfluorinated compounds (Fuertes et al., 2017; Liu & Mejia Avendaño, 2013). High levels of PFASs did not exhibit toxicity to microbial communities in the biodegradation process (Liu & Mejia Avendaño, 2013). A study by O'Carroll et al. (2020) observed the microbial communities of groundwater samples, which was polluted with PFOS. The findings showed that PFOS could be decomposed by microbial communities, especially for 6:2 fluorotelomer sulfonate (FTS) (O'Carroll et al., 2020). A strong association between PFOS concentrations and the relative abundance of over 100 individual genera was observed. The *Desulfococcus* genus was found to be significantly positively correlated with PFOS, while the *Oxalobacteraceae* family was significantly negatively correlated with PFOS. In addition, the results suggest the inhibition effects were found in certain genera at high concentrations of PFOS, while some genera, such as *Acidimicrobium* and *Gordonia* were simulated from low to medium-range concentrations (O'Carroll et al., 2020).

It is reported that the aerobic bacterium *Pseudomonas aeruginosa* (strain HJ4) biodegrades PFOS in the concentration range of 1400–1800 $\mu g\ L^{-1}$ (Kwon et al., 2014). The decomposition rate of PFOS was estimated to be about 67% within 48 hours of incubation, while the residual levels of PFOS were kept constant. *Pseudomonas aeruginosa* can be separated from the municipal communities of sludge in a wastewater treatment plant (Schultz et al., 2006). The addition of 0.1% of glucose by volume could help the growth of the HJ4 strain in the degradation experiments (Fuertes et al., 2017; Kwon et al., 2014). During the degradation process, the concentration of fluoride ions was determined with ion chromatography. The results indicated that no fluoride ions were detected, supporting that no defluorination of PFOS occurred (Kwon et al., 2014). In contrast, the analysis from tandem mass spectrometry (MS) demonstrated that the products of perfluorohexane sulfonate and perfluorobutanesulfonate (PFBS) were detected and the concentrations were in the range of 4–26 ng L^{-1}. Furthermore, several unknown products were formed and detected in aqueous solution. These findings suggested that the degradation mechanism of PFOS by *Pseudomonas aeruginosa* tends to break down the C–C bonds of PFOS rather than C–F bonds (Kwon et al., 2014).

Ochoa-Herrera et al. (2016) and Ochoa-Herrera et al. (2008) studied the toxicity sensitivity of microbial communities to the biodegradability of PFOS, as well as a short chain of perfluoroalkyl sulfonic acid, polyfluoroalkyl sulfonic acid, and carboxylic acid. They used municipal wastewater with PFOS as inoculants. They did not observe the toxicity in response to the methanogenic activity of anaerobic wastewater of the chemical compounds at concentrations as high as 500 mg L^{-1}

(Ochoa-Herrera et al., 2008; Ochoa-Herrera et al., 2016). These compounds are PFOS, 6:2 FTSA, PFBS, TFA (trifluoroacetic acid), and C3- and C5-PFAS compounds. All investigated PFASs have high microbial degradation resistance. In addition, Ochoa-Herrera et al. reported that the defluorination efficiency of PFOS and PFOS fluoride branching-chain isomers was estimated to be approximately 18% and 71%, respectively, when the reduction dehalogenation was combined with vitamin B12 as a biocatalyst and titanium (III) citrate as a bulk reducing agent in anaerobic wastewater of microbial communities (Ochoa-Herrera et al., 2008; Ochoa-Herrera et al., 2016). Some studies also have documented the aerobic degradation of some PFAS, including 6:2 FTSA, $F(CF_2)_6(CH_2)_2SO_3H$, and 8:2 FTOH ($F(CF_2)_8$(CH2)2OH) (Liou et al., 2010; Liu et al., 2010; Wang et al., 2009).

In these aerobic degradation reactions, functional groups of -CF_2- are oxidized to form short-chain perfluorocarboxylic acids and carbon dioxide (Russell et al., 2008). The biodegradation of the polyfluorocarbon chain initially focused on the hydrocarbon functional groups connected by the polyfluorocarbon portion (Wackett, 2022; Wang et al., 2009). Additional functional groups (e.g., chlorine, benzene ring, sulfonate, carboxylate, or phosphonate) offer a mild entry point of reactions, which could initiate the bond of C-F and lead to its breakup (Wackett, 2022; Wang et al., 2009). If additional functions are linked with a chain or ring with highly fluorinated or perfluorinated chains, the chemical and metabolic pathways of achieving defluorination are complicated, and differ from the pathway described above (Wackett, 2022; Wang et al., 2009).

Nevertheless, conflicting evidence on the biodegradation metabolism of PFOS is also documented. Liou et al. (2010) conducted experiments on five different microbial communities for microbial metabolism of PFOA, which are collected from sediment at industrial sites, soil at urban sewage treatment plants, agricultural soil, and soil at two fire training regions. In 259 days of anaerobic cultivation, PFOA at concentrations of 100 ppm and 100 ppb was not in defluorination processes in combination with some electron donors, such as acetate, lactate, and ethanol (Liou et al., 2010). They also examined PFOA at concentrations of 100 ppm and 100 ppb with trichloroethylene (TCE) in a 5% inoculation dose (microbial communities) obtained from a sewage treatment plant. The anaerobic cultivation was carried out within 65 days, and no reductive defluorination metabolism of PFOA was found (Liou et al., 2010). In the third scenario, the authors checked a potential defluorination conversion of PFOA at a concentration of 100 ppm and 100 ppb with five electron acceptors in a 5% inoculation dose (microbial communities) collected from the wastewater treatment plant (Liou et al., 2010). Over an incubation period of 110 days, five electron acceptors including aerobic, nitrate, iron, sulfate, and methane are incapable of removing fluoride from PFOA. The concentrations of PFOA did not vary significantly (Liou et al., 2010). These experiments did not attempt to find metabolites, to elucidate possible processes. Based on the prior studies, it is inferred that certain PFAS species may be susceptible to biodegradation under certain conditions. Further endeavors are thus encouraged to create a biodegradation technology of PFAS at an industrial scale.

5.4.2 *Biodegradation and Transformation of PFAS Using Fungi*

Approximately 75% of the microbial communities in soil are composed of fungi, which are essential for ecosystems make a great contribution to the he biotransformation of PFAS (Dickman & Aga, 2022; Liu & Mejia Avendaño, 2013). Prior researchers have documented that other pollutants (e.g., synthetic dyes, aliphatic hydrocarbons, polycyclic aromatic hydrocarbons) can be biodegraded by fungi through basic metabolic reactions (Liu & Mejia Avendaño, 2013). These results suggest that fungi are able to bio-transform PFAS in the environment (Brunn et al., 2023). The most commonly used fungi for degrading toxic pollutants are brown rot fungus *Aspergillus niger* and white rot fungus *Aspergillus flavus*. And *A. niger* dioxygenase, xylanases, cellulases, and hemicellulases (Grgas et al., 2023; Kurwadkar et al., 2022).

Previous studies have shown that the use of enzyme catalyzed oxidation humification reaction (ECOHR) is feasible for the remediation and treatment of perfluorooctanoic acid, as well as the effective decomposition of PFOA (Colosi et al., 2009). ECOHR is performed by natural extracellular enzymes obtained from white and brown rot fungi (*Aspergillus flavus.*and *Aspergillus niger*), including laccase (LaC), lignin peroxidases (LiP), manganese peroxidases (MnP), etc. (Weber & Huang, 2003). ECOHR reactions are usually derived from soil systems and are featured by a series of oxidation reactions in the process of humification (Naidja et al., 1998). At the beginning process, the enzymes reacted with PFAS by producing reactive radicals through the conversion of chemical compounds (Naidja et al., 1998). Then, the reactive radicals degrade PFAS as an inactive molecule. Degradation experiment of PFOA using horseradish peroxidase (HRP), hydrogen peroxide and 4-methoxyphenol with ECOHR were conducted by Colosi et al. (2009). The levels of PFOA and 4-methoxyphenol were determined by high-performance liquid chromatography, while the levels of fluoride ions were measured using ion chromatography. The by-products were characterized with gas chromatography-mass spectrometry (GC-MS). The toxicity tests during the degradation process were monitored by the Microtox toxicity testing system. Within 6 h, they found a degradation rate of 68% in the PFOA and a reduction of 98% in aquatic toxicity. Without the presence of hydrogen peroxide and 4-methoxyphenol, the removal of PFOA was negligible. Thus, the authors proposed that degradation mechanism include the formation of radical intermediates during the reaction of 4-methoxyphenol and HRP, and the non-specific degradation of PFOA accomplished by conversing the bond of C–F in PFOA with radical intermediates to CO_2 and fluoride ions (Colosi et al., 2009).

The fungal biotransformation experiments of PFAS by Merino et al. used 6:2 FTOH as a shorter-chain substituent and 8:2 FTOH as a longer-chain substituent (Merino et al., 2016; Merino et al., 2018). The 6:2 FTOH could be converted into PFCAs and other compounds in the environment through biological and physicochemical processes. Compared to bacterial biotransformation, 6:2 FTOH is more easily biotransformation by fungal (Tseng et al., 2014). The experiments studied the biotransformation of 6:2 FTOH with two fungal strains of wood decay, i.e., *Gloephyllum trabeum* and *Trametes versicolor*, as well as six fungal isolates, i.e., *Fusarium* sp., *Aspergillus* sp., *Aspergillus* sp., *Fusarium* sp., *Penicillium* sp., and *Penicillium* sp. (Merino et al., 2016, 2018). The chemical products of biotransformation metabolites were characterized using liquid chromatography-tandem mass spectrometry. The results indicated that six chemical transformation products were determined by *Gloephyllum trabeum* and *Trametes versicolor*. All six fungal isolates are capable of converting 6:2 FTOH to 5–9 chemical products during a period of 28 days under experimental conditions (Merino et al., 2016, 2018). Supplementing the culture medium may increase the biotransformation rate of 6:2 FTOH. The experiments also assessed the tolerance of six fungal isolates to PFAS in the concentration from 0 to 1000 mg L^{-1}, and found that six fungal isolates could live with the enhanced levels of PFOA and PFOS (Merino et al., 2016; Merino et al., 2018).

Luo et al. (2018) employed *Pleurotus ostreatus* isolated from laccase to degrade PFOS in a period of 162 days. In the degradation of PFOS, 1-hydroxybenzotriazole (HBT) is introduced as a mediator in the humification process of the ECOHR reaction. Laccase is a type of phenoloxidases that is capable of catalyzing ECOHR reaction, while HBT has the characteristics of high reaction efficiency in the laccase application systems. In the decomposition process, the level of fluoride was measured by ion chromatography, while the chemical products were characterized by electrospray ionization-high-resolution mass spectrometry (Luo et al., 2018). The defluorination ratio with the presence of Cu^{2+} (47%) was comparable with that having Mg^{2+} (47%). In contrast, the release rate of fluoride was slower in the system with Mg^{2+} than with Cu^{2+}. The findings indicated that degradation of PFOS in the system with Mg^{2+} and with Cu^{2+} were performed in similar mechanisms but at different reaction rates (Luo et al., 2018). Theoretical model results derived from density functional theory demonstrated that a complex of PFOS and metal ions could lead to a decrease in the energy of the C-C bond through the helical configuration. This complex could help the radicals of HBT to attack the skeleton of C-C in PFOS. The direct attack of radicals triggered a chain reaction of the free radical process, resulting in the products of fluorides and other organic fluorinated compounds. Based on the prior studies, ECOHR may provide a sustainable way to degrade contamination PFOS in the environment with the addition of appropriate enzymes and mediators. Yet, degradation outcomes of PFAS using fungi require more reaction time and are dependable on the applied incubation matrix (Luo et al., 2018). Research on biodegradation and defluorination of PFAS may center on the increase in the reaction efficiencies and decrease in the formation of varying organic products under mild operating conditions (Luo et al., 2018).

5.5 Conclusions

In summary, a further step strives to focus on the regulation of high-risk PFAS in drinking water with the implementation of effective ban at a national scale. Thus, an uninterrupted effort to monitor the frequency of high-risk PFAS in drinking water could be achieved, which may a more balanced and focused approach to the management of high-risk PFAS across several continents with different levels of socioeconomic development. Finally, the research on safer remediation alternatives and technologies of PFAS with lost cost and short time may minimize the levels of PFAS in drinking water and improve the health and quality of life of global citizens.

References

Ahrens, L., & Bundschuh, M. (2014). Fate and effects of poly- and perfluoroalkyl substances in the aquatic environment: A review. *Environmental Toxicology and Chemistry, 33*(9), 1921–1929. https://doi.org/10.1002/etc.2663

Blake, B. E., Pinney, S. M., Hines, E. P., Fenton, S. E., & Ferguson, K. K. (2018). Associations between longitudinal serum perfluoroalkyl substance (PFAS) levels and measures of thyroid hormone, kidney function, and body mass index in the Fernald community cohort. *Environmental Pollution, 242*, 894–904. https://doi.org/10.1016/j.envpol.2018.07.042

Brunn, H., Arnold, G., Körner, W., Rippen, G., Steinhäuser, K. G., & Valentin, I. (2023). PFAS: Forever chemicals—Persistent, bioaccumulative and mobile. Reviewing the status and the need for their phase out and remediation of contaminated sites. *Environmental Sciences Europe, 35*(1), 20. https://doi.org/10.1186/s12302-023-00721-8

Brusseau, M. L., & Van Glubt, S. (2021). The influence of molecular structure on PFAS adsorption at air-water interfaces in electrolyte solutions. *Chemosphere, 281*, 130829. https://doi.org/10.1016/j.chemosphere.2021.130829

Buck, R. C., Franklin, J., Berger, U., Conder, J. M., Cousins, I. T., de Voogt, P., Jensen, A. A., Kannan, K., Mabury, S. A., & van Leeuwen, S. P. J. (2011). Perfluoroalkyl and polyfluoroalkyl substances in the environment: Terminology, classification, and origins. *Integrated Environmental Assessment and Management, 7*(4), 513–541. https://doi.org/10.1002/ieam.258

Bunting, S. Y., Lapworth, D. J., Crane, E. J., Grima-Olmedo, J., Koroša, A., Kuczyńska, A., Mali, N., Rosenqvist, L., van Vliet, M. E., Togola, A., & Lopez, B. (2021). Emerging organic compounds in European groundwater. *Environmental Pollution, 269*, 115945. https://doi.org/10.1016/j.envpol.2020.115945

Butt, C. M., Muir, D. C. G., & Mabury, S. A. (2014). Biotransformation pathways of fluorotelomer-based polyfluoroalkyl substances: A review. *Environmental Toxicology and Chemistry, 33*(2), 243–267. https://doi.org/10.1002/etc.2407

Cantoni, B., Turolla, A., Wellmitz, J., Ruhl, A. S., & Antonelli, M. (2021). Perfluoroalkyl substances (PFAS) adsorption in drinking water by granular activated carbon: Influence of activated carbon and PFAS characteristics. *Science of the Total Environment, 795*, 148821. https://doi.org/10.1016/j.scitotenv.2021.148821

Cheng, J., Vecitis, C. D., Park, H., Mader, B. T., & Hoffmann, M. R. (2008). Sonochemical degradation of perfluorooctane sulfonate (PFOS) and perfluorooctanoate (PFOA) in landfill groundwater: Environmental matrix effects. *Environmental Science & Technology, 42*(21), 8057–8063. https://doi.org/10.1021/es8013858

Colosi, L. M., Pinto, R. A., Huang, Q., & Weber, W. J., Jr. (2009). Peroxidase-mediated degradation of perfluorooctanoic acid. *Environmental Toxicology and Chemistry, 28*(2), 264–271. https://doi.org/10.1897/08-282.1

Cousins, I. T., DeWitt, J. C., Glüge, J., Goldenman, G., Herzke, D., Lohmann, R., Ng, C. A., Scheringer, M., & Wang, Z. (2020). The high persistence of PFAS is sufficient for their management as a chemical class. *Environmental Science: Processes & Impacts, 22*(12), 2307–2312. https://doi.org/10.1039/D0EM00355G

Cousins, I. T., Johansson, J. H., Salter, M. E., Sha, B., & Scheringer, M. (2022). Outside the safe operating space of a new planetary boundary for per- and polyfluoroalkyl substances (PFAS). *Environmental Science & Technology, 56*(16), 11172–11179. https://doi.org/10.1021/acs.est.2c02765

Crawford, N. M., Fenton, S. E., Strynar, M., Hines, E. P., Pritchard, D. A., & Steiner, A. Z. (2017). Effects of perfluorinated chemicals on thyroid function, markers of ovarian reserve, and natural fertility. *Reproductive Toxicology, 69*, 53–59. https://doi.org/10.1016/j.reprotox.2017.01.006
Dickman, R. A., & Aga, D. S. (2022). A review of recent studies on toxicity, sequestration, and degradation of per- and polyfluoroalkyl substances (PFAS). *Journal of Hazardous Materials, 436*, 129120. https://doi.org/10.1016/j.jhazmat.2022.129120
Domingo, J. L., & Nadal, M. (2017). Per- and polyfluoroalkyl substances (PFASs) in food and human dietary intake: A review of the recent scientific literature. *Journal of Agricultural and Food Chemistry, 65*(3), 533–543. https://doi.org/10.1021/acs.jafc.6b04683
ECHA. (2023). *Annex XV restriction report: Proposal for a restriction*. European Chemicals Agency. https://echa.europa.eu/documents/10162/f605d4b5-7c17-7414-8823-b49b9fd43aea
EPA. (2023a). *PFAS master list of PFAS substances*. U.S. Environmental Protection Agency. https://comptox.epa.gov/dashboard/chemical-lists/pfasmaster
EPA. (2023b). *Risk management for per- and polyfluoroalkyl substances (PFASs) under TSCA*. U.S. Environmental Protection Agency. https://www.epa.gov/assessing-and-managing-chemicals-under-tsca/risk-management-and-polyfluoroalkyl-substances-pfas
Feng, X., Chen, X., Yang, Y., Yang, L., Zhu, Y., Shan, G., Zhu, L., & Zhang, S. (2021). External and internal human exposure to PFOA and HFPOs around a mega fluorochemical industrial park, China: Differences and implications. *Environment International, 157*, 106824. https://doi.org/10.1016/j.envint.2021.106824
Fuertes, I., Gómez-Lavín, S., Elizalde, M. P., & Urtiaga, A. (2017). Perfluorinated alkyl substances (PFASs) in northern Spain municipal solid waste landfill leachates. *Chemosphere, 168*, 399–407. https://doi.org/10.1016/j.chemosphere.2016.10.072
Gomis, M. I., Vestergren, R., MacLeod, M., Mueller, J. F., & Cousins, I. T. (2017). Historical human exposure to perfluoroalkyl acids in the United States and Australia reconstructed from biomonitoring data using population-based pharmacokinetic modelling. *Environment International, 108*, 92–102. https://doi.org/10.1016/j.envint.2017.08.002
Grgas, D., Petrina, A., Štefanac, T., Bešlo, D., & Landeka Dragičević, T. (2023). A review: Per- and polyfluoroalkyl substances—Biological degradation. *Toxics, 11*(5), 446. https://doi.org/10.3390/toxics11050446
Hori, H., Nagaoka, Y., Yamamoto, A., Sano, T., Yamashita, N., Taniyasu, S., Kutsuna, S., Osaka, I., & Arakawa, R. (2006). Efficient decomposition of environmentally persistent perfluorooctanesulfonate and related fluorochemicals using zerovalent iron in subcritical water. *Environmental Science & Technology, 40*(3), 1049–1054. https://doi.org/10.1021/es0517419
Kurwadkar, S., Dane, J., Kanel, S. R., Nadagouda, M. N., Cawdrey, R. W., Ambade, B., Struckhoff, G. C., & Wilkin, R. (2022). Per- and polyfluoroalkyl substances in water and wastewater: A critical review of their global occurrence and distribution. *Science of the Total Environment, 809*, 151003. https://doi.org/10.1016/j.scitotenv.2021.151003
Kwon, B. G., Lim, H. J., Na, S. H., Choi, B. I., Shin, D. S., & Chung, S. Y. (2014). Biodegradation of perfluorooctanesulfonate (PFOS) as an emerging contaminant. *Chemosphere, 109*, 221–225. https://doi.org/10.1016/j.chemosphere.2014.01.072
Lampic, A., & Parnis, J. M. (2020). Property estimation of per- and polyfluoroalkyl substances: A comparative assessment of estimation methods. *Environmental Toxicology and Chemistry, 39*(4), 775–786. https://doi.org/10.1002/etc.4681
Lee, Y.-C., Lo, S.-L., Kuo, J., & Lin, Y.-L. (2012). Persulfate oxidation of perfluorooctanoic acid under the temperatures of 20–40°C. *Chemical Engineering Journal, 198–199*, 27–32. https://doi.org/10.1016/j.cej.2012.05.073
Liou, J. S. C., Szostek, B., DeRito, C. M., & Madsen, E. L. (2010). Investigating the biodegradability of perfluorooctanoic acid. *Chemosphere, 80*(2), 176–183. https://doi.org/10.1016/j.chemosphere.2010.03.009
Liu, J., & Mejia Avendaño, S. (2013). Microbial degradation of polyfluoroalkyl chemicals in the environment: A review. *Environment International, 61*, 98–114. https://doi.org/10.1016/j.envint.2013.08.022

Liu, J., Wang, N., Szostek, B., Buck, R. C., Panciroli, P. K., Folsom, P. W., Sulecki, L. M., & Bellin, C. A. (2010). 6-2 Fluorotelomer alcohol aerobic biodegradation in soil and mixed bacterial culture. *Chemosphere, 78*(4), 437–444. https://doi.org/10.1016/j.chemosphere.2009.10.044

Liu, L., Li, D., Li, C., Ji, R., & Tian, X. (2018). Metal nanoparticles by doping carbon nanotubes improved the sorption of perfluorooctanoic acid. *Journal of Hazardous Materials, 351*, 206–214. https://doi.org/10.1016/j.jhazmat.2018.03.001

Liu, S., Liu, Z., Tan, W., Johnson, A. C., Sweetman, A. J., Sun, X., Liu, Y., Chen, C., Guo, H., Liu, H., Wan, X., & Zhang, L. (2023). Transport and transformation of perfluoroalkyl acids, isomer profiles, novel alternatives and unknown precursors from factories to dinner plates in China: New insights into crop bioaccumulation prediction and risk assessment. *Environment International, 172*, 107795. https://doi.org/10.1016/j.envint.2023.107795

Luo, Q., Yan, X., Lu, J., & Huang, Q. (2018). Perfluorooctanesulfonate degrades in a laccase-mediator system. *Environmental Science & Technology, 52*(18), 10617–10626. https://doi.org/10.1021/acs.est.8b00839

Merino, N., Qu, Y., Deeb, R. A., Hawley, E. L., Hoffmann, M. R., & Mahendra, S. (2016). Degradation and removal methods for perfluoroalkyl and polyfluoroalkyl substances in water. *Environmental Engineering Science, 33*(9), 615–649. https://doi.org/10.1089/ees.2016.0233

Merino, N., Wang, M., Ambrocio, R., Mak, K., O'Connor, E., Gao, A., Hawley, E. L., Deeb, R. A., Tseng, L. Y., & Mahendra, S. (2018). Fungal biotransformation of 6:2 fluorotelomer alcohol. *Remediation Journal, 28*(2), 59–70. https://doi.org/10.1002/rem.21550

Miner, K. R., Clifford, H., Taruscio, T., Potocki, M., Solomon, G., Ritari, M., Napper, I. E., Gajurel, A. P., & Mayewski, P. A. (2021). Deposition of PFAS 'forever chemicals' on Mt. Everest. *Science of the Total Environment, 759*, 144421. https://doi.org/10.1016/j.scitotenv.2020.144421

Mohd Azmi, L. H., Williams, D. R., & Ladewig, B. P. (2021). Polymer-assisted modification of metal-organic framework MIL-96 (Al): Influence of HPAM concentration on particle size, crystal morphology and removal of harmful environmental pollutant PFOA. *Chemosphere, 262*, 128072. https://doi.org/10.1016/j.chemosphere.2020.128072

Moriwaki, H., Takagi, Y., Tanaka, M., Tsuruho, K., Okitsu, K., & Maeda, Y. (2005). Sonochemical decomposition of perfluorooctane sulfonate and perfluorooctanoic acid. *Environmental Science & Technology, 39*(9), 3388–3392. https://doi.org/10.1021/es040342v

Muir, D. C. G., & Howard, P. H. (2006). Are there other persistent organic pollutants? A challenge for environmental chemists. *Environmental Science & Technology, 40*(23), 7157–7166. https://doi.org/10.1021/es061677a

Murray, C. C., Vatankhah, H., McDonough, C. A., Nickerson, A., Hedtke, T. T., Cath, T. Y., Higgins, C. P., & Bellona, C. L. (2019). Removal of per- and polyfluoroalkyl substances using super-fine powder activated carbon and ceramic membrane filtration. *Journal of Hazardous Materials, 366*, 160–168. https://doi.org/10.1016/j.jhazmat.2018.11.050

Naidja, A., Huang, P. M., & Bollag, J. M. (1998). Comparison of reaction products from the transformation of catechol catalyzed by birnessite or tyrosinase. *Soil Science Society of America Journal, 62*(1), 188–195. https://doi.org/10.2136/sssaj1998.03615995006200010025x

Norén, E., Lindh, C., Glynn, A., Rylander, L., Pineda, D., & Nielsen, C. (2021). Temporal trends, 2000–2017, of perfluoroalkyl acid (PFAA) concentrations in serum of Swedish adolescents. *Environment International, 155*, 106716. https://doi.org/10.1016/j.envint.2021.106716

O'Carroll, D. M., Jeffries, T. C., Lee, M. J., Le, S. T., Yeung, A., Wallace, S., Battye, N., Patch, D. J., Manefield, M. J., & Weber, K. P. (2020). Developing a roadmap to determine per- and polyfluoroalkyl substances-microbial population interactions. *Science of the Total Environment, 712*, 135994. https://doi.org/10.1016/j.scitotenv.2019.135994

Ochoa-Herrera, V., Field, J. A., Luna-Velasco, A., & Sierra-Alvarez, R. (2016). Microbial toxicity and biodegradability of perfluorooctane sulfonate (PFOS) and shorter chain perfluoroalkyl and polyfluoroalkyl substances (PFASs). *Environmental Science: Processes & Impacts, 18*(9), 1236–1246. https://doi.org/10.1039/C6EM00366D

Ochoa-Herrera, V., Sierra-Alvarez, R., Somogyi, A., Jacobsen, N. E., Wysocki, V. H., & Field, J. A. (2008). Reductive defluorination of perfluorooctane sulfonate. *Environmental Science & Technology, 42*(9), 3260–3264. https://doi.org/10.1021/es702842q

OECD. (2021). *Reconciling terminology of the universe of per- and polyfluoroalkyl substances (PFASs)* (OECD environment, health and safety publications series on risk management No. 61). OECD Publishing.

OECD. (2023). *What are PFASs and what are they used for?* https://www.oecd.org/chemicalsafety/portal-perfluorinated-chemicals/aboutpfass/

Olsen, G. W., & Ley, C. A. (2015). Prostate cancer and PFOA. *Journal of Occupational and Environmental Medicine, 57*(6). https://doi.org/10.1097/JOM.0000000000000446

Petriello, M. C., Mottaleb, M. A., Serio, T. C., Balyan, B., Cave, M. C., Pavuk, M., Birnbaum, L. S., & Morris, A. J. (2022). Serum concentrations of legacy and emerging per- and polyfluoroalkyl substances in the Anniston community health surveys (ACHS I and ACHS II). *Environment International, 158*, 106907. https://doi.org/10.1016/j.envint.2021.106907

Pokhrel, N., Vabbina, P. K., & Pala, N. (2016). Sonochemistry: Science and engineering. *Ultrasonics Sonochemistry, 29*, 104–128. https://doi.org/10.1016/j.ultsonch.2015.07.023

Polishchuk, P. (2017). Interpretation of quantitative structure–activity relationship models: Past, present, and future. *Journal of Chemical Information and Modeling, 57*(11), 2618–2639. https://doi.org/10.1021/acs.jcim.7b00274

Russell, M. H., Berti, W. R., Szostek, B., & Buck, R. C. (2008). Investigation of the biodegradation potential of a fluoroacrylate polymer product in aerobic soils. *Environmental Science & Technology, 42*(3), 800–807. https://doi.org/10.1021/es0710499

Schultz, M. M., Higgins, C. P., Huset, C. A., Luthy, R. G., Barofsky, D. F., & Field, J. A. (2006). Fluorochemical mass flows in a municipal wastewater treatment facility. *Environmental Science & Technology, 40*(23), 7350–7357. https://doi.org/10.1021/es061025m

Scott, B. F., Spencer, C., Lopez, E., & Muir, D. C. G. (2009). Perfluorinated alkyl acid concentrations in Canadian rivers and creeks. *Water Quality Research Journal, 44*(3), 263–277. https://doi.org/10.2166/wqrj.2009.028

Shahsavari, E., Rouch, D., Khudur, L. S., Thomas, D., Aburto-Medina, A., & Ball, A. S. (2021). Challenges and current status of the biological treatment of PFAS-contaminated soils. *Frontiers in Bioengineering and Biotechnology, 8*, 602040. https://doi.org/10.3389/fbioe.2020.602040

Spyrakis, F., & Dragani, T. A. (2023). The EU'S per- and polyfluoroalkyl substances (PFAS) ban: A case of policy over science. *Toxics, 11*(9), 721. https://doi.org/10.3390/toxics11090721

Stratton, G. R., Dai, F., Bellona, C. L., Holsen, T. M., Dickenson, E. R. V., & Mededovic Thagard, S. (2017). Plasma-based water treatment: Efficient transformation of perfluoroalkyl substances in prepared solutions and contaminated groundwater. *Environmental Science & Technology, 51*(3), 1643–1648. https://doi.org/10.1021/acs.est.6b04215

Su, H., Dou, X., Xu, D., Feng, L., Liu, Y., Du, Z., & Zhang, L. (2022). Fe0-loaded superfine powdered activated carbon prepared by ball milling for synergistic adsorption and persulfate activation to remove aqueous carbamazepine. *Chemosphere, 293*, 133665. https://doi.org/10.1016/j.chemosphere.2022.133665

Tarapore, P., & Ouyang, B. (2021). Perfluoroalkyl chemicals and male reproductive health: Do PFOA and PFOS increase risk for male infertility? *International Journal of Environmental Research and Public Health, 18*(7), 3794. https://doi.org/10.3390/ijerph18073794

Thapa, B. S., Pandit, S., Mishra, R. K., Joshi, S., Idris, A. M., & Tusher, T. R. (2024). Emergence of per- and poly-fluoroalkyl substances (PFAS) and advances in the remediation strategies. *Science of the Total Environment, 916*, 170142. https://doi.org/10.1016/j.scitotenv.2024.170142

Tseng, N., Wang, N., Szostek, B., & Mahendra, S. (2014). Biotransformation of 6:2 fluorotelomer alcohol (6:2 FTOH) by a wood-rotting fungus. *Environmental Science & Technology, 48*(7), 4012–4020. https://doi.org/10.1021/es4057483

Veciana, M., Bräunig, J., Farhat, A., Pype, M. L., Freguia, S., Carvalho, G., Keller, J., & Ledezma, P. (2022). Electrochemical oxidation processes for PFAS removal from contaminated water and wastewater: Fundamentals, gaps and opportunities towards practical implementation. *Journal of Hazardous Materials, 434*, 128886. https://doi.org/10.1016/j.jhazmat.2022.128886

Vecitis, C. D., Park, H., Cheng, J., Mader, B. T., & Hoffmann, M. R. (2008). Kinetics and mechanism of the sonolytic conversion of the aqueous perfluorinated surfactants, perfluorooctanoate (PFOA), and perfluorooctane sulfonate (PFOS) into inorganic products. *The Journal of Physical Chemistry A, 112*(18), 4261–4270. https://doi.org/10.1021/jp801081y

Wackett, L. P. (2022). Strategies for the biodegradation of polyfluorinated compounds. *Microorganisms, 10*(8), 1664. https://doi.org/10.3390/microorganisms10081664

Wang, F., Zhuang, Y., Dong, B., & Wu, J. (2022). Review on per- and poly-fluoroalkyl substances' (PFASs') pollution characteristics and possible sources in surface water and precipitation of China. *Water, 14*(5), 812. https://doi.org/10.3390/w14050812

Wang, N., Liu, J., Buck, R. C., Korzeniowski, S. H., Wolstenholme, B. W., Folsom, P. W., & Sulecki, L. M. (2011). 6:2 Fluorotelomer sulfonate aerobic biotransformation in activated sludge of waste water treatment plants. *Chemosphere, 82*(6), 853–858. https://doi.org/10.1016/j.chemosphere.2010.11.003

Wang, N., Szostek, B., Buck, R. C., Folsom, P. W., Sulecki, L. M., & Gannon, J. T. (2009). 8-2 Fluorotelomer alcohol aerobic soil biodegradation: Pathways, metabolites, and metabolite yields. *Chemosphere, 75*(8), 1089–1096. https://doi.org/10.1016/j.chemosphere.2009.01.033

Wang, Z., Buser, A. M., Cousins, I. T., Demattio, S., Drost, W., Johansson, O., Ohno, K., Patlewicz, G., Richard, A. M., Walker, G. W., White, G. S., & Leinala, E. (2021). A new OECD definition for per- and polyfluoroalkyl substances. *Environmental Science & Technology, 55*(23), 15575–15578. https://doi.org/10.1021/acs.est.1c06896

Weber, W. J., & Huang, Q. (2003). Inclusion of persistent organic pollutants in humification processes: Direct chemical incorporation of phenanthrene via oxidative coupling. *Environmental Science & Technology, 37*(18), 4221–4227. https://doi.org/10.1021/es030330u

Wu, C., Klemes, M. J., Trang, B., Dichtel, W. R., & Helbling, D. E. (2020). Exploring the factors that influence the adsorption of anionic PFAS on conventional and emerging adsorbents in aquatic matrices. *Water Research, 182*, 115950. https://doi.org/10.1016/j.watres.2020.115950

Yang, Y., Zheng, Z., Ji, W., Xu, J., & Zhang, X. (2020). Insights to perfluorooctanoic acid adsorption micro-mechanism over Fe-based metal organic frameworks: Combining computational calculation with response surface methodology. *Journal of Hazardous Materials, 395*, 122686. https://doi.org/10.1016/j.jhazmat.2020.122686

Yassin, J. M., Taddesse, A. M., & Sánchez-Sánchez, M. (2022). Sustainable synthesis of a new semiamorphous Ti-BDC MOF material and the photocatalytic performance of its ternary composites with Ag3PO4 and g-C3N4. *Applied Surface Science, 578*, 151996. https://doi.org/10.1016/j.apsusc.2021.151996

Chapter 6
Advances in Mycoremediation: Effective Strategies for Heavy Metal Removal from Wastewater

N. K. Praful, Binaya Kumar Pattnaik, Bishnu Prasad Sahoo, and Rasmita Behera

Abstract The increasing incidences of heavy metal contamination in aquatic ecosystems, driven by swift industrialization, present a significant risk to both ecosystem and public health. Conventional remediation techniques like chemical precipitation, ion exchange, and membrane filtration frequently encounter challenges concerning cost, effectiveness at low metal concentrations, and the potential for secondary pollution. The application of fungi for the detoxification or removal of heavy metals, known as mycoremediation, has surfaced as a sustainable and effective alternative. This chapter provides an in-depth examination of the principles, mechanisms, and applications of mycoremediation in the context of wastewater treatment. This study delves into significant fungal species such as *Aspergillus*, *Penicillium*, *Phanerochaete chrysosporium*, and *Pleurotus*, which exhibit robust metal-binding abilities via biosorption, bioaccumulation, biotransformation, and bioleaching processes. This chapter examines the incorporation of fungi throughout different phases of wastewater treatment, as well as the application of agricultural and industrial waste as substrates for fungal growth. In light of the obstacles associated with field scalability, microbial competition, and biomass disposal, recent advancements in genetic engineering, immobilization techniques, and bioreactor design present considerable potential. This chapter highlights mycoremediation as

N. K. Praful
Symbiosis Institute of Technology, Symbiosis International (Deemed University), Pune, Maharashtra, India

B. K. Pattnaik (✉)
Institute of Environment Education and Research, Bharati Vidyapeeth (Deemed to be University), Pune, Maharashtra, India

B. P. Sahoo
Forest Ecology and Climate Change Division, Forest Research Institute, Dehradun, India

R. Behera
Symbiosis Centre for Management Studies, Symbiosis International (Deemed University), Pune, Maharashtra, India

C. M. Hussain et al. (eds.), *Fungi in Waste Water Treatment*, Springer Water,
https://doi.org/10.1007/978-3-032-12693-1_6

an economical and environmentally sustainable approach to addressing heavy metal contamination in wastewater systems.

Keywords Mycoremediation · Heavy metal removal · Fungal bioremediation · Wastewater treatment · Biosorption mechanisms

6.1 Introduction

The rapid industrialization of the last century has led to a concerning increase in environmental pollution, especially the contamination of aquatic systems with hazardous heavy metals. These contaminants, Chromium (Cr), lead (Pb), nickel (Ni), zinc (Zn), arsenic (As), cadmium (Cd), and mercury (Hg) are non-biodegradable and tend to accumulate in aquatic environments, presenting significant hazards to both human health and the environment (Singh et al., 2023). Traditional methods for addressing heavy metal-contaminated effluents, such as Ion exchange, chemical precipitation, and membrane filtration, frequently encounter elevated operational expenses, inefficiency at low concentrations, and the production of secondary pollutants (Verma et al., 2017).

Considering these constraints, mycoremediation, the application of fungi to eliminate or neutralize environmental contaminants, has surfaced as a sustainable and economical alternative. Fungal species demonstrate exceptional capacities to bind, accumulate, and alter heavy metals via mechanisms including biosorption, bioaccumulation, and enzymatic biotransformation (Ayele et al., 2021). Filamentous fungi have elevated surface-area-to-volume ratios, several functional groups in their cell walls, and an array of extracellular enzymes, rendering them highly effective for environmental remediation purposes (Ferreira et al., 2020).

Recent study has validated the metal-binding efficacy of extensively examined fungus such as *Aspergillus*, *Penicillium*, and *Pleurotus* spp., while also revealing proteomic and metabolic changes that improve their performance under multi-metal stress conditions (Dey et al., 2024). Additionally, fungal biomass is frequently available as an industrial waste, improving the economic viability of large-scale implementation (Thiyur & Samuel, 2020).

This chapter critically assesses contemporary advances in mycoremediation approaches for heavy metal extraction from wastewater. This study will investigate the biological principles of fungal remediation, enumerate effective fungal species, analyze mechanistic techniques, and assess practical applications and potential advancements in the field. This chapter emphasizes the potential of fungus as environmentally sustainable bioremediation agents to combat the escalating global issue of heavy metal contamination.

6.2 Fundamentals of Mycoremediation

6.2.1 Definition and Overview

Mycoremediation is a fungi-based bioremediation strategy that leverages metabolic processes to detoxify or extract heavy metals and co-contaminants from wastewater, sediments, and soil. It utilizes the inherent metabolic and biosorptive abilities of fungi to convert harmful compounds into less toxic or nontoxic forms (Ayele et al., 2021). Fungi, particularly filamentous species, exhibit various structural and physiological characteristics that enhance their effectiveness as remediators (Ghosh et al., 2023). Their extensive mycelial networks offer significant surface areas for engagement with pollutants. The main constituents of the fungal cell wall are glycoproteins, glucans, and chitin. It features functional groups such as Amine, phosphate, carboxyl, and hydroxyl, which facilitate the binding of heavy metal ions through biosorption mechanisms (Shourie & Vijayalakshmi, 2022).

6.2.2 Advantages Over Other Bioremediation Agents

Fungi present notable advantages compared to other bioremediation agents, including bacteria and algae, particularly for cleaning up ecosystems contaminated with heavy metals. The primary advantage is their capacity to endure and prosper in extreme environmental circumstances, including wide variations in temperature, salinity, and pH. This adaptability enables fungi to operate efficiently in industrial wastewater environments where other microorganisms may not survive (Roy et al., 2025).

Fungi have a distinctive range of extracellular enzymes, including peroxidases, laccases, and reductases, which help the degradation of organic pollutants and engage in redox reactions that detoxify heavy metals via biotransformation and immobilization processes (Dinakarkumar et al., 2024). A significant benefit is the saprophytic characteristic of numerous fungal species, enabling their growth on diverse organic waste materials, including straw, sawdust, and agricultural residues. This feature renders extensive fungal cultivation both economically viable and environmentally sustainable (Mastranzo-Pérez et al., 2025). Additionally, numerous fungi employed in mycoremediation are non-pathogenic and safe for handling, presenting minimal risk to both workers and the surrounding environment (Sen et al., 2023).

Fungi generally produce a significant biomass yield, thereby increasing their metal-binding surface area and biosorption capacity. This characteristic, along with their resilience and enzyme production, establishes fungi as more efficient and scalable bioremediation agents in comparison to bacteria and algae (Shourie & Vijayalakshmi, 2022).

6.3 Heavy Metals in Wastewater: Sources, Impacts, and Regulatory Concerns

Heavy metals are amongst the most persistent and hazardous pollutants found in industrial effluent. Various industrial processes, such as textile production, tanning, mining, electroplating, manufacturing, battery production, and the manufacture of paints and pigments, all contribute to the release of these contaminants into aquatic systems (Binkley & Simpson, 2003). Metals frequently identified include chromium (Cr), zinc (Zn), nickel (Ni), cadmium (Cd), copper (Cu), mercury (Hg), and lead (Pb). Due to their non-biodegradable nature and ability to bioaccumulate in the food chain, these metals pose significant risks (Akpor et al., 2014). The toxicological impacts of these metals on the environment are significant and complex. Heavy metals in aquatic ecosystems disrupt microbial communities, induce oxidative stress in aquatic fauna, and lead to habitat degradation, including algal blooms and sedimentation problems. In plants, their presence can hinder nutrient uptake, photosynthesis, and growth, ultimately diminishing crop quality and yield (Rafique et al., 2022). A frequently neglected yet essential effect pertains to the wastewater treatment infrastructure itself. Heavy metals impede microbial activity in biological treatment systems, consequently diminishing the effectiveness of processes including nitrification, sludge digestion, and overall chemical oxygen demand (COD) removal (Gupta & Diwan, 2017). This not only raises treatment costs but also heightens the risk of untreated discharges contaminating natural water bodies.

Various international regulatory frameworks have established permissible limits for heavy metal concentrations in industrial effluents to mitigate these risks. Adherence to these standards exhibits global inconsistency, especially in developing nations where industrial growth frequently surpasses environmental regulation. The increasing concern regarding the limitations of the traditional treatment methods has given rise to a heightened interest in cost-effective and ecologically sustainable technologies, such as mycoremediation, for addressing the heavy metal contaminations in wastewater (Khan, 2021).

6.3.1 Major Industrial Sources of Heavy Metal Pollution

Heavy metal pollution in wastewater largely arises from several industrial sectors, where metals are consumed, processed, or produced as by-products (Barakat, 2011). These industries include mining, metallurgy, chemical production, textiles, tanneries, electroplating, and petroleum refining, among others. Figure 6.1 demonstrate the different sources of heavy metals entering the aquatic environment.

Fig. 6.1 Source of Heavy metal contamination in water

6.3.1.1 Chemical Production Industry (CPI)

The chemical production industry comprises a variety of manufacturing industries, such as petroleum, fertilizers, insecticides, pulp and paper, paints, dyes, pigments, and soaps. The key sources of heavy metals in this sector include chemical plant emissions, subterranean pipeline leaks, incorrect waste disposal, raw materials, and chemical processes (Rahman et al., 2022). Common heavy metals encountered in this sector include are Cadmium (Cd), iron (Fe), arsenic (As), nickel (Ni), aluminum (Al), lead (Pb), chromium (Cr), mercury (Hg), zinc (Zn), and copper (Cu). The textile industry is particularly recognized for employing synthetic dyes and chemicals that commonly include heavy metals and other compounds resistant to degradation (Adane et al., 2021; Kant, 2012). Wastewater generated during petroleum production often contains dissolved heavy metals for example mercury and cadmium, which present significant risks to the environment (Peng et al., 2020). Similarly, wastewater produced during wood digestion, pulp bleaching, washing, and dewatering in paper production contains both heavy metals and organic substances (Singh & Chandra, 2019).

6.3.1.2 Textile Industry

The textile industry significantly contributes to heavy metal pollution, mostly due to the use of dyes, finishing chemicals, and other chemical additives that often include heavy metals. Toxic metals like copper (Cu), chromium (Cr), and arsenic (As) can be found in textile dyes, particularly those used in synthetic materials, along with other dangerous substances that are not biodegradable and linger in the environment (Velusamy et al., 2021). These metals and other toxins may be present in textile effluent, which is frequently released into water systems untreated, causing serious pollution and harm to ecosystems (Khan & Malik, 2014). Additionally, effluents from the textile industry may also contain high concentrations of chemicals from processes such as bleaching, scouring, and dyeing, further contributing to environmental toxicity (Yaseen & Scholz, 2019).

6.3.1.3 Tanneries

The tanning industry, which processes animal skins to generate leather items, is another important contributor to heavy metal pollution, primarily owing to the application of salts of chromium in the tanning method. The wastewater from these operations contains significant concentrations of hazardous chromium compounds, and chromium (Cr) is frequently employed as a tanning agent. Improper disposal of tannery effluents typically results in the pollution of adjacent water bodies with these toxic metals, as well as other pollutants like sulfides and salts. The release of untreated or insufficiently treated tannery wastewater can result in substantial environmental harm, damaging both water quality and aquatic life (Yaseen & Scholz, 2019).

6.3.1.4 Mining and Metallurgy Industry (MOPI)

Mining and ore processing industries are major contributors to heavy metal pollution. These sectors extract and refine metals used in various applications, including construction, electronics, energy, and manufacturing. Metals such as Silver (Ag), copper (Cu), lead (Pb), gold (Au), iron (Fe), and others are commonly processed in these industries (Fashola et al., 2016). Main heavy metals released during mining and metallurgical operations include Cadmium (Cd), copper (Cu), manganese (Mn), arsenic (As), nickel (Ni), zinc (Zn), chromium (Cr), aluminum (Al), lead (Pb) and mercury (Hg) (Angon et al., 2024). Pollution arises from critical activities such as ore crushing, grinding, chemical leaching, smelting, refining, and acid mine drainage (Niu et al., 2023). For instance, a study investigating steel industry effluent showed elevated levels of zinc (Zn) and cadmium (Cd) in urban waterways due to metal accumulation in sediments. Notably, up to 90% of the wastewater generated during smelting, rolling, and other operations is discharged without adequate treatment, severely impacting surrounding water systems (Das et al., 2021).

6.3.1.5 Electroplating Industry

The electroplating industry applies metallic coatings—often using chromium, nickel, zinc, silver, and gold—to various substrates to enhance durability, appearance, and corrosion resistance. Common in sectors such as automotive, aerospace, electronics, and jewelry, electroplating processes wastewater heavily laden with toxic heavy metals (Rahman et al., 2021; Rajoria et al., 2022). Processes contributing to HM contamination include water demineralization, aqueous cleaning, solvent recovery, and acid pickling (Babu et al., 2009). Without effective treatment systems, the discharge from electroplating facilities poses a significant hazard to aquatic ecosystems and public health.

6.3.2 Toxic Effects of Heavy Metals on Human Health and Ecosystems

6.3.2.1 Lead (Pb)

Lead is a highly toxic metal that primarily enters the human body by ingesting contaminated food or water, by inhaling dust, or by dermal contact with lead-laden materials (Wani et al., 2015). It is a well-known neurotoxin and has a particularly damaging effect on brain development in children. Lead exposure in children is associated with cognitive deficits, lower IQ, and behavioral problems (Wani et al., 2015). In adults, lead toxicity can lead to high blood pressure, reproductive problems, and kidney damage related issues. Lead also disrupts hematopoiesis, leading to anemia and other blood disorders. The U.S. Environmental Protection Agency (EPA) has set a maximum permissible concentration of 0.015 mg/L for lead in drinking water, while the World Health Organization (WHO) advises keeping children's blood lead levels below 10 μg/dL to avoid harmful health impacts (Wani et al., 2015). Lead accumulates in the environment, especially in water and soil, where it can persist for long durations due to its non-biodegradable nature. This contamination can disrupt plant growth, affect wildlife, and bioaccumulate in food chains, leading to the poisoning of both aquatic and terrestrial organisms (Wani et al., 2015).

6.3.2.2 Cadmium (Cd)

Cadmium is released into the environment through various industrial activities, such as fossil fuel combustion, metal extraction, and the agricultural use of sewage sludge. It can also be found in cigarette smoke. Cadmium exposure primarily results from the ingestion of contaminated water. Chronic exposure to cadmium can cause kidney failure, bone disease, and cardiovascular problems. It is also a carcinogen, linked to an increased risk of lung cancer and prostate cancer (Matebese et al.,

2024). The WHO recommends a limit of 0.003 mg/L for cadmium in drinking water, while the EPA has set the maximum permissible limit concentration at 0.005 mg/L. Cadmium contamination degrades soil fertility and hampers plant growth, while its accumulation in the food chain poses a lasting risk to both human and wildlife health (Kubier et al., 2019).

6.3.2.3 Mercury (Hg)

Mercury is a naturally occurring element that is discharged into the environment by industrial processes such as chemical manufacture, coal combustion, and mining. Particularly in its methylmercury form, which bioaccumulates in aquatic environments, mainly in fish, it is extremely dangerous. Inhaling vapors, eating tainted fish and seafood, or being exposed at work are the main ways that humans get exposed to mercury (Kubier et al., 2019). Mercury poisoning can harm the nervous system, resulting in cognitive impairment, tremors, and memory loss. It is particularly detrimental to the development of the fetus and pregnant women, affecting brain development and causing long-term cognitive and motor issues in offspring (Fernandes Azevedo et al., 2012). The WHO suggests a mercury limit of 0.006 mg/L for drinking water, but the EPA has set a stricter maximum contaminant level at 0.002 mg/L. Additionally, exposure to mercury disrupts entire food chains, harms aquatic ecosystems, and negatively impacts fish and other aquatic organisms (Kortei et al., 2020).

6.3.2.4 Chromium (Cr)

Chromium is a hazardous metal that is utilized in many industrial processes, including metal plating, tanning leather, and pigment manufacturing. It is especially useful in its hexavalent form (Cr(VI)). Inhaling chromium dust, coming into contact with chromium-containing goods on the skin, or consuming tainted food or water can all expose one to chromium (Sharma et al., 2022). Prolonged exposure to Cr(VI) can lead to kidney and liver damage, skin ulcers, gastrointestinal problems, and lung cancer. The WHO suggests a limit of 0.05 mg/L for chromium in drinking water, but the EPA has set the maximum permissible content at 0.1 mg/L. Environmental chromium pollution hinders photosynthesis, stunts plant development, and interferes with aquatic life's ability to reproduce (Georgaki et al., 2023).

6.3.2.5 Zinc (Zn)

Zinc is an essential trace element required for numerous physiological processes, such as immune function, protein production, and DNA synthesis. Excessive zinc exposure can lead to stomach pain, nausea, and vomiting, along with other gastrointestinal issues (Hussain et al., 2022). Extended exposure to elevated zinc levels can disrupt the body's capacity to absorb copper, resulting in a copper shortage and adversely impacting numerous enzyme systems. The WHO recommends limiting zinc levels in drinking water to a maximum of 5 mg/L to prevent adverse health effects. Excessive zinc can also disrupt aquatic ecosystems by altering water chemistry, lowering oxygen levels, and compromising the health of fish and invertebrates. Zinc pollution of soil can diminish soil fertility, decreasing agricultural productivity (Singh et al., 2022).

6.3.2.6 Nickel (Ni)

Nickel is a commonly found metal that enters the environment through industrial activities such as metal production, fossil fuel burning, and mining. It can cause a range of health problems, particularly when exposure occurs through inhalation of nickel dust or direct contact with nickel-containing products. Chronic exposure to nickel can cause respiratory problems, skin allergies, and lung cancer. Nickel toxicity has also been linked to cardiovascular diseases and kidney damage (Genchi et al., 2020). The EPA has set a maximum permissible concentration of 0.1 mg/L for nickel in drinking water, while the WHO recommends a limit of 0.02 mg/L. Nickel contamination in the environment can damage plant health, reduce soil fertility, and harm aquatic organisms by accumulating in their tissues (Genchi et al., 2020).

6.3.2.7 Arsenic (As)

Arsenic is a naturally occurring metalloid that is found in water and can be released into the environment through industrial discharges, mining, and agricultural activities. Chronic exposure to arsenic, especially from consuming contaminated water, has been associated with gastrointestinal problems, skin lesions, and a higher potential risk of cancers of the skin, liver, lungs, and bladder (Jaishankar et al., 2014). The EPA has set the maximum permissible concentration of arsenic in drinking water at 0.01 mg/L, and the WHO recommends the same limit. Arsenic contamination of the environment can damage aquatic ecosystems, disrupt photosynthesis, inhibit plant growth, and result in ecological imbalance and a loss of biodiversity.

6.3.3 Regulatory Standards for Heavy Metal Contamination

Regulatory agencies like the U.S. EPA and WHO have established standards for the maximum allowable concentrations of heavy metals in drinking water, soil, and air to mitigate the harmful effects of heavy metal exposure (Table 6.1). These standards are critical in limiting human exposure to toxic metals and preventing acute

Table 6.1 Heavy metal regulatory limits

Heavy metal	EPA limit (mg/L)	WHO limit (mg/L)
Arsenic (As)	0.01	0.01
Cadmium (Cd)	0.003	0.005
Chromium (Cr)	0.05	0.1
Lead (Pb)	0.01	0.015
Mercury (Hg)	0.006	0.002
Zinc (Zn)	5.0	5.0

poisoning and chronic diseases. For instance, the EPA sets the maximum allowable levels for lead and mercury in drinking water at 0.015 mg/L and 0.002 mg/L, respectively, while the WHO advises maximum levels of 0.01 mg/L for arsenic and 0.05 mg/L for chromium (Shashank & Velayudhannair, 2024).

These guidelines are crucial for protecting public health, but they also underscore the necessity for effective remediation technologies to treat contaminated water and soil. Considering the persistence of heavy metals in the environment, rigorous monitoring, regulation, and remediation efforts are essential to prevent these contaminants from accumulating to dangerous levels in ecosystems and human communities.

6.3.4 The Need for Remediation of Heavy Metal Contamination

The persistence and bioaccumulative nature of heavy metals in the environment necessitate the development of effective remediation strategies. Traditional methods, such as ion exchange, chemical precipitation, and reverse osmosis, are frequently used to remove heavy metals from contaminated water. However, these methods can be costly and may generate secondary waste that requires further treatment. Mycoremediation is an emerging and eco-friendly solution to heavy metal contamination, a bioremediation approach that utilizes fungi to degrade, immobilize, or detoxify harmful metals. Certain fungi, mainly white rot fungi like Phanerochaete chrysosporium, have shown a remarkable ability to absorb and transform heavy metals into less toxic forms. These fungi produce extracellular enzymes, such as laccases and peroxidases, to break down environmental metal contaminants. Additionally, fungi can immobilize metals by binding them to their cell walls, preventing them from spreading further into the environment. Other bioremediation methods, such as phytoremediation (using plants) and bacterial remediation, also promise to remove heavy metals from contaminated soil and water. These approaches are sustainable, environmentally friendly, and cost-effective, offering a viable alternative to traditional chemical-based remediation approaches.

6.4 Fungal Species Used in Mycoremediation

6.4.1 Overview of Key Fungi

A variety of fungal species has been identified for their ability to remediate environments contaminated with heavy metals. These fungi employ biosorption, bioaccumulation, and enzymatic transformation to immobilize or detoxify metals. Species of the genus Pleurotus, commonly known as oyster mushrooms, are among the most extensively studied due to their widespread cultivation, rapid growth, and tolerance to metals including cobalt (Co), copper (Cu), and nickel (Ni) (Chiu et al., 1998). Research indicates that Pleurotus spp. can thrive in metal-stressed environments and generate antioxidant enzymes such as superoxide dismutase and catalase to mitigate oxidative stress, thereby improving their bioremediation capabilities (Mohamadhasani & Rahimi, 2022).

White-rot fungi, specifically *Trametes versicolor and Phanerochaete chrysosporium*, represent a significant category utilized in mycoremediation. These fungi exhibit a strong ligninolytic enzyme system that can decompose complex pollutants and convert metal ions. The capacity to generate extracellular enzymes such as peroxidases and laccases enables the alteration of metal oxidation states, thereby reducing toxicity or facilitating sequestration (Fragoeiro & Magan, 2005; Gugel et al., 2024).

Filamentous fungi, including *Aspergillus niger*, *Aspergillus fumigatus*, and *Penicillium rubens*, exhibit notable biosorptive capacities (Iskandar et al., 2011). *A. niger* can effectively eliminate up to 43% of chromium (Cr) and 98% of cadmium (Cd)) from contaminated soil, attributed to its dense mycelial networks and functional groups present on the cell wall that enhance metal binding (Khan et al., 2019).

Penicillium species have been documented to biosorb iron (Fe) and copper (Cu) from contaminated environments, facilitated by their capacity to thrive in low-pH conditions and synthesize chelating compounds (Tagyan et al., 2023). Additional notable genera encompass Fusarium, Trichoderma, Curvularia, and Rhizopus, all exhibiting differing levels of metal tolerance and removal efficacy contingent upon environmental conditions and the specific type of metal involved (Zafar et al., 2007). *Curvularia lunata* exhibits significant biosorption potential for lead (Pb) and manganese (Mn), indicating its applicability in contaminated aquatic environments (Sugitha et al., 2024).

The findings indicate that selecting a suitable fungal species for mycoremediation is contingent upon the specific contaminant and the environmental matrix, such as soil, water, or sludge. The availability of fungal biomass, frequently a byproduct of food or pharmaceutical industries, renders fungi a practical and cost-effective choice for large-scale remediation programs (Ayele et al., 2021).

6.4.2 Traits Supporting Heavy Metal Tolerance and Accumulation in Fungi

Fungi have diverse biological adaptations that facilitate their survival in metal-contaminated habitats and their capacity to absorb heavy metals, rendering them especially useful for mycoremediation. The composition of the fungal cell wall, characterized by a high concentration of chitin, glucans, and glycoproteins, is one of the most essential features. The structural components exhibit several functional groups—including carboxyl, hydroxyl, amino groups, and phosphate—that function as active binding sites for metal ions. This characteristic enables both live and deceased fungal biomass to passively sequester metals via biosorption (Viraraghavan & Srinivasan, 2011).

Besides their chemical composition, fungi—particularly filamentous varieties—develop enormous mycelial networks that offer a substantial surface area for interaction with pollutants. This anatomical characteristic improves their ability to uptake heavy metals from aquatic or terrestrial habitats (Ramya et al., 2021). Upon entering the cell, heavy metals are frequently sequestered in vacuoles or associated with intracellular chelators such as metallothioneins and glutathione. These sequestration systems mitigate metal toxicity and safeguard vital cellular functions (Manisha et al., 2016).

Such an adaptation is the synthesis of melanin by some fungi. Melanin serves as a natural defense against oxidative stress and chelates metal ions, hence augmenting the fungi's resilience to hazardous conditions. Melanized fungal species exhibit enhanced survivability and remediation efficacy in severely contaminated environments (Tamjidi et al., 2022). The amalgamation of structural robustness, metabolic flexibility, and metal-chelating capacities highlights the significant role fungi can assume in sustained heavy metal cleanup.

6.5 Mechanistic Strategies for Metal Removal

Fungal species demonstrate an impressive ability to engage with heavy metals via various biochemical and physiological mechanisms. The mechanisms exhibit a range of complexity, from passive adsorption on cell surfaces to active intracellular accumulation and enzymatic transformation. Comprehending each of these processes is crucial for enhancing fungal performance across various environmental conditions. The key mechanisms involved in the remediation of heavy metals through fungal processes are detailed below:

6.5.1 Biosorption

Biosorption is a passive process, independent of metabolism, where heavy metal ions bind to the surfaces of fungal cells. The fungal cell wall comprises polysaccharides, including chitin and glucans, along with proteins abundant in functional groups (e.g., Amino, hydroxyl, phosphate, carboxyl) that engage with metal ions by mechanisms such as Chelation, ion exchange, physical adsorption, and complexation. Both viable and non-viable fungal biomass (e.g., Aspergillus, Penicillium, Ganoderma) have significant biosorption efficacy for metals such as Cd^{2+}, Cu^{2+}, and Pb^{2+} (Ayele et al., 2021; Muraleedharan & Venkobachar, 1994).

6.5.2 Bioleaching

Bioleaching is a metabolic process in which fungi excrete organic acids (e.g., oxalic, gluconic, citric) that solubilize metal ions from solid substrates such as ores, fly ash, or polluted soil. These acids lower the pH and chelate metals, hence improving their solubility and mobilization. *Aspergillus niger* and *A. nomius* are effective bioleachers for metals such as Zn, Cd, and Pb in acidic conditions (Chatterjee et al., 2020; Yang et al., 2009).

6.5.3 Bioprecipitation

Bioprecipitation involves fungi releasing ligands or enzymes that facilitate the conversion of metal ions into insoluble salts through precipitation. Fungi generate oxalates, phosphates, or sulfides that interact with free metal ions to create metal oxalates (e.g., Ca–Cd–oxalate), phosphates, or sulfides, thus immobilizing the metals. This method is very effective in diminishing the mobility and bioavailability of metals in situ (Liang & Gadd, 2017; Zheng et al., 2024).

6.5.4 Bioaccumulation

Bioaccumulation is a metabolism-driven process in which living fungal cells actively translocate metal ions via the cell membrane. The metals are then complexed with intracellular ligands, including metallothioneins, phytochelatins, and glutathione, and sequestered within organelles such as vacuoles. This technique is exceptionally precise and efficacious at low metal concentrations. Indigenous strains such as *Aspergillus niger* and *Penicillium rubens* have demonstrated significant bioaccumulation of Cd and Cr in soil rehabilitation initiatives (Bibbins-Martínez et al., 2023; Khan et al., 2019).

6.5.5 Biotransformation

Biotransformation denotes the enzymatic modification of metal oxidation states, hence decreasing their toxicity. Fungi generate reductases and oxidases that transform metals into less reactive states; for instance, they reduce hexavalent chromium (Cr^{6+}), a carcinogen, to trivalent chromium (Cr^{3+}), which is comparatively less detrimental. This method is essential for metal detoxification and enhancing environmental safety (Refaey et al., 2021; Shourie & Vijayalakshmi, 2022).

6.5.6 Biomineralization

Fungi perform a vital role in biomineralization by modifying environmental pH and producing enzymes, including phosphatases, which contribute to the formation of stable mineral compounds. The specified conditions facilitate the precipitation of metal phosphates or carbonates, resulting in the permanent immobilization of metals in stable crystalline structures. *Penicillium simplicissimum* can precipitate uranium in the form of uranyl phosphate, which makes it environmentally inert (Liang & Gadd, 2017).

Figure 6.2 explains the major mechanism governing the mycroremediation of heavy metals from wastewater. These four methods frequently function concurrently

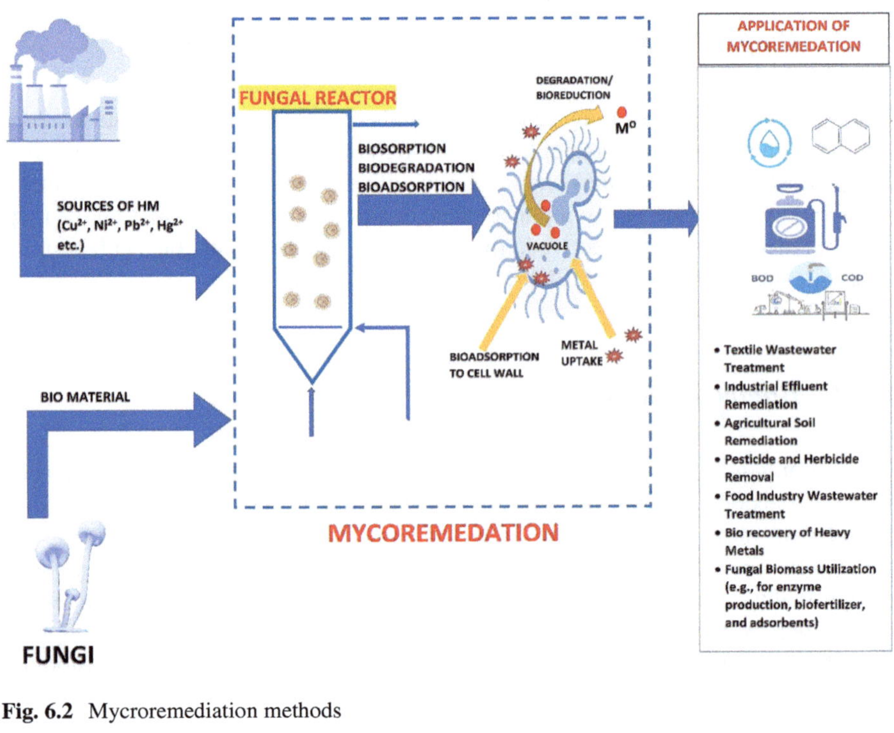

Fig. 6.2 Mycroremediation methods

within fungal systems. *Aspergillus niger* and *Phanerochaete chrysosporium* initially adsorb metals via biosorption, subsequently transport and sequester them intracellularly through bioaccumulation, and ultimately convert or precipitate them through enzymatic biotransformation or organic acid excretion (Fortier et al., 2015). These mechanisms elucidate the strong efficacy of fungi in metal-contaminated environments and provide a framework for developing integrated and scalable fungal-based remediation systems.

6.6 Application of Fungi in Different Stages of Wastewater Treatment

Fungi can be strategically integrated at different stages of wastewater treatment to enhance the efficiency of heavy metal removal (Fig. 6.3). Their adaptability, high metal-binding capacity, and ability to grow in harsh conditions make them suitable for multiple roles across the treatment process.

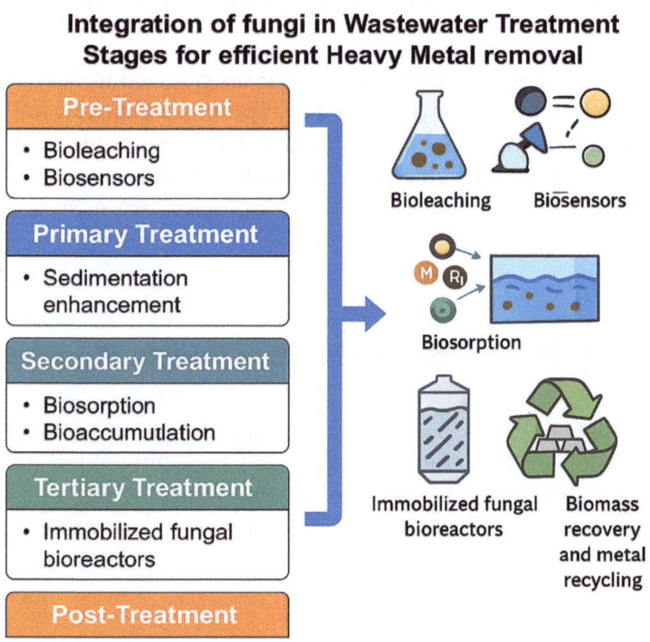

Fig. 6.3 Integration of fungi in different stages of wastewater treatment

6.6.1 Pre-treatment: Bioleaching and Biosensors

Fungi such as Aspergillus and Penicillium secrete organic acids (e.g., citric, oxalic), which solubilize heavy metals from suspended solids. This bioleaching process increases the mobility of metals, making them easier to remove in downstream treatments. Additionally, fungal metabolic activity can be used in biosensor systems to detect metal concentrations in influent wastewater (Manisha et al., 2016).

6.6.2 Primary Treatment: Sedimentation Enhancement

Though fungi are not typically involved in sedimentation, their filamentous mycelia can assist by forming networks that trap metal particulates and enhance floc formation. This increases the sedimentation rate and efficiency of heavy metal removal during primary clarification (Wang et al., 2024).

6.6.3 Secondary Treatment: Biosorption and Bioaccumulation

Fungi like *Aspergillus niger* and *Rhizopus arrhizus* can be introduced into aeration basins or bioreactors where they perform biosorption via their cell walls and bioaccumulation through active metal uptake. Fungal cells are capable of binding metals such as Pb, Cu, and Cd with high efficiency. The use of immobilized fungal biomass improves reusability and operational control in biological treatment systems (Kumar et al., 2019).

6.6.4 Tertiary Treatment: Immobilized Fungal Bioreactors

In advanced treatment stages, packed-bed and fluidized-bed bioreactors loaded with immobilized fungi are used for continuous and high-throughput metal removal. These systems maintain high biosorption capacity over time and are ideal for treating low-concentration effluents before discharge or reuse (Kapoor & Viraraghavan, 1998).

6.6.5 Post-Treatment: Biomass Recovery and Metal Recycling

Following remediation, metal-laden fungal biomass can be subjected to acid elution or thermal processing to recover valuable metals or render the biomass safe for disposal. This step supports resource recovery and sustainability in the treatment process (Sharma et al., 2022).

6.7 Process Optimization and Environmental Parameters

The efficacy of fungal-based heavy metal cleanup is heavily influenced by environmental and operating circumstances. Enhancing essential parameters such as pH, fungal biomass dosage, temperature, contact duration, and initial heavy metal concentration is vital for attaining optimal biosorption effectiveness.

6.7.1 pH

The pH is crucial in influencing the availability of metal ions and the surface charge of fungal cells. Numerous studies indicate that optimal fungal biosorption occurs within a mildly acidic to neutral pH range. Trichoderma viride exhibited optimal chromium removal efficiency at pH 6, so affirming the biosorption process's sensitivity to hydrogen ion concentration (Tello-Galarreta et al., 2023). Cadmium biosorption by Actinomucor sp. was determined to be optimum at pH 7–8 (Masoudi et al., 2017).

6.7.2 Temperature

Temperature additionally affects fungal metabolism and biosorption kinetics. Optimal temperature ranges typically lie between 25–45 °C, with research indicating enhanced cadmium absorption by *Actinomucor* sp. at 45 °C (Masoudi et al., 2017), while *Aspergillus niger* exhibits maximum activity at 35 °C (Bhateria & Dhaka, 2019).

6.7.3 Dosage

The dosage of biomass directly influences the quantity of accessible metal-binding sites. Elevated biosorbent concentration often results in enhanced removal efficiency until equilibrium is attained. Aspergillus flavus demonstrated optimal zinc and nickel removal at the biomass concentration level of 2 g/L (Raut & Wani, 2010).

6.7.4 Contact Time

Contact duration is essential, as fungal biosorption often attains equilibrium within 30–120 min. Trichoderma viride accomplished approximately 98% chromium removal in merely 45 min (El-Kassas & El-Taher, 2009), but Actinomucor sp. necessitated about 120 min for optimal cadmium absorption (Masoudi et al., 2017).

6.7.5 Initial Concentration

The initial metal concentration influences uptake dynamics by saturating binding sites. Elevated concentrations may enhance absorption per unit biomass while diminishing percentage elimination. For instance, Penicillium sp. demonstrated optimum biosorption of nickel and lead at metal concentrations of 123–133 mg/L (Aytar et al., 2014).

6.7.6 Immobilization

Immobilization strategies have been extensively utilized to improve fungal performance and durability in diverse situations. Fungal biomass can be immobilized on substrates such as alginate beads, charcoal, or polyurethane foam. This method safeguards fungal cells while facilitating recurrent utilization, ease of separation, and stability in flow systems (Ayele et al., 2021).

The utilization of agricultural and industrial wastes as growth substrates for fungi in mycoremediation has gained prominence. These waste materials—such as straw, sawdust, sugarcane bagasse, or leftover brewing grains—diminish expenses while promoting sustainable circular economy frameworks. These substrates are nutrient-dense and facilitate substantial fungal biomass generation, essential for large-scale use.

6.8 Challenges and Limitations

6.8.1 Scale-Up and Commercialization Issues

A primary obstacle in mycoremediation is the difficulty of translating laboratory techniques to practical outside applications. Fungi have significant biosorption and transformation efficiency in controlled conditions; however, these efficiencies frequently diminish in field situations due to environmental variability, including fluctuating moisture, pH, temperature, and pollutant dispersion (Roy et al., 2025). Moreover, there is an absence of commercially accessible, standardized bioreactor designs specifically designed for fungal treatment, hindering the advancement of automated or continuous mycoremediation systems. Cost-effectiveness is a significant obstacle, especially in relation to traditional treatment modalities that are firmly embedded within the current infrastructure.

6.8.2 Competition with Native Microbes

In both natural and contaminated environments, introduced fungal strains must compete with native microbial communities, which include bacteria, other fungi, and actinomycetes. Microbial competition may diminish the efficacy of inoculated fungi by restricting nutrient access and available surface colonization sites. The effectiveness of mycoremediation frequently relies on the ecological compatibility of the introduced fungi and their capacity to either dominate or synergize with indigenous organisms (Nielsen et al., 2010). In the absence of effective competition management, there may be a rapid decline in fungal activity, which could diminish long-term remediation potential.

6.8.3 Metal Recovery and Disposal of Fungal Biomass

An essential yet frequently neglected concern in mycoremediation is the secure management and disposal of metal-laden fungal biomass. Following biosorption or bioaccumulation, the fungal material harbors concentrated heavy metals, which may present secondary pollution hazards if inadequately managed (Arikan et al., 2020). Methods include incineration, composting, and chemical extraction are employed for biomass treatment and metal recovery (Wang & Chen, 2009); however, these techniques can be expensive, energy-intensive, or generate toxic byproducts (Gadd, 2009). It is imperative to develop cost-effective and sustainable solutions for the disposal of post-treatment biomass to ensure practical applicability.

6.8.4 Genetic and Metabolic Constraints

Despite the inherent resistance of numerous fungi to heavy metals, genetic and metabolic constraints may hinder their remediation efficacy, particularly in environments with mixed metals or elevated toxicity levels. Certain fungi are deficient in the enzyme systems required to convert specific metals or decompose intricate co-contaminants. Moreover, even advantageous strains may demonstrate sluggish growth rates, variable enzyme production, or susceptibility to metal concentrations above specific thresholds (Chan et al., 2016). Despite progress in genetic engineering—such as CRISPR-Cas9 and transcriptome profiling—regulatory, biosafety, and ethical issues with genetically modified organisms (GMOs) continue to impede their deployment in open ecosystems.

These issues underscore the necessity for comprehensive, interdisciplinary research and regulatory cooperation to fully realize the potential of mycoremediation. Resolving these challenges via innovation and standardization will be essential for its future scalability and worldwide acceptance.

6.9 Conclusion

This chapter presents a comprehensive examination of the application of fungi in the remediation of wastewater contaminated with heavy metals. The text underscores the increasing environmental issues stemming from industrial pollution, especially due to hazardous metals such as mercury, lead, arsenic, cadmium, and chromium, while also pointing out the shortcomings of conventional remediation techniques like ion exchange and chemical precipitation. Mycoremediation presents a compelling alternative, attributed to the distinctive physiological and biochemical characteristics of fungi. Filamentous fungi like *Aspergillus niger*, *Phanerochaete chrysosporium*, and *Pleurotus* spp. demonstrate the ability to eliminate heavy metals via various mechanisms such as bioaccumulation, bioleaching, biotransformation, and biosorption. This group of fungi demonstrates remarkable resilience in harsh environmental conditions that typically challenge bacteria and algae, positioning them as highly effective candidates for various industrial applications.

This chapter outlines the application of fungi throughout different phases of wastewater treatment, ranging from pre-treatment bioleaching to tertiary treatment with immobilized fungal bioreactors, showcasing their adaptability and effectiveness. Utilizing agricultural and industrial waste as substrates for fungi significantly boosts both the economic viability and environmental sustainability of the process. Fungal biomass, frequently generated as a byproduct in various industries, serves as an accessible and economical resource for extensive remediation efforts. Nonetheless, the practical application encounters obstacles including environmental variability, competition with indigenous microbes, the lack of standardized bioreactor systems, and the secure disposal of metal-laden fungal biomass. Moreover, specific fungi encounter genetic and metabolic limitations, especially in intricate or highly toxic settings.

In light of these constraints, the future outlook for mycoremediation appears to be quite encouraging. Progress in genetic engineering, immobilization methods, and the combination of fungal treatment with additional remediation technologies is anticipated to improve performance and scalability. The chapter concludes that fungi present a sustainable, adaptable, and effective approach to tackling heavy metal contamination. With ongoing innovation and interdisciplinary collaboration, mycoremediation holds significant promise for practical environmental cleanup efforts.

References

Adane, T., Adugna, A. T., & Alemayehu, E. (2021). Textile industry effluent treatment techniques. *Journal of Chemistry*. https://doi.org/10.1155/2021/5314404

Akpor, O. B., Ohiobor, G. O., & Olaolu, T. D. (2014). Heavy metal pollutants in wastewater effluents: Sources, effects and remediation. *Advances in Bioscience and Bioengineering, 2*(4), 37–43. https://doi.org/10.11648/j.abb.20140204.11

Angon, P. B., Islam, M. S., KC, S., Das, A., Anjum, N., Poudel, A., & Suchi, S. A. (2024). Sources, effects and present perspectives of heavy metals contamination: Soil, plants and human food chain. *Heliyon, 10*(7), e28357. https://doi.org/10.1016/j.heliyon.2024.e28357

Arikan, E. B., Bouchareb, E. M., & Dizge, N. (2020). Investigation of fungal treatment potential for bulgur cooking process wastewater. *Bioresource Technology Reports, 11*, 100468. https://doi.org/10.1016/j.biteb.2020.100468

Ayele, A., Haile, S., Alemu, D., Tesfaye, T., & Kamaraj, M. (2021). Mycoremediation: Fungal-based technology for biosorption of heavy metals—A review. In *Strategies and tools for pollutant mitigation* (pp. 355–373). Springer International Publishing. https://doi.org/10.1007/978-3-030-63575-6_17

Aytar, P., Gedikli, S., Buruk, Y., Cabuk, A., & Burnak, N. (2014). Lead and nickel biosorption with a fungal biomass isolated from metal mine drainage: Box–Behnken experimental design. *International journal of Environmental Science and Technology, 11*(6), 1631–1640. https://doi.org/10.1007/s13762-013-0354-5

Babu, B. R., Bhanu, S. U., & Meera, K. S. (2009). Waste minimization in electroplating industries: A review. *Journal of Environmental Science and Health, Part C, 27*(3), 155–177. https://doi.org/10.1080/10590500903124158

Barakat, M. A. (2011). New trends in removing heavy metals from industrial wastewater. *Arabian Journal of Chemistry, 4*(4), 361–377. https://doi.org/10.1016/j.arabjc.2010.07.019

Bhateria, R., & Dhaka, R. (2019). Optimization and statistical modelling of cadmium biosorption process in aqueous medium by *Aspergillus niger* using response surface methodology and principal component analysis. *Ecological Engineering, 135*, 127–138. https://doi.org/10.1016/j.ecoleng.2019.05.010

Bibbins-Martínez, M., Juárez-Hernández, J., López-Domínguez, J. Y., Nava-Galicia, S. B., Martínez-Tozcano, L. J., Juárez-Atonal', R., Cortés-Espinosa, D., & Díaz-Godinez, G. (2023). Potential application of fungal biosorption and/or bioaccumulation for the bioremediation of wastewater contamination: A review. *Journal of Environmental Biology, 44*(2), 135–145.

Binkley, J., & Simpson, J. A. (2003). Heavy metals in wastewater treatment processes. In D. Mara & W. M. Horan (Eds.), *Handbook of water and wastewater microbiology* (pp. 597–610). Academic. https://doi.org/10.1016/B978-012470100-7/50036-4

Chan, W. K., Wildeboer, D., Garelick, H., & Purchase, D. (2016). Mycoremediation of heavy metal/metalloid-contaminated soil: Current understanding and future prospects. In D. Purchase (Ed.), *Fungal applications in sustainable environmental biotechnology* (pp. 249–272). Springer International Publishing. https://doi.org/10.1007/978-3-319-42852-9_10

Chatterjee, A., Das, R., & Abraham, J. (2020). Bioleaching of heavy metals from spent batteries using aspergillus nomius JAMK1. *International Journal of Environmental Science and Technology, 17*(1), 49–66. https://doi.org/10.1007/s13762-019-02255-0

Chiu, S. W., Chan, Y. H., Law, S. C., Cheung, K. T., & Moore, D. (1998). Cadmium and manganese in contrast to calcium reduce yield and nutritional values of the edible mushroom *Pleurotus pulmonarius*. *Mycological Research, 102*(4), 449–457. https://doi.org/10.1017/S0953756297005728

Das, P. P., Anweshan, Mondal, P., Sinha, A., Biswas, P., Sarkar, S., & Purkait, M. K. (2021). Integrated ozonation assisted electrocoagulation process for the removal of cyanide from steel industry wastewater. *Chemosphere, 263*, 128370. https://doi.org/10.1016/j.chemosphere.2020.128370

Dey, P., Malik, A., Singh, D. K., Haange, S.-B., von Bergen, M., & Jehmlich, N. (2024). Unveiling fungal strategies: Mycoremediation in multi-metal pesticide environment using proteomics. *Scientific Reports, 14*(1), 23171. https://doi.org/10.1038/s41598-024-74517-y

Dinakarkumar, Y., Ramakrishnan, G., Gujjula, K. R., Vasu, V., Balamurugan, P., & Murali, G. (2024). Fungal bioremediation: An overview of the mechanisms, applications and future perspectives. *Environmental Chemistry and Ecotoxicology, 6*, 293–302. https://doi.org/10.1016/j.enceco.2024.07.002

El-Kassas, H. Y. & El-Taher, E. M. (2009). Optimization of Batch Process Parameters by Response Surface Methodology for Mycoremediation of Chrome-VI by a Chromium Resistant Strain of Marine Trichoderma Viride Optimization of Batch Process Parameters by Response Surface Methodology for Mycoremediat. American-Eurasian Journal of Agricultural & Environmental Sciences 5, 676–681.

Fashola, M., Ngole-Jeme, V., & Babalola, O. (2016). Heavy metal pollution from gold mines: Environmental effects and bacterial strategies for resistance. *International Journal of Environmental Research and Public Health, 13*(11), 1047. https://doi.org/10.3390/ijerph13111047

Fernandes Azevedo, B., Barros Furieri, L., Peçanha, F. M., Wiggers, G. A., Frizera Vassallo, P., Ronacher Simões, M., Fiorim, J., Rossi de Batista, P., Fioresi, M., Rossoni, L., Stefanon, I., Alonso, M. J., Salaices, M., & Valentim Vassallo, D. (2012). Toxic effects of mercury on the cardiovascular and central nervous systems. *Journal of Biomedicine and Biotechnology, 2012*, 1–11. https://doi.org/10.1155/2012/949048

Ferreira, J. A., Varjani, S., & Taherzadeh, M. J. (2020). A critical review on the ubiquitous role of filamentous fungi in pollution mitigation. *Current Pollution Reports, 6*(4), 295–309. https://doi.org/10.1007/s40726-020-00156-2

Fortier, J., Truax, B., Gagnon, D., & Lambert, F. (2015). Biomass carbon, nitrogen and phosphorus stocks in hybrid poplar buffers, herbaceous buffers and natural woodlots in the riparian zone on agricultural land. *Journal of Environmental Management, 154*, 333–345. https://doi.org/10.1016/j.jenvman.2015.02.039

Fragoeiro, S., & Magan, N. (2005). Enzymatic activity, osmotic stress and degradation of pesticide mixtures in soil extract liquid broth inoculated with Phanerochaete chrysosporium and Trametes versicolor. *Environmental Microbiology, 7*(3), 348–355. https://doi.org/10.1111/j.1462-2920.2005.00699.x

Gadd, G. M. (2009). Biosorption: Critical review of scientific rationale, environmental importance and significance for pollution treatment. *Journal of Chemical Technology & Biotechnology, 84*(1), 13–28. https://doi.org/10.1002/jctb.1999

Genchi, G., Carocci, A., Lauria, G., Sinicropi, M. S., & Catalano, A. (2020). Nickel: Human health and environmental toxicology. *International Journal of Environmental Research and Public Health, 17*(3), 679. https://doi.org/10.3390/ijerph17030679

Georgaki, M.-N., Charalambous, M., Kazakis, N., Talias, M. A., Georgakis, C., Papamitsou, T., & Mytiglaki, C. (2023). Chromium in water and carcinogenic human health risk. *Environments, 10*(2), 33. https://doi.org/10.3390/environments10020033

Ghosh, S., Rusyn, I., Dmytruk, O. V., Dmytruk, K. V., Onyeaka, H., Gryzenhout, M., & Gafforov, Y. (2023). Filamentous fungi for sustainable remediation of pharmaceutical compounds, heavy metal and oil hydrocarbons. *Frontiers in Bioengineering and Biotechnology, 11*, 1106973.

Gugel, I., Summa, D., Costa, S., Manfredini, S., Vertuani, S., Marchetti, F., & Tamburini, E. (2024). Mycoremediation of synthetic azo dyes by white-rot fungi grown on diary waste: A step toward sustainable and circular bioeconomy. *Fermentation, 10*(2), 80. https://doi.org/10.3390/fermentation10020080

Gupta, P., & Diwan, B. (2017). Bacterial exopolysaccharide mediated heavy metal removal: A review on biosynthesis, mechanism and remediation strategies. *Biotechnology Reports, 13*, 58–71. https://doi.org/10.1016/j.btre.2016.12.006

Hussain, S., Khan, M., Sheikh, T. M. M., Mumtaz, M. Z., Chohan, T. A., Shamim, S., & Liu, Y. (2022). Zinc essentiality, toxicity, and its bacterial bioremediation: A comprehensive insight. *Frontiers in Microbiology, 13*. https://doi.org/10.3389/fmicb.2022.900740

Iskandar, N. L., Zainudin, N. A. I. M., & Tan, S. G. (2011). Tolerance and biosorption of copper (cu) and lead (Pb) by filamentous fungi isolated from a freshwater ecosystem. *Journal of Environmental Sciences, 23*(5), 824–830. https://doi.org/10.1016/S1001-0742(10)60475-5

Jaishankar, M., Tseten, T., Anbalagan, N., Mathew, B. B., & Beeregowda, K. N. (2014). Toxicity, mechanism and health effects of some heavy metals. *Interdisciplinary Toxicology, 7*(2), 60–72. https://doi.org/10.2478/intox-2014-0009

Kant, R. (2012). Textile dyeing industry an environmental hazard. *Natural Science, 04*(01), 22–26. https://doi.org/10.4236/ns.2012.41004

Kapoor, A., & Viraraghavan, T. (1998). Removal of heavy metals from aqueous solutions using immobilized fungal biomass in continuous mode. *Water Research, 32*(6), 1968–1977. https://doi.org/10.1016/S0043-1354(97)00417-X

Khan, I., Ali, M., Aftab, M., Shakir, S., Qayyum, S., Haleem, K. S., & Tauseef, I. (2019). Mycoremediation: A treatment for heavy metal-polluted soil using indigenous metallotolerant fungi. *Environmental Monitoring and Assessment, 191*(10), 622. https://doi.org/10.1007/s10661-019-7781-9

Khan, S., & Malik, A. (2014). Environmental and health effects of textile industry wastewater. In *Environmental deterioration and human health* (pp. 55–71). Springer. https://doi.org/10.1007/978-94-007-7890-0_4

Khan, S. H. (2021). Advanced approaches for heavy metals removal from industrial wastewater. In M. P. Shah, S. Rodriguez Couto, & I. W. Kumar (Eds.), *New trends in removal of heavy metals from industrial wastewater* (pp. 403–440). Elsevier. https://doi.org/10.1016/B978-0-12-822965-1.00017-9

Kortei, N. K., Heymann, M. E., Essuman, E. K., Kpodo, F. M., Akonor, P. T., Lokpo, S. Y., Boadi, N. O., Ayim-Akonor, M., & Tettey, C. (2020). Health risk assessment and levels of toxic metals in fishes (Oreochromis noliticus and Clarias anguillaris) from Ankobrah and Pra basins: Impact of illegal mining activities on food safety. *Toxicology Reports, 7*, 360–369. https://doi.org/10.1016/j.toxrep.2020.02.011

Kubier, A., Wilkin, R. T., & Pichler, T. (2019). Cadmium in soils and groundwater: A review. *Applied Geochemistry, 108*, 104388. https://doi.org/10.1016/j.apgeochem.2019.104388

Kumar, A., Kumar, V., & Singh, J. (2019). Role of fungi in the removal of heavy metals and dyes from wastewater by biosorption processes. In A. N. Yadav, S. Singh, S. Mishra, & A. Gupta (Eds.), *Recent advancement in white biotechnology through fungi: Volume 3: Perspective for sustainable environments* (pp. 397–418). Springer International Publishing. https://doi.org/10.1007/978-3-030-25506-0_16

Liang, X., & Gadd, G. M. (2017). Metal and metalloid biorecovery using fungi. *Microbial Biotechnology, 10*(5), 1199–1205. https://doi.org/10.1111/1751-7915.12767

Manisha, S., Pratibha, S., Shaima, M. S., Qaisar, M., & Arun, K. (2016). Heavy metal removal from industrial wastewater using fungi: Uptake mechanism and biochemical aspects. *Journal of Environmental Engineering, 142*(9), C6015001. https://doi.org/10.1061/(ASCE)EE.1943-7870.0000983

Masoudi, R., Basiri, M., & Moghimi, H. (2017). Mycoremediation of cadmium from aqueous solution using newly isolated Actinomucor sp.: Isotherm and kinetic studies. *Desalination and Water Treatment, 80*, 193–202. https://doi.org/10.5004/dwt.2017.20867

Mastranzo-Pérez, L. A., Hernández-Domínguez, E. M., Falcón-León, M. P., & Álvarez-Cervantes, J. (2025). Pleurotus spp: A cosmopolitan fungi of biotechnological importance. *Scientia Agropecuaria, 16*(1), 79–91. https://doi.org/10.17268/sci.agropecu.2025.008

Matebese, F., Mosai, A. K., Tutu, H., & Tshentu, Z. R. (2024). Mining wastewater treatment technologies and resource recovery techniques: A review. *Heliyon, 10*(3), e24730. https://doi.org/10.1016/j.heliyon.2024.e24730

Mohamadhasani, F., & Rahimi, M. (2022). Growth response and mycoremediation of heavy metals by fungus Pleurotus sp. *Scientific Reports, 12*(1), 19947. https://doi.org/10.1038/s41598-022-24349-5

Muraleedharan, T. R., & Venkobachar, L. I. (1994). Further insight into the mechanism of biosorption of heavy metals by Ganoderma lucidum. *Environmental Technology, 15*(11), 1015–1027. https://doi.org/10.1080/09593339409385510

Nielsen, D. R., Amarasiriwardena, G. S., & Prather, K. L. J. (2010). Predicting the adsorption of second generation biofuels by polymeric resins with applications for in situ product recovery (ISPR). *Bioresource Technology, 101*(8), 2762–2769. https://doi.org/10.1016/j.biortech.2009.12.003

Niu, S., Xia, Y., Yang, C., & Liu, C. (2023). Impacts of the steel industry on sediment pollution by heavy metals in urban water system. *Environmental Pollution, 335*, 122364. https://doi.org/10.1016/j.envpol.2023.122364

Peng, B., Yao, Z., Wang, X., Crombeen, M., Sweeney, D. G., & Tam, K. C. (2020). Cellulose-based materials in wastewater treatment of petroleum industry. *Green Energy & Environment, 5*(1), 37–49. https://doi.org/10.1016/j.gee.2019.09.003

Rafique, M., Hajra, S., Tahir, M. B., Gillani, S. S. A., & Irshad, M. (2022). A review on sources of heavy metals, their toxicity and removal technique using physico-chemical processes from wastewater. *Environmental Science and Pollution Research, 29*(11), 16772–16781. https://doi.org/10.1007/s11356-022-18638-9

Rahman, M. L., Wong, Z. J., Sarjadi, M. S., Soloi, S., Arshad, S. E., Bidin, K., & Musta, B. (2021). Heavy metals removal from electroplating wastewater by waste Fiber-based poly(amidoxime) ligand. *Water, 13*(9), 1260. https://doi.org/10.3390/w13091260

Rahman, S. U., Yasin, G., Nawaz, M. F., Cheng, H., Azhar, M. F., Riaz, L., Javed, A., & Lu, Y. (2022). Evaluation of heavy metal phytoremediation potential of six tree species of Faisalabad city of Pakistan during summer and winter seasons. *Journal of Environmental Management, 320*, 115801. https://doi.org/10.1016/j.jenvman.2022.115801

Rajoria, S., Vashishtha, M., & Sangal, V. K. (2022). Treatment of electroplating industry wastewater: A review on the various techniques. *Environmental Science and Pollution Research, 29*(48), 72196–72246. https://doi.org/10.1007/s11356-022-18643-y

Ramya, D., Jennifer Michellin Kiruba, N., & Thatheyus, A. J. (2021). Biosorption of heavy metals using fungal biosorbents—A review. In V. K. Sharma, M. P. Shah, S. Parmar, & S. A. Kumar (Eds.), *Fungi bio-prospects in sustainable agriculture, environment and nano-technology volume 2: extremophilic fungi and myco-mediated environmental management* (pp. 331–352). Academic. https://doi.org/10.1016/B978-0-12-821925-6.00015-0

Raut, P. S., & Wani, P. P. (2010). Removal of heavy metals from industrial effluent by biomass of aspergillus flavus. *Nature, Environment and Pollution Technology, 9*(3), 639–642.

Refaey, M., Abdel-Azeem, A. M., Abo Nahas, H. H., Abdel-Azeem, M. A., & El-Saharty, A. A. (2021). Role of fungi in bioremediation of soil contaminated with heavy metals. In A. M. Abdel-Azeem, A. N. Yadav, N. Yadav, & Z. Usmani (Eds.), *Industrially important fungi for sustainable development: Volume 1: Biodiversity and ecological perspectives* (pp. 509–540). Springer International Publishing. https://doi.org/10.1007/978-3-030-67561-5_16

Roy, A., Gogoi, N., Haider, F., & Farooq, M. (2025). Mycoremediation for sustainable remediation of environmental pollutants. *Biocatalysis and Agricultural Biotechnology, 64*, 103526. https://doi.org/10.1016/j.bcab.2025.103526

Sen, K., Llewellyn, M., Taheri, B., Turner, R. J., Berglund, T., & Maloney, K. (2023). Mechanism of fungal remediation of wetland water: Stropharia rugosoannulata as promising fungal species for the development of biofilters to remove clinically important pathogenic and antibiotic resistant bacteria in contaminated water. *Frontiers in Microbiology, 14*, 1234586.

Sharma, P., Singh, S. P., Parakh, S. K., & Tong, Y. W. (2022). Health hazards of hexavalent chromium (Cr (VI)) and its microbial reduction. *Bioengineered, 13*(3), 4923–4938. https://doi.org/10.1080/21655979.2022.2037273

Shashank, A. S., & Velayudhannair, K. (2024). Comprehensive assessment of heavy metal pollution and ecological risk in lakes of Bangalore, Karnataka, India. *Environment and Ecology, 42*(2), 479–491. https://doi.org/10.60151/envec/ZLAO4807

Shourie, A., & Vijayalakshmi, U. (2022). Fungal diversity and its role in Mycoremediation. *Geomicrobiology Journal, 39*(3–5), 426–444. https://doi.org/10.1080/01490451.2022.2032883

Singh, A., Sharma, A., Verma, R., Chopade, R., Pandit, P., Nagar, V., Aseri, V., Choudhary, S., Awasthi, G., Awasthi, K., & Sankhla, M. (2022). Heavy metal contamination of water and their toxic effect on living organisms. In *The toxicity of environmental pollutants*. IntechOpen. https://doi.org/10.5772/intechopen.105075

Singh, A. K., & Chandra, R. (2019). Pollutants released from the pulp paper industry: Aquatic toxicity and their health hazards. *Aquatic Toxicology, 211*, 202–216. https://doi.org/10.1016/j.aquatox.2019.04.007

Singh, V., Singh, N., Rai, S. N., Kumar, A., Singh, A. K., Singh, M. P., Sahoo, A., Shekhar, S., Vamanu, E., & Mishra, V. (2023). Heavy metal contamination in the aquatic ecosystem: Toxicity and its remediation using eco-friendly approaches. *Toxics, 11*(2), 147. https://doi.org/10.3390/toxics11020147

Sugitha, S., Vishnu Priya, P., Kavya Kanishka, T., Duraimurugan, A., Suganthi, M., Ashok Kumar, K., Jayanthi, M., Durgadevi, R., Ramprasath, C., & Abirami, G. (2024). Mycoremediation of heavy metals by Curvularia lunata from Buckingham Canal, Neelankarai, Chennai. *World Journal of Microbiology and Biotechnology, 41*(1), 1. https://doi.org/10.1007/s11274-024-04218-1

Tagyan, A. I., Yasser, M. M., Mousa, A. M., Alkhalifah, D. H. M., Hozzein, W. N., & Marzouk, M. A. (2023). A feasibility study on the recall of metallophilic fungi from Fe(III)-contaminated soil and evaluating their mycoremediation capacity: Experimental and theoretical study. *Journal of Fungi, 9*(3), 382). https://doi.org/10.3390/jof9030382

Tamjidi, S., Ameri, A., & Esmaeili, H. (2022). A review of the application of fungi as an effective and attractive bio-adsorbent for biosorption of heavy metals from wastewater. *Environmental Monitoring and Assessment, 195*(1), 91. https://doi.org/10.1007/s10661-022-10687-4

Tello-Galarreta, F. A., Durand-Paz, J. H., Rojas-Villacorta, W., Cabanillas-Chirinos, L., De La Cruz-Noriega, M., Nazario-Naveda, R., Benites, S. M., & Rojas-Flores, S. (2023). In vitro effect of molasses concentration, pH, and time on chromium removal by Trichoderma spp. from the effluents of a Peruvian tannery. *Processes, 11*(5), 1557. https://doi.org/10.3390/pr11051557

Thiyur, S., & Samuel, J. (2020). Fungi: An invaluable tool for the valorization of industrial waste. In J. Singh & B. Gehlot (Eds.), *New and future developments in microbial biotechnology and bioengineering. Recent advances in application of fungi and fungal metabolites: Environmental and industrial aspects* (pp. 121–131). Elsevier. https://doi.org/10.1016/B978-0-12-821007-9.00011-5

Velusamy, S., Roy, A., Sundaram, S., & Kumar Mallick, T. (2021). A review on heavy metal ions and containing dyes removal through graphene oxide-based adsorption strategies for textile wastewater treatment. *The Chemical Record, 21*(7), 1570–1610. https://doi.org/10.1002/tcr.202000153

Verma, T., Maurya, A., Tripathi, M., & Garg, S. K. (2017). In T. Satyanarayana, S. K. Deshmukh, & B. N. Johri (Eds.), *Mycoremediation: An alternative treatment strategy for heavy metal-laden wastewater BT - Developments in fungal biology and applied mycology* (pp. 315–340). Springer. https://doi.org/10.1007/978-981-10-4768-8_17

Viraraghavan, T., & Srinivasan, A. (2011). Fungal biosorption and biosorbents. In P. Kotrba, M. Mackova, & T. Macek (Eds.), *Microbial biosorption of metals* (pp. 143–158). Springer. https://doi.org/10.1007/978-94-007-0443-5_6

Wang, J., & Chen, C. (2009). Biosorbents for heavy metals removal and their future. *Biotechnology Advances, 27*(2), 195–226. https://doi.org/10.1016/j.biotechadv.2008.11.002

Wang, L., Zhang, K., Shareen, Yin, Z., Yu, L., Qiu, X., Zhou, S., Chen, R., & Wang, Q. (2024). Exploring the potential of *Aspergillus terreus* 2021, WLL-ISO: A dual-functional agent for heavy metal removal and self-aggregation in wastewater treatment. *Separation and Purification Technology, 350*, 127964. https://doi.org/10.1016/j.seppur.2024.127964

Wani, A. L., Ara, A., & Usmani, J. A. (2015). Lead toxicity: A review. *Interdisciplinary Toxicology, 8*(2), 55–64. https://doi.org/10.1515/intox-2015-0009

Yang, J., Wang, Q., Luo, Q., Wang, Q., & Wu, T. (2009). Biosorption behavior of heavy metals in bioleaching process of MSWI fly ash by aspergillus Niger. *Biochemical Engineering Journal, 46*(3), 294–299. https://doi.org/10.1016/j.bej.2009.05.022

Yaseen, D. A., & Scholz, M. (2019). Textile dye wastewater characteristics and constituents of synthetic effluents: A critical review. *International Journal of Environmental Science and Technology, 16*(2), 1193–1226. https://doi.org/10.1007/s13762-018-2130-z

Zafar, S., Aqil, F., & Ahmad, I. (2007). Metal tolerance and biosorption potential of filamentous fungi isolated from metal contaminated agricultural soil. *Bioresource Technology, 98*(13), 2557–2561. https://doi.org/10.1016/j.biortech.2006.09.051

Zheng, X., Lin, H., Du, D., Li, G., Alam, O., Cheng, Z., Liu, X., Jiang, S., & Li, J. (2024). Remediation of heavy metals polluted soil environment: A critical review on biological approaches. *Ecotoxicology and Environmental Safety, 284*, 116883. https://doi.org/10.1016/j.ecoenv.2024.116883

Chapter 7
Fungal Treatment of Textile Wastewater for Fungi in Wastewater Treatment

Sudipta Choudhury, Umamaheshwari S, and Kanthesh M. Basalingappa

Abstract The textile industry generates large quantities of effluent contaminated with heavy metals, dyes, and other hazardous chemicals, and thus, it is one of the leading sources of worldwide water pollution. Using fungi as bioremediation agents, which has been of great interest recently, proves to be an efficient substitute for treating textile wastewater. Because of their distinct enzymatic nature, fungi can break down all pollutants, encompassing toxic chemicals and recalcitrant dyes with resistance to degradation properties, making this process sustainable and a long-lasting source of wastewater remediation. This study investigates the role of fungi in treating wastewater produced from textile manufacturing with clear indications of benefits over traditional methods, mechanisms and scope of uses.

This chapter explains the efficiency of fungal-based treatment systems in reducing pollutant concentrations and enhancing water quality using case studies and empirical data. In addition, it discusses new methods and future trends for developing fungal treatment technologies. This chapter provides an overall evaluation of the ability of fungi to revolutionise wastewater treatment in the textile industry by discussing the opportunities and challenges of fungal bioremediation.

Keywords Fungi · Textile wastewater · Bioremediation · Metals · Toxicity

---

The original version of this chapter has been revised. The correction to this chapter can be found at https://doi.org/10.1007/978-3-032-12693-1_14

S. Choudhury · K. M. Basalingappa (✉)
Division of Molecular Biology, School of Life Sciences, JSS Academy of Higher Education and Research, Mysuru, Karnataka, India
e-mail: kantheshmb@jssuni.edu.in

U. S
Department of Microbiology, School of Life Sciences, JSS Academy of Higher Education and Research, Mysuru, Karnataka, India
e-mail: umamaheshwari@jssuni.edu.in

C. M. Hussain et al. (eds.), *Fungi in Waste Water Treatment*, Springer Water,
https://doi.org/10.1007/978-3-032-12693-1_7

7.1 Introduction

7.1.1 Wastewater Treatment

The textile industry is among the significant industries causing water pollution, producing quantities of wastewater, often laden with a vast array of pollutants. About 80–150 l of water is consumed to produce a kilogram of fabric, consequently leading to enormous volumes of wastewater production (Jakhar & Styszko, 2025). The constituents of this wastewater comprise synthetic dyes, heavy metals, organic solvents, and an array of chemicals employed during production processes. The presence of synthetic dyes significantly challenges owing to their considerable stability and resilience against biological degradation, with estimates suggesting that up to 200,000 tons of dyes may be lost each year if not subjected to effective treatment (Yin et al., 2019). Chromium and cadmium, some heavy metals, pose an important issue, imposing substantial environmental and health risks (Selim et al., 2021).

Adequate wastewater treatment from textile sources is a matter of environmental security and a requirement by onerous regulations. Various treatment techniques have been established, including physical, chemical, and biological methods (Fig. 7.1). The common approaches include screening, coagulation-flocculation, activated sludge processes, and advanced oxidation techniques (Amanuel, 2019; Bello, 2017). Nonetheless, the complexity and diversity of contaminants necessitate an integrated strategy combining these approaches to achieve the highest level of treatment efficiency (Adane et al., 2021).

Textile wastewater is characterised by various pollutants that are hard to treat. The major pollutants found in the effluent of textile industries include synthetic dyes, heavy metals, organic chemicals, and many toxins. Synthetic dyes, including azo and anthraquinone types, are a big challenge as they do not biodegrade quickly and thus cause much damage to the environment when left untreated and allowed to flow into bodies of water (Pundir et al., 2024). Some heavy metals like chromium, cadmium, lead, and mercury find their way into the effluent systems due to the

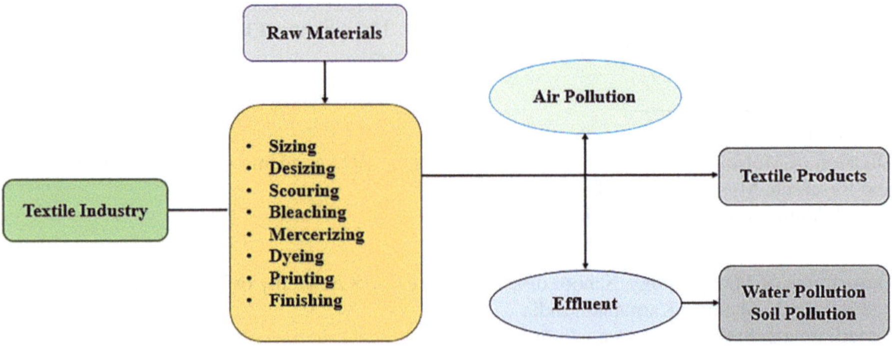

Fig. 7.1 Different stages of product formation in the textile industry with corresponding wastewater (Pundir et al., 2024)

dyeing and finishing processes and create health hazards for humans and organisms in the aquatic environment (Selim et al., 2021). In addition, wastewater from the textile industry contains high COD levels, suspended solids, and many organic solvents and surfactants used during manufacturing, making it quite challenging to clean up (Vasconcelos et al., 2022).

Such pollutants not only deteriorate the aesthetic quality of aquatic water bodies but also cause eutrophication and decrease the rate of photosynthesis in the marine environment (Ogunwumi et al., 2025). The traditional removal techniques fail to remove these pollutants because of their complex nature and high resistance (Gallo et al., 2023). Hence, new and effective wastewater treatment methods that can overcome the problems related to textile wastewater treatment are urgently needed (Fig. 7.2).

Fungi play a highly significant role in the cleaning of textile wastewater, most especially for their ability to biodegrade complex polluting materials like synthetic dyes and heavy metals. Inefficiency is attributed to the secretion of extracellular enzymes that can carry on Biosorption. Fungi, including the white rot fungi such as Trametes versicolor and Phanerochaete chrysosporium, produce ligninolytic enzymes such as laccases, manganese peroxidases, and lignin peroxidases. These are essential for degrading resistant dye compounds (Pundir et al., 2024). These enzymes catalyse the oxidation of various substrates and allow for the degradation of phenolic and non-phenolic compounds in textile effluents. Besides the enzymatic degradation process, fungi can adsorb dyes and other impurities on their biomass, dramatically reducing the harmful substances in wastewater. For instance, studies show that Aspergillus flavus and Fusarium oxysporum can achieve remarkable decolourisation percentages of up to 78% while treating real textile effluents (Selim

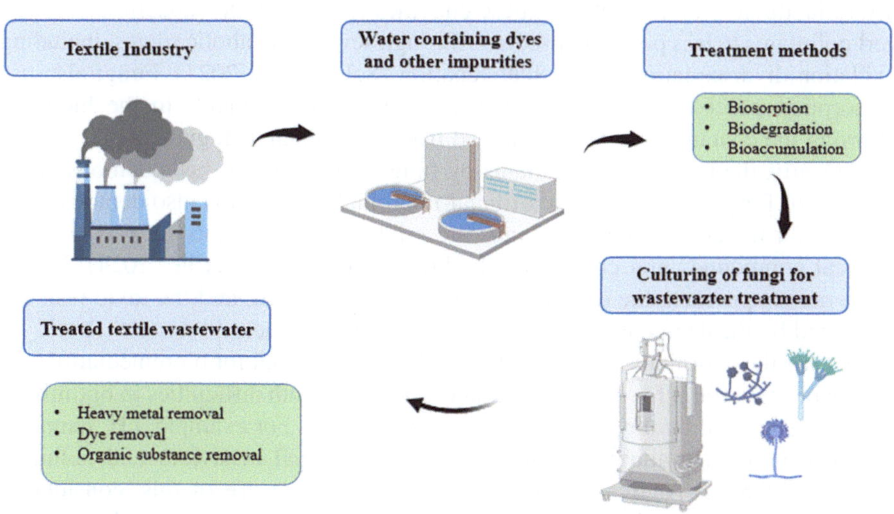

Fig. 7.2 Flow chart showing the involvement of fungi in the treatment methods to remove dyes, pigments, and heavy metals from textile wastewater using Biosorption, bioaccumulation, and biodegradation (Pundir et al., 2024)

et al., 2021). Moreover, the adaptability of fungal species allows them to thrive in various environmental settings, making them suitable for a variety of applications in the treatment of wastewater (Andressa Chicatto et al., 2018). Fungi is used in some procedures like bioremediation, and it is applied in combination with bacteria to improve the efficiency of removing pollutants.

This collaborative methodology harnesses the advantages inherent in both microorganisms, resulting in enhanced degradation and wastewater detoxification rates (Efremenko et al., 2024). Nonetheless, obstacles persist in scaling these processes for industrial utilisation, which requires additional investigation into optimising growth conditions and creating effective bioreactor systems (Latif et al., 2023). In summary, fungal treatment signifies a viable option for tackling the possible issues associated with the environment and textile wastewater, providing a sustainable alternative to conventional chemical approaches.

7.1.1.1 Fungi as Bioremediating Agents

Recently, fungi have emerged as potential bioremediation capable of removing significant loads of environmental contaminants, including those implicated in causing causal contamination of industrial wastewater effluents from the textile industry. Depending on their metabolic characteristics, they can be utilised as complex organic pollutant detoxifying agents and, thus, must be involved in bioremediation—fungi (particularly *Trametes versicolor*, *Phanerochaete chrysosporium*, and *Aspergillus* sp.

Producing ligninolytic metalloenzymes that can oxidise and cleave a broad range of dye structures (laccases and peroxidases) is among the ecological strategies fungi use to facilitate bioremediation. These enzymes have catalytic capabilities to convert pollutants to less pollutant products through several metabolic routes, including oxidation, hydroxylation, and dehalogenation (Selim et al., 2021). Fungi also use biosorption processes. These organisms can bind contaminants to the biomass, allowing the contaminants to be removed from the effluent (Magnoli et al., 2023).

Recently, the removal of pollutants by fungal consortia has been demonstrated. The use of fungi and bacteria in artificial microbial consortia is also demonstrated to work for the deconstruction of intricate mixtures of pollutants such as pharmaceutical compounds and chlorinated herbicides (Efremenko et al., 2024). In this cooperative approach, the degradation yields are calibrated, and the toxic residues generated by the degradation process are rendered nontoxic.

Despite the enormous promise in the application of fungi for bioremediation, the applicability of fungi for bioremediation is limited by both difficulties in optimising fungal growth and by the treatment process scalability. For example, pH, temperature, and nutrient availability significantly affect fungal behaviour and pollution removal potential (Andressa Chicatto et al., 2018). Because of this, considering these issues in conjunction with the developments in science and technology is crucial to gaining the evolved capacity of the fungi to have a higher level of bioremediation ability (Latif et al., 2023).

7.1.1.2 Fungal Enzymatic Capabilities

The fungi can degrade complex organic molecules, and this property makes them suitable for treating textile wastewater. Their various extracellular enzymes can specifically cleave unions of large organic molecules to degrade dyes and other chemicals found in textile wastewater. Such enzymes bring about several oxidative reactions, transforming the old, complex and poisoned substrates into much smaller, relatively harmless compounds (Vaksmaa et al., 2023; Deshmukh et al., 2016).

7.1.1.3 Diversity of Enzymes

Fungi produce different extracellular enzymes, such as lignin peroxidases, manganese peroxidases, and laccases; these enzymes profoundly influence the degradation of complex organic compounds such as dyes and other synthetic chemicals in textile wastewater. These enzymes bring about oxidative reactions, which can finally break down a resistant substance into a much simpler and less harmful compound (Vaksmaa et al., 2023; Deshmukh et al., 2016).

7.1.1.4 Modes of Biodegradation

Several mechanisms initiate biodegradation:

1. Oxidative degradation: Fungi utilise oxidative enzymes to catalyse the breakdown of complex polymers into smaller molecules that can be assimilated or further degraded. This is important in removing colour and toxicity from dye-laden wastewater (Okal et al., 2023; Srikanth et al., 2022).
2. Hydrolytic activity: Fungi utilise some hydrolytic enzymes to contribute to the cleavage of different chemical bonds within the polymer molecule, further enhancing the degradation process. This dual enzymatic action allows for multi-spectrum pollutant degradation by fungi (Vaksmaa et al., 2023; Okal et al., 2023).

Fungal bioremediation is a promising approach toward textile wastewater treatment for several reasons.

1. Considerable efficiency: Studies have shown that selective fungal strains can degrade several of the dyes used in textiles, including well-known azo dyes, which are recalcitrant to being eliminated using conventional treatment methods. For instance, fungi like *Aspergillus* and *Trichoderma* have been mentioned for their ability to decolourise and detoxify dye effluents (Okal et al., 2023; Srikanth et al., 2022).
2. Resilience: They flourish in various environmental conditions, allowing them to deduce optimum efficiency in pH values and temperatures that meet industrial wastewater conditions (Vaksmaa et al., 2023; Deshmukh et al., 2016).

7.1.1.5 Advantages of Fungal Treatment

Fungal treatment is gradually gaining credence as a promising alternative to traditional wastewater treatment methods involving physical and chemical processes, mainly due to recent work's key advantages.

1. Effective degradation of pollutants.

   Fungi, especially white rot fungi, possess the unique potential for a wide range of enzyme productions, like laccases and peroxidases, which can degrade complex organic pollutants in wastewater-from dyes to pharmaceuticals. This enzyme oxidises and degrades the recalcitrant compounds otherwise resistant to conventional treatments (Latif et al., 2023; Pundir et al., 2024).
2. Environmental sustainability

   Fungal treatment is environmentally more sustainable than chemical methods that may produce secondary pollutants. Fungal bioremediation uses natural metabolic pathways to detoxify contaminants, which is very much in line with and supportive of sustainability to limit the use of chemicals and decrease waste (Pundir et al., 2024; Thirumalaivasan et al., 2024).
3. Economic viability.

   The operational costs for fungal treatment are generally lower than those of other traditional methods. Fungi grow well on several possible substrates, such as agricultural waste, thereby lessening the need for chemical reagents or energy inputs and thus further reducing the cost (Ghosh et al., 2023; Malik et al., 2023).
4. Flexibility and adaptability.

   Fungal bioremediation can be applied in different settings and easily integrated with other bioremediation techniques for enhanced efficiency in pollutant removal. This versatility in their adaptation allows them to effectively treat diverse types of wastewater (Latif et al., 2023; Thirumalaivasan et al., 2024).
5. Potential byproduct utilisation

   In addition to pollutant degradation, the enhancement of fungal treatments can result in the generation of some valuable byproducts, such as enzymes and bioactive compounds, which can find wide applications in other fields, thereby enhancing the economic aspects of the bioremediation process (Ghosh et al., 2023; Malik et al., 2023).

Table 7.1 summarise some of the most effective fungal species used in wastewater treatment:

7.1.2 Overview

This chapter intends to provide current research evidence on how different fungal species efficiently break down complex pollutants in textile wastes, including heavy metals and dyes. This chapter describes how fungi can metabolise these toxic compounds using enzymes and presents research examples and data suggesting that

Table 7.1 Summarise some of the most effective fungal species used in wastewater treatment

Fungal species	Pollutants treated	Key mechanisms	References
Aspergillus piperis	Chloride ions	Ion sequestration and enzymatic conversion of Cl^-	(Han et al., 2023)
Penicillium sp.	Various organic pollutants, heavy metals	Enzyme production (e.g., laccases) for biodegradation	(Buratti et al., 2022; Ghosh et al., 2023)
Trichoderma sp.	Dyes, hydrocarbons	Ligninolytic enzymes for dye decolorization and degradation	(Pundir et al., 2024)
Fusarium sp.	Pharmaceuticals, dyes	Enzymatic breakdown of complex organic compounds	(Ghosh et al., 2023)
Phanerochaete chrysosporium	Lignin, dyes	Lignin peroxidases and manganese peroxidases for lignin degradation	(Ghosh et al., 2023)
Geotrichum sp.	Organic pollutants, nutrients	Biodegradation and nutrient removal capabilities	(Buratti et al., 2022; Pundir et al., 2024)

these organisms could be biodegradable. This chapter also fills significant research gaps, particularly expanding the fungal treatment processes for industrial applications. It highlights the need for further studies on the conditions enhancing operation, such as pH and temperature, and the culture's performance stability in practical wastewater treatment environments.

Additionally, future studies' directions on improving fungal treatment technologies are noted. These are the analysis of genetic engineering to enhance the fungal comprehensive application, the study of the integrated treatment systems containing fungi and other microorganisms, and the examination of the economic effectiveness of utilising these technologies in the industry. Further, it highlights the need to focus on typical regulatory issues to improve the use of fungal bioremediation in wastewater treatment procedures.

7.1.3 Pollutants in Textile Effluents

7.1.3.1 Synthetic Dyes

Cationic synthetic dyes are some of the most severe pollutants in textile wastewater. The textile industry has many synthetic dyes because of the brightness and fastness of colours. Common types include:

- Azo dyes: These dyes are the most used synthetic dyes, containing about 70% of all the dyes used in textile industries. It can decompose into toxic amines, which are carcinogenic (Kishor et al., 2020).
- Reactive dyes: These dyes are suitable for bonding with fibres, which makes them mainly used for cotton and other natural fibres. The high solubility of these compounds in water makes them challenging to eliminate during water treatment (Bhatia et al., 2017).

- Disperse dyes: Of all the categories of dyes, disperse dyes are pretty challenging to treat, mainly polyester, because of their hydrophobic characteristic (Khan & Malik, 2018).

The release of these dyes into water bodies causes enormous aesthetic pollution and toxicity to aquatic life and human beings. Reports indicate that these dyes released directly into water courses without prior treatment can produce severe environmental damage, aesthetic degradation, and toxicity within the aquatic biome (Kallawar & Bhanvase, 2024). It was observed that the concentration of dyes in textile effluents varied between 10 and 7000 mg/L due to differences in the manufacturing processes. Those wastewater bodies lead to significant aesthetic pollution and pose toxicity risks to aquatic life and humans. Studies report dye concentrations in textile effluents ranging from 10 mg/L to over 7000 mg/L, depending on the manufacturing processes used (Kishor et al., 2020).

7.1.3.2 Heavy Metals

Another composition of textile wastewater is heavy metals, which are sourced from the dyes or chemical supplements used in the textile lines. The employment of diverse chemicals in textile processing brings heavy metals, like chromium, cadmium, lead, and mercury, into water. These are toxic heavy metals that can accumulate within the ecosystem, thereby seriously threatening humans and wildlife with health dangers. Generally, heavy metals originate in the dyeing processes and the chemical treatments used to finish textiles (Selim et al., 2021).

Common heavy metals found include:

- Chromium: Another helpful element in dyes is chromium, which is dangerous and accumulates in living tissues and can lead to critical health problems (Khan & Malik, 2018).
- Lead: Current in some dye formulations, lead results in neuropathologic effects and developmental problems.
- Cadmium and arsenic: These metals are used in many industrial processes and are very dangerous to health because they are carcinogenic (Kishor et al., 2020).

Some heavy metals are not biodegradable; hence, they find their way into the food chain and may cause long-term impacts on the ecosystem (Andressa Chicatto et al., 2018).

7.1.3.3 Hazardous Chemicals

Textile wastewater comprises numerous dangerous chemicals employed during the processing and manufacturing of textiles. These include:

- Acids and alkalis: Employed as a buffer in dyeing processes; may be toxic to water.

- Surfactants and detergents: These compounds aid in dye dissolution but carry toxic effects to water-accommodating life forms and high BOD to the water receiving them (Andressa Chicatto et al., 2018).
- Binders and dispersants: Employed to improve the dyeing properly; they are sometimes refractory and cannot be quickly metabolised biologically.
- Other toxic substances: Besides dyes and heavy metals, textile process wastewater contains a variety of other toxic substances like acids, alkalis, detergents, and auxiliary chemicals used in the dyeing and finishing stage. The presence of these substances makes textile effluents complex and thus demands appropriate treatment techniques (Deshmukh et al., 2025). These chemicals not only enhance the toxicity of the effluent discharges but also add to the high BOD and COD proportional to oxygen consumption in the water bodies (Khan & Malik, 2018; Kishor et al., 2020).

7.1.3.4 Hazards of the Natural and Artificial Environment and Health

The discharge of untreated or inadequately treated textile wastewater has dire consequences for both the environment and human health:

- Aquatic ecosystems: The introduction of synthetic dyes reduces light penetration to water bodies, thereby affecting the process of photosynthesis in water-suspending plants and the dissolved oxygen necessary for fish survival in water bodies. This leads to the spoliation of biological diversity and an ecosystem's disturbed and unpredicted dynamics (Bhatia et al., 2017).
- Soil contamination: It is shown that when textile effluents enter the soil system, soil chemistry is changed, plant growth is affected, and the groundwater gets polluted.
- Human health risks: Through contact with infected water sources, one can suffer from skin and respiratory illnesses and, in the long term, be afflicted by carcinogens in dyes and heavy metals (Kishor et al., 2020).
- Suspended solids: The suspended solids in textile processes usually contain fibres and particulate matter. Hence, their removal is required before the onset of subsequent treatment processes. Such solids cause difficulties in the treatment operations and increase the associated costs of the operation (Ogunwumi et al., 2025).

7.2 Current Treatment Methods

Textile wastewater requires proper treatment because it contains very high levels of pollutants, which include synthetic dyes, heavy metals and some dangerous chemicals. The current treatment methods employed and the limitations of the physical, chemical and biological methods in this Table 7.2 are also shown.

Table 7.2 Summarise for the current treatment methods and limitations of the physical, chemical and biological methods

Treatment method	Description	Effectiveness	Limitations	References
Membrane filtration	Utilizes semi-permeable membranes to separate contaminants from water	High removal efficiency (up to 99% for turbidity and color)	Membrane fouling and blockage; requires regular maintenance; high operational costs.	(Ceretta et al., 2021; Kathing & Saini, 2022)
Coagulation-flocculation	Involves adding coagulants to aggregate and remove suspended particles	Effective for removing suspended solids and some dyes (up to 93% removal)	Generates sludge that requires disposal; may not effectively degrade all dye compounds.	(Kathing & Saini, 2022; Pattnaik et al., 2018)
Activated sludge	A biological treatment method using microorganisms to degrade organic matter	Can achieve significant COD reduction (>90%)	Sensitive to toxic compounds; produces large volumes of sludge; requires aeration and regular monitoring	(Andressa Chicatto et al., 2018; Castillo-Suárez et al., 2023)
Advanced oxidation processes (AOPs)	Chemical processes that generate hydroxyl radicals to oxidize pollutants	High efficiency in degrading complex organic compounds (over 98% removal) (Lebron et al., 2021)	High operational costs; generates secondary pollutants; requires careful handling of chemicals	(Castillo-Suárez et al., 2023; Ceretta et al., 2021)
Electrocoagulation	Uses electrical current to destabilize and remove contaminants from wastewater	Effective for color removal and heavy metal precipitation (up to 100% in some cases)	High energy consumption; equipment maintenance can be costly; effectiveness can vary with wastewater composition	(Kathing & Saini, 2022; Pattnaik et al., 2018)
Biological treatments	Utilizes fungi, bacteria, or algae for biodegradation of pollutants	Effective in reducing dye concentration and toxicity (up to 78% in real wastewater)	Slower treatment rates; may require pre-treatment of wastewater; effectiveness can be inhibited by toxic substances	(Andressa Chicatto et al., 2018; Kathing & Saini, 2022)

7.3 Analysis of Treatment Methods

1. Membrane filtration: This method is widely used for efficient cleaning due to its ability to remove many kinds of interference, like dyes and suspended solids. Nonetheless, membrane fouling poses a significant problem that lowers efficacy and raises costs (Pattnaik et al., 2018).
2. Coagulation-flocculation: Compared to the previous methods, this method is quite effective in removing turbidity and some dyes but produces sludge, which must be disposed of. Further, it does not remove all dye compounds to their highest possible degree of decomposition—thus, it may pose further environmental risks (Castillo-Suárez et al., 2023).
3. Activated sludge: This commonly applied biological technique can attain good COD elimination but is less tolerant to toxic substances in textile waste. Moreover, sludge produced in this amount is becoming a delicacy to dispose of (Kathing & Saini, 2022).
4. Advanced oxidation processes: The key advantages of AOPs are their ability to break down highly structured mini pollutants like complex organic compounds; however, they have a high operational cost and a high potential to form secondary pollutants (Kathing & Saini, 2022).
5. Electrocoagulation: Despite its applicability in treating highly contaminated wastewater, this method suffers from high energy consumption and variable efficiency depending on the type of water (Pattnaik et al., 2018).
6. Biological Treatments: Microorganisms' usage for degradation is eco-friendly; nevertheless, the contextual contamination of toxic components hinders microbial growth; further, advancements in enhancing these systems regarding their full-fledged industrial acceptability are still required (Ceretta et al., 2021).

7.4 Problem in Treatment

Managing wastewater retrieval from textile sources is pretty challenging due to the following reasons:

1. Chemical resistance: Some chemical dyes are considerably non-biodegradable. Hence, advanced treatment methods are used to break the substances further (Cao et al., 2021).
2. Composition variability: The composition of wastewater differs significantly depending on the textile being processed and the specific chemicals used in its production (Halepoto et al., 2022).
3. Regulatory compliance: Textile manufacturers are bound by strict environmental regulations on effluent quality. This requires that treatment systems are in place to ensure compliance with limits for various pollutants (Amanuel, 2019)

7.4.1 Mechanisms of Fungal Bioremediation

7.4.1.1 Fungal Species Involved in Wastewater Treatment (Fig. 7.3)

Aspergillus sp.

Some of these species are useful in breaking down synthetic dyes and other organic materials ordinarily contained in textile wastewater. Some of these fungi are *Aspergillus flavus* and *Aspergillus niger*.

- Role in degradation: It has been discovered that when *Aspergillus flavus* is applied in combination with other fungi, such as *Fusarium oxysporum*, decolourisation rates of up to 78% of azo dyes can be obtained (Selim et al., 2021).
- Besides the dye decolourisation, the fungal consortium also minimised other physicochemical pollutants, which include total dissolved solids and chemical oxygen demand (COD), displaying versatility in the treatment system.

Trametes versicolor

The second most researched white rot fungus is *Trametes versicolor*, known for its efficient action towards most environmental pollutants, such as pharmacological compounds and micropollutants, more than any other group of fungi.

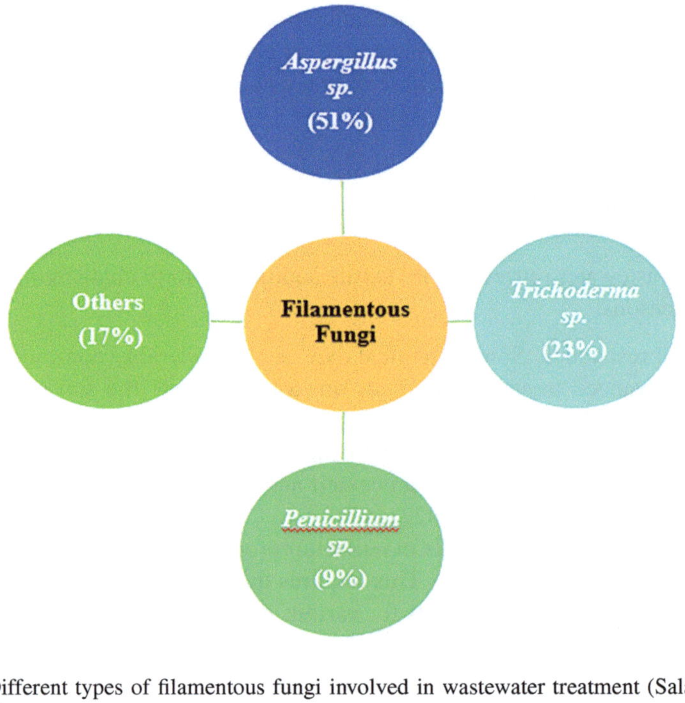

Fig. 7.3 Different types of filamentous fungi involved in wastewater treatment (Salazar-Cerezo et al., 2023)

- Role in degradation: Aquilegia has been very effective in eradicating pharmaceutical compounds in water; the following is the efficiency of this fungus. In one study, up to 69% of the elimination of pharmaceutical contaminants in hospital effluents was achieved (Halepoto et al., 2022).
- *T. versicolor* has been immobilised on rotating biological contactors for continuous flow operations, and the authors noted the high removal efficiency of the organic carbon loading in a recent study by Cruz et al. (Del Álamo et al., 2020). This fungus benefits from the enhanced bio-oxidation to degrade large, complicated chemical compounds into less hazardous products.

Penicillium chrysogenum

He et al. (2018) examined the ability of *Penicillium chrysogenum* to treat azo dyes in textile wastewater. This study proved that such species of this fungus could efficiently decolourise the Direct Black 22 (DB22) dye at a bench scale. It was further revealed that not only *did P. chrysogenum* decolourise the dye, but it was also able to deactivate the phytotoxicity of treated effluents and thus can be used for irrigation. The work highlighted the benefits of immobilising fungal biomass, resulting in better treatment performance and the possibility of biomass recovery for reuse for several cycles (Lanfranconi et al., 2022).

Phanerochaete chrysosporium

A white rot fungus named *Phanerochaete chrysosporium* is one of the most studied fungi specialising in the degradation of lignin. It comes out with various extracellular enzymes like lignin peroxidases and manganese peroxidases that play crucial roles in degrading complex organic hazardous wastes.

- Role in degradation: This fungus has been documented to possess the capacity to biodegrade different environmental contaminants such as phenols and azo dyes. For instance, Altes tests show that *P. chrysosporium* can effectively lower the levels of organic matter in pulping wastewater. For instance, in one study, the authors identified only seven organic compounds after 1 week of treatment of *P. chrysosporium*, compared to 43 before treatment (Yang et al., 2019).

Further, we have shown that this fungus can metabolise complex xenobiotics using the ligninolytic enzyme system, which initiates oxidation reactions that are mandatory for a wide variety of xenobiotic degradation.

Emmia latemarginata

In the work of Silva et al., *E. latemarginata* was one of the six fungal strains that were isolated from textile industry effluents, and this proved to be potentially helpful in decolourising azo, indigo, and anthraquinone dyes. The research proved that this strain produced high amounts of ligninolytic enzymes, including lignin peroxidase and manganese peroxidase, needed to degrade the dye under enzyme induction conditions (Juárez-Hernández et al., 2021).

Other Notable Fungi

In addition to the species above, several other fungi have been identified for their wastewater treatment capabilities:

- *Fusarium oxysporum*: Usually used in conjunction with *Aspergillus species*, it helps improve the efficiency of various dye degradation processes.
- *Penicillium* sp.: They discharge their capacity to break numerous organic pollutants and heavy metals in bioremedial processes (Selim et al., 2021).

7.4.1.2 Enzymatic Pathways in Fungal Bioremediation

Fungi are increasingly recognised because they secrete several extracellular enzymes to break down heterogeneous organic compounds, including synthetic dyes. Some of these enzymes are lignin peroxidase, manganese peroxidase and laccase. The current analysis describes these fungal enzymes and how they enable the degradation of textile dyes and other intricate organic structures concerning biotransformation and Biosorption.

Lignin Peroxidase (LiP)

Lignin peroxidase is an extracellular heme enzyme secreted by certain white rot fungi, especially Phanerochaete chrysosporium. The biodegradation process is intensely involved in the oxidation breakdown of lignin and other large molecular organic structures.

- Mechanism of action: LiP enhances the oxidation process of phenolic and non-phenolic substrates through the involvement of Hydrogen. The enzyme produces free radicals that can catalyse the reduction of complex dye structures like azo dyes into simpler and non-hazardous forms (Andressa Chicatto et al., 2018).

For example, the ability of LiP to compounds of azo dyes and cleave azo bonds (-N=N-) to form less toxic metabolites has been confirmed by several authors (Selim et al., 2021).

Manganese Peroxidase (MnP)

Another important enzyme that is secreted by fungi such as *Phanerochaete chrysosporium* and *Trametes versicolor* is manganese peroxidase. Although primarily analogous to ligninase, it is also associated with the degradation of dyes.

- Mechanism of action: MnP is an enzyme that incubates manganese ions (Mn^{2+}) into manganese oxides (Mn^{3+}), oxidising phenolic substrates and other substrates. This process gives rise to reactive oxygen species, which are involved in the degradation of large molecular structures of the applied dyes (Zhou et al., 2019). The enzyme can decolorise dyes, such as Reactive Red 120, at relatively high decolorisation rates (Del Álamo et al., 2020).

7.4.1.3 Laccase

Laccase is a multi-copper oxidase enzyme produced by many fungi, such as *Trametes versicolor* and *Aspergillus* species. It is also capable of oxidising phenolic compounds and aromatic amines.

- Mechanism of action: Laccase catalyses the oxidation of substrates by phenolic compounds that donate electrons to molecular oxygen, producing water as a byproduct. This may result in the polymerisation of the dye molecules or more complete mineralisation of the dye into harmless end products (Kishor et al., 2020). Their use for decolourising various synthetic dyes, Indigos, Methylene blue, etc., showed superior performance even in laboratory conditions (Selim et al., 2021).

7.4.1.4 Biotransformation and Biosorption Mechanisms

Two processes described when treating textile water have undergone fungal treatment: biotransformation and Biosorption.

Biotransformation

Biotransformation is the chemical alteration an organism makes on a drug or chemical compound. In the case of using fungi for bioremediation:

- Enzymatic degradation: Upon exposure to pollutants, fungi produce enzymes that destroy complex organic pollutants into simpler forms through oxidation. Laccases, for example, can oxidise the phenolic structures of dyes, allowing dye degradation into less toxic forms (Andressa Chicatto et al., 2018).
- Formation of intermediate metabolite: In the process of degradation, there are intermediate metabolites formed that may still exert some toxicity, but these can

be further treated enzymatically for complete mineralisation into consecutive innoxious final products, water and Carbon dioxide (Halepoto et al., 2022).

Biosorption

Biosorption is the process of passive bearing on biopolymers composed of fungal biomass, which absorbs pollutants.

- Mechanism: Since fungal mycelia are filamentous and have a high aspect ratio, they can encounter and adsorb dye molecules in wastewater. This encounter is also mediated by dose pH, temperature and dye concentration (Bibbins-Martínez et al., 2023). It has also been observed that optimal pH ranges enhance the adsorption potential of the fungal mycelia as the charged dye molecules and the surface of the fungi interact electrostatically.
- Role in treatment: Though Biosorption may not fully degrade the dyes, it removes them from wastewater before the enzymatic treatment. Therefore, combining both strategies improves the overall treatment (Jadhav & Kumar, 2020).

7.4.2 Comparison of Fungal Treatment with Other Biological Methods

The treatment of textile wastewater using biological methods has gained significant attention, e.g., with fungi and bacteria. Table 7.3 compares fungi treatment methods to bacterial procedures with a focus on the ability of these methods to handle complex pollutants, including azo dyes. In addition, the cost-effectiveness or the economisation that can be formulated using the fungal treatment method in industrial applications is discussed, along with the research conducted.

7.4.2.1 Economic Feasibility of Fungal Treatment

Recent studies have explored the cost-benefit analysis of fungal applications in wastewater treatment:

- A study by He et al. (2018) reported that the application of *Penicillium chrysogenum* biocatalyst for dye degradation offered a promising economic feasibility owing to relatively low operational costs of immobilising technologies, which facilitated biomass recycling and thereby decreased sludge formation (Lanfranconi et al., 2022).
- Research by Khan et al. highlighted that the consortium of *Aspergillus flavus* and Fusarium oxysporum demonstrated significant pollutant removal efficiencies while suggesting that the overall operational costs could be competitive with

Table 7.3 Comparison between fungal treatment methods and bacterial procedures

Aspect	Fungal treatment	Bacterial treatment (activated sludge/anaerobic digestion)	References
Mechanism of action	Utilizes enzymes (e.g., laccase, lignin peroxidase) to degrade complex organic pollutants through oxidation and biotransformation	Relies on microbial metabolism to break down organic matter; anaerobic digestion produces biogas from organic waste	(Lanfranconi et al., 2022; Selim et al., 2021)
Pollutant degradation	Effective in degrading synthetic dyes (e.g., azo dyes) and reducing heavy metals through biosorption and enzymatic degradation	Effective for organic matter reduction; however, bacterial degradation of azo dyes can produce toxic byproducts (aromatic amines)	(Halepoto et al., 2022; Thirumalaivasan et al., 2024)
Treatment efficiency	High efficiency in laboratory settings; varying results in large-scale applications due to environmental conditions and fungal growth challenges.	Generally high efficiency in removing BOD/COD; performance can be affected by toxic compounds and operational conditions.	(Del Álamo et al., 2020; Dell'Anno et al., 2022)
Time Requirements	Longer treatment times are often required for effective degradation, with some studies indicating several days for significant results.	Faster treatment times compared to fungi, typically within hours to a few days depending on conditions.	(Del Álamo et al., 2020; Selim et al., 2021)
Sludge Production	Minimal sludge production; immobilized fungi can be reused, reducing disposal issues.	Significant sludge production requiring management and disposal, leading to additional costs and environmental concerns.	(Lanfranconi et al., 2022; Thirumalaivasan et al., 2024)
Operational Costs	Potentially lower operational costs when using immobilized fungi; however, initial setup costs can be high.	Generally lower initial costs but can incur higher operational costs due to energy requirements for aeration and sludge handling.	(Del Álamo et al., 2020; Halepoto et al., 2022)
Economic Feasibility	Studies suggest that fungal treatment can be economically viable, especially when considering the reuse of biomass and reduced sludge disposal costs.	Established technology with known costs; however, efficiency may decrease with complex pollutants requiring additional treatments.	(Dell'Anno et al., 2022; Lanfranconi et al., 2022)
Limitations	Challenges include long hydraulic retention times, maintaining fungal cultures in reactors, and potential need for pre-treatment of wastewater to enhance fungal activity.	Limitations include sensitivity to toxic compounds, the need for regular monitoring, and the potential formation of toxic byproducts during dye degradation.	(Selim et al., 2021; Sudha et al., 2014)

traditional bacterial methods when accounting for reduced sludge management needs (Selim et al., 2021).

7.4.3 *Innovations and Advances in Fungal Wastewater Treatment*

7.4.3.1 Immobilisation Techniques

Immobilisation of fungal cells has become a new attractive technology in wastewater treatment, especially in increasing the efficacy and reusability of fungal bioremediation (Table 7.4).

Table 7.4 List of immobilization techniques in the innovation of fungal wastewater treatment

Aspect	Description	Examples of innovations	References
Immobilization techniques	Methods to confine fungal cells to a specific area while maintaining biological activity	– Alginate beads: Fungal cells are encapsulated in alginate beads, providing protection and ease of handling – Loofa Sponges: A natural support that allows for effective fungal attachment and retention	(Iqbal & Saeed, 2005; Lanfranconi et al., 2022)
Advantages of immobilization	Enhances treatment efficiency, allows for easy separation of biomass from treated water, and enables reuse of fungal cultures	– Studies show that immobilized fungi have higher decolorization rates for azo dyes compared to free cells – Immobilization reduces the hydraulic retention time required for effective treatment	(Lanfranconi et al., 2022; Selim et al., 2021)
Fungal species used	Various fungal species have been successfully immobilized for wastewater treatment	– *Penicillium chrysogenum* has been immobilized in alginate beads and loofa sponges for heavy metal removal – *Aspergillus flavus* and *Fusarium oxysporum* were used in a consortium to treat textile effluents	(Ali et al., 2021; Selim et al., 2021)
Pollutants treated	Immobilized fungi have shown effectiveness in degrading synthetic dyes and removing heavy metals	– *Penicillium chrysogenum* immobilized on loofa sponges achieved significant removal of Cd(II) from wastewater – Azo dye degradation using immobilized *Trametes versicolor* resulted in high decolorization percentages	(Ali et al., 2021; Lanfranconi et al., 2022)

Table 7.4 (continued)

Aspect	Description	Examples of innovations	References
Limitations of immobilization	Some challenges include potential mass transfer limitations, reduced metabolic activity under immobilization conditions, and the need for optimal operational parameters	– Immobilization can lead to decreased diffusion rates of substrates to fungal cells – The cost of materials for immobilization may be a concern in large-scale applications	(Ali et al., 2021; Selim et al., 2021)
Economic feasibility	The use of immobilized fungi can be economically viable due to reduced sludge production and the ability to reuse biomass	– Cost-benefit analyses indicate that while initial setup costs may be higher, long-term savings from reduced sludge management can offset these costs – Studies suggest that using natural supports like loofa can lower operational costs compared to synthetic materials	(Ali et al., 2021; Selim et al., 2021)

7.5 Advances in Genetic Engineering and Metabolic Pathway Optimization

With recent developments in genetically engineered tools, fungal strains have shown enhanced pollutant degradation ability for wastewater applications. The process by which scientists create fungal strains is altered by modifying the metabolic pathways, thus optimising the enzyme synthesis that could quite efficiently degrade complex organic pollutants, such as synthetic dyes and heavy metals (Table 7.5).

Table 7.5 Genetic engineering and metabolic pathway optimisation in fungal wastewater treatment

Aspect	Description	Examples of innovations	References
Genetic modification techniques	Various methods have been developed to enhance the genetic traits of fungi	– CRISPR–Cas9: A versatile tool for precise gene editing in filamentous fungi, allowing targeted mutations to enhance enzyme production – Agrobacterium-mediated transformation (AMT): This method allows for stable integration of exogenous genes into fungal genomes	(Li et al., 2017; Nødvig et al., 2015)
Enhanced Enzyme Production	Genetic modifications have led to increased production of enzymes critical for pollutant degradation	– *Aspergillus oryzae* strains engineered to overexpress laccase and manganese peroxidase showed improved dye decolorization capabilities – *Trametes versicolor* modified through CRISPR technology exhibited enhanced ligninolytic enzyme activity, leading to better degradation of lignin and associated dyes	(Jin et al., 2021; Salazar-Cerezo et al., 2023)
Metabolic Pathway Optimization	Genetic engineering has enabled the re-routing of metabolic pathways for improved degradation efficiency	– The disruption of non-homologous end-joining pathways in *A. oryzae* increased homologous recombination efficiency, facilitating the introduction of multiple beneficial traits – Metabolic engineering approaches have been applied to enhance the production of secondary metabolites involved in dye degradation	(Oliveira et al., 2024; Rajhans et al., 2021)
Field Applications	Engineered fungal strains are being tested in real-world wastewater treatment scenarios	– Field trials using genetically modified *Aspergillus* strains demonstrated significant reductions in COD and dye concentrations in textile effluents – Studies indicated that modified *Trametes versicolor* strains could effectively treat industrial wastewater with high dye loads	(Salazar-Cerezo et al., 2023; Zhao et al., 2016)
Limitations and Challenges	Despite advancements, challenges remain in the application of genetically engineered fungi	– Regulatory hurdles for the use of genetically modified organisms in environmental applications can limit deployment – Potential ecological impacts and stability of genetically modified strains in natural environments require further investigation	(Oliveira et al., 2024; Rajhans et al., 2021)

7.6 Combined Treatment Approaches in Wastewater Purification (Table 7.6)

Table 7.6 List of combined treatment approaches in wastewater purification

Aspect	Description	Examples of Innovations	References
Fungal–bacterial consortia	The combination of fungi and bacteria can leverage the strengths of both groups for improved pollutant degradation	A study by Badia-Fabregat et al. (2021) utilized a fungal/bacterial consortium that included *Trametes versicolor* and *Pseudomonas fluorescens*, achieving significant reductions in COD and dye concentrations	(Mawad et al., 2020)
Microalgae integration	Incorporating microalgae into fungal–bacterial systems can enhance nutrient uptake and further degrade pollutants	Research by Puentes-Morales et al. (2021) demonstrated that a fungal/bacterial consortium followed by *Chlorella* sp. treatment effectively reduced pollutants in non-domestic wastewater (Puentes-Morales et al., 2021)	(Mawad et al., 2020)
Physical treatments	Combining biological treatments with physical methods such as membrane filtration or adsorption can improve overall efficiency	A study showed that pre-treatment with activated carbon followed by a fungal consortium significantly improved dye removal rates compared to biological treatment alone (Sahu et al., 2023)	(Thirumalaivasan et al., 2024)
Biological fenton process	Utilizing fungi in conjunction with Fenton's reagent can enhance the degradation of recalcitrant compounds	Research indicated that combining *Phanerochaete chrysosporium* with Fenton's reagent resulted in higher decolorization rates for azo dyes (Wibowo et al., 2024)	(Thirumalaivasan et al., 2024)
Synergistic Effects	The interaction between fungi and bacteria can lead to enhanced degradation pathways and improved pollutant removal efficiencies	Studies have shown that fungal–bacterial consortia can increase the removal efficiency of heavy metals by up to 45% compared to individual cultures (Rudakova et al., 2024)	(Yu et al., 2024)
Limitations	Challenges include the need for optimizing conditions for each organism, potential antagonistic interactions, and scaling up processes for industrial applications	The complexity of maintaining optimal conditions for both fungal and bacterial growth can complicate treatment systems (Khan et al., 2021)	(Mandolini et al., 2021)

7.6.1 Environmental Impact of Fungal Treatment

7.6.1.1 Benefits

Reduced Chemical Usage

Fungal treatments often rely on biological processes to degrade pollutants, which can significantly decrease the need for harsh chemicals typically used in conventional wastewater treatment. For example, *Penicillium chrysogenum* has successfully degraded azo dyes without using toxic substances to reduce secondary pollution further. This amount of chemical reduction reduces operating costs and enhances the environmental impact of wastewater treatment processes (Lanfranconi et al., 2022).

Lower Sludge Production

Fungal treatments yield fewer sludges on average than bacterial treatments such as activated sludge processes. This is due to the ability of fungi to use a mixture of complex organic molecules as carbon sources for growth, thereby facilitating an increased efficiency of degradation. Reduced sludge production solves disposal problems commonly encountered with traditional treatments and, as a result, minimises environmental footprint (Selim et al., 2021).

Biodegradation of Complex Pollutants

Fungi possess unique enzymatic capabilities that enable them to degrade complex organic pollutants that are resistant to degradation by bacteria. White rot fungi (WRF), such as *Trametes versicolor*, can decompose lignin and related dyes via the secretion of ligninolytic enzymes (Latif et al., 2023). This capability improves the systemwide effectiveness of wastewater treatment systems.

7.6.1.2 Risks

Introduction of Nonnative Species

Introducing nonnative fungal species in wastewater treatment presents ecological risks if escape to natural habitats can happen. Nonnative fungi can displace endemic species, alter established ecological stability, and convey pathogens (Mandolini et al., 2021). When fungal treatments are implemented in "real-life" settings, this risk requires thoughtful and careful consideration and management.

Potential Pathogenicity

Certain fungi used in bioremediation may harbour pathogenic or relatives that may be unhealthy to humans and wildlife when released into the environment (Mawad et al., 2020). The occurrence of mycotoxins from some fungal species constitutes another worry about food safety and ecosystem integrity.

7.6.1.3 Regulatory Standards

The regulatory oversight framework of wastewater treatment is critical for assuring that effluent quality protects human health before it is discharged offsite to the environment. In most areas, especially in the textile sector, high effluent quality standards are regulated:

Effluent Quality Standards

Environmental Protection Agency (EPA) is one of the Regulatory agencies in the United States and the rest of the world that have established limits for different pollutants in wastewater discharges. These limits usually contain indices like Biological Oxygen Demand (BOD), Chemical Oxygen Demand (COD), total suspended solids (TSS) and particular pollutants such as heavy metals and dyes (Selim et al., 2021).

Guidelines for Bioremediation Technologies

Although there are well-established regulatory frameworks for traditional chemical treatments, there are still developments for guidelines on biological treatments, specifically those derived from fungi. Therapeutic interventions in fungi must be effective in achieving predefined effluent quality standards to be approved for general use (Mawad et al., 2020).

Environmental Impact Assessments

Environmental impact assessments are often needed before large-scale application of fungal bioremediation technologies to assess potential risks of releasing genetically engineered or nonnative fungal species into the environment. These evaluations support the requirement for environmental impact to be considered and mitigated (Yu et al., 2024).

7.6.2 Research Gaps in Fungal Wastewater Treatment

7.6.2.1 Improving Efficiency of Dye Degradation

Even with the promising potential of fungi in the textile industry for degrading synthetic dyes, more research is still needed to optimise fungi performance in degrading synthetic dyes. Key areas include:

- Enzyme optimisation: Studies on ligninolytic enzyme production and activity under different conditions to achieve the highest dye degradation per unit of time (Latif et al., 2023).
- Strain selection: Identification and genetic modification of fungal strains that have enhanced dye degradation ability or tolerance of wastewater-borne toxic substances (Selim et al., 2021).

7.6.2.2 Developing Scalable Fungal Bioreactors

There is a need to move from lab-scale experiments to large-scale applications, and the bioreactor design requires optimal growth conditions for fungi to be maintained during the efficient removal of pollutants. Research should focus on:

- Bioreactor design: An innovative arrangement of bioreactor configurations is proposed by which mass transfer and nutrient distribution can be efficiently maximised for fungal growth (Yu et al., 2024).
- Operational conditions: [Relevance of factors influencing fungal performance and environmental conditions] Understanding the impacts of changing pH, temperature, and nutrient availability on fungal performance in natural settings (Mandolini et al., 2021).

7.6.2.3 Overcoming Operational Challenges

Treatments for fungal diseases are often beset with problems due to operating conditions which may impair performance.

- Hydraulic retention time: Investigating techniques to decrease the hydraulic retention times with good treatment performance (Lanfranconi et al., 2022).
- Toxicity management: Strategies to reduce toxic compound effects on fungal growth and enzymatic activity (Mawad et al., 2020).

7.6.2.4 Potential for Large-Scale Applications

Integration into Existing Wastewater Treatment Facilities

The prospective application of combining fungal treatment with current wastewater treatment systems is attractive:

- Pilot projects: Current pilot projects are investigating the possibility of integrating fungal treatment with other traditional techniques (e.g., activated sludge), which could significantly improve overall treatment performance (Thirumalaivasan et al., 2024).
- Cost-benefit analysis: Research has shown that embedding fungi into current systems can lower chemical usage and sludge production, rendering it economically rational due to low-cost benefits (Selim et al., 2021).

Barriers to Widespread Adoption

Many barriers have to be overcome for the broader use of ITS:

- Regulatory hurdles: The introduction of genetically modified or non-indigenous fungal species can be subject to regulatory review through requirements for thorough risk analysis and safety assessment (Mandolini et al., 2021).
- Public perception: Educating stakeholders about the safety and efficacy of fungal treatments is crucial for gaining acceptance in industrial applications (Yu et al., 2024).

7.6.2.5 Sustainability Considerations

Environmental Footprint Reduction

Fungal treatments can reduce the environmental impact of the textile industry in several ways:

- Minimising chemical use: Fungi can act for contaminant degradation without strong chemicals and reduced secondary pollution potential (Lanfranconi et al., 2022).
- Lowering sludge production: Treatment with fungi generally results in lower amounts of sludge than are caused by traditional approaches, thereby reducing problems caused by surplus biomass (Selim et al., 2021). Treated effluents can be repurposed for agricultural use, promoting water conservation in regions facing scarcity (Mawad et al., 2020).
- Biomass utilisation: The biomass generated by fungal treatment can be used as a fertiliser or as a part of other biotechnological applications, which has a positive consequence for sustainability (Latif et al., 2023).

7.7 Conclusion

The search for treatment options for fungal applications for wastewater from the textile industry has shown great promise towards solving the multifaceted problems of this sector. Primary findings emphasise the remarkable potential of fungi, especially their enzymes, which can successfully degrade industrial synthetic dyes and heavy metals, which are majorly present in textile effluents. Research has demonstrated that fungi such as *Trametes species*, *Penicillium chrysogenum*, and various Aspergillus species can achieve high rates of pollutant removal tendency. Also, they produce less sludge than traditional bacterial methods. The blending of treatments with fungi has spectacularly increased the efficacy of the entire treatment into an auspicious system for contemporary wastewater treatment. Fungi have an enormous potential to innovate wastewater treatment processes (Fig. 7.4).

By harnessing their natural abilities for bioremediation, we can develop more sustainable and effective treatment solutions that mitigate pollution and contribute to cleaner water systems. The adoption of fungal treatments in the treatment of conventional wastewater facilities is a potential mechanism for significant reductions in chemical inputs and operational costs, in addition to its potential to reduce the environmental impacts of sludge disposal. In order to realise this potential, however, several policy and technological developments are required. Regulatory frameworks must be moved to reflect the deployment of different biological remedial methods, such as fungi, by maintaining safety and efficacy standards.

The focus on future research and development mainly leads to different procedures that can be performed through gene technology and metabolic pathways, which may enhance the fungal strains likewise. By applying another stainable practice and embracing innovative solutions, we can pave the way toward a future where wastewater treatment protects ecosystems and supports environmental sustainability and public health.

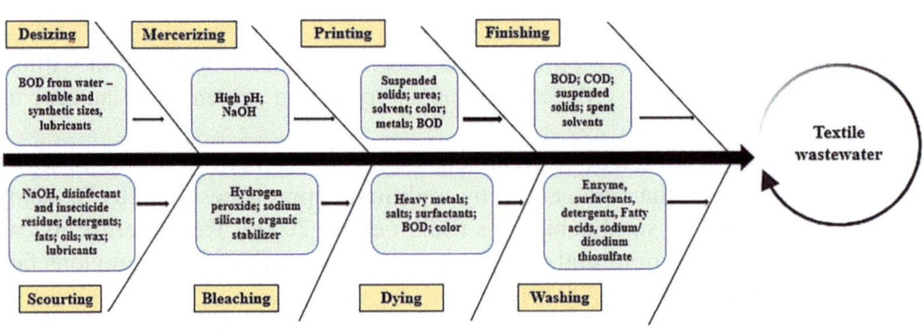

Fig. 7.4 The process of Textile wastewater production and the primary pollutants in the wastewater streams (Castillo-Suárez et al., 2023)

References

Adane, T., Adugna, A. T., & Alemayehu, E. (2021). Textile industry effluent treatment techniques. *Journal of Chemistry*. https://doi.org/10.1155/2021/5314404

Ali, E. A. M., Sayed, M. A., Abdel-Rahman, T. M., & Hussein, R. (2021). Fungal remediation of Cd(II) from wastewater using immobilization techniques. *RSC Advances*. https://doi.org/10.1039/D0RA08578B

Amanuel, L. (2019). Wastewater treatment plant and its design for textile industry. *Current Trends in Fashion Technology & Textile Engineering*. https://doi.org/10.19080/CTFTTE.2019.05.555663

Andressa Chicatto, J., Jefferson Gonçalves, M., Altmajer-Vaz, D., & Ballod Tavares, L. B. (2018). Treatment of the textile wastewater through fungi: A sustainable alternative. *Sustainability in Debate*. https://doi.org/10.18472/SustDeb.v9n1.2018.26460

Bello, I. A. (2017). Challenges in textile wastewater and current palliative methods: An overview. *IIUM Engineering Journal*. https://doi.org/10.31436/iiumej.v18i2.742

Bhatia, D., Sharma, N. R., Singh, J., & Kanwar, R. S. (2017). Biological methods for textile dye removal from wastewater: A review. *Critical Reviews in Environmental Science and Technology*. https://doi.org/10.1080/10643389.2017.1393263

Bibbins-Martínez, M., Juárez-Hernández, J., López-Domínguez, J. Y., Nava-Galicia, S. B., Martínez-Tozcano, L. J., Juárez-Atonal', R., & Díaz-Godinez, G. (2023). Potential application of fungal biosorption and/or bioaccumulation for the bioremediation of wastewater contamination: A review. *Journal of Environmental Biology*. https://doi.org/10.22438/jeb/44/2/MRN-5093

Buratti, S., Girometta, C. E., Baiguera, R. M., Barucco, B., Bernardi, M., De Girolamo, G., & Savino, E. (2022). Fungal diversity in two wastewater treatment plants in North Italy. *Microorganisms, 10*(6). https://doi.org/10.3390/microorganisms10061096

Cao, N., Yue, C., Lin, Z., Li, W., Zhang, H., Pang, J., & Jiang, Z. (2021). Durable and chemical resistant ultra-permeable nanofiltration membrane for the separation of textile wastewater. *Journal of Hazardous Materials*. https://doi.org/10.1016/j.jhazmat.2021.125489

Castillo-Suárez, L. A., Sierra-Sánchez, A. G., Linares-Hernández, I., Martínez-Miranda, V., & Teutli-Sequeira, E. A. (2023). A critical review of textile industry wastewater: Green technologies for the removal of indigo dyes. *International Journal of Environmental Science and Technology*. https://doi.org/10.1007/s13762-023-04810-2

Ceretta, M. B., Nercessian, D., & Wolski, E. A. (2021). Current trends on role of biological treatment in integrated treatment technologies of textile wastewater. *Frontiers in Microbiology*. https://doi.org/10.3389/fmicb.2021.651025

Del Álamo, A. C., Pariente, M. I., Martínez, F., & Molina, R. (2020). Trametes versicolor immobilized on rotating biological contactors as alternative biological treatment for the removal of emerging concern micropollutants. *Water Research*. https://doi.org/10.1016/j.watres.2019.115313

Dell'Anno, F., Rastelli, E., Buschi, E., Barone, G., Beolchini, F., & Dell'Anno, A. (2022). Fungi can be more effective than bacteria for the bioremediation of marine sediments highly contaminated with heavy metals. *Microorganisms*. https://doi.org/10.3390/microorganisms10050993

Deshmukh, R., Khardenavis, A. A., & Purohit, H. J. (2016). Diverse metabolic capacities of fungi for bioremediation. *Indian Journal of Microbiology*. https://doi.org/10.1007/s12088-016-0584-6

Deshmukh, S. M., Dhokpande, S. R., Sankhe, A., & Khandekar, A. (2025). Effluent wastewater technologies for textile industry: A review. *Reviews in Inorganic Chemistry*. https://doi.org/10.1515/revic-2024-0046

Efremenko, E., Stepanov, N., Senko, O., Aslanli, A., Maslova, O., & Lyagin, I. (2024). Using fungi in artificial microbial consortia to solve bioremediation problems. *Microorganisms*. https://doi.org/10.3390/microorganisms12030470

Gallo, G., Arancio, W., Palazzotto, E., Capri, F. C., & Alduina, R. (2023). Microbial biotechnology for wastewater treatment into circular economy. *Current Developments in Biotechnology and Bioengineering*. https://doi.org/10.1016/B978-0-323-99920-5.00003-2

Ghosh, S., Rusyn, I., Dmytruk, O. V., Dmytruk, K. V., Onyeaka, H., Gryzenhout, M., & Gafforov, Y. (2023). Filamentous fungi for sustainable remediation of pharmaceutical compounds, heavy metal and oil hydrocarbons. *Frontiers in Bioengineering and Biotechnology*. https://doi.org/10.3389/fbioe.2023.1106973

Halepoto, H., Gong, T., & Memon, H. (2022). Current status and research trends of textile wastewater treatments—A bibliometric-based study. *Frontiers in Environmental Science*. https://doi.org/10.3389/fenvs.2022.1042256

Han, Z., Moh, E. S., Santos, A. L., Barcellos, I. C., Peng, Y., Huang, W., & Ye, J. (2023). Dechlorination of wastewater from shell-based glucosamine processing by mangrove wetland-derived fungi. *Frontiers in Microbiology*. https://doi.org/10.3389/fmicb.2023.1271286

Iqbal, M., & Saeed, A. (2005). Novel method for cell immobilization and its application for production of organic acid. *Letters in Applied Microbiology*. https://doi.org/10.1111/j.1472-765X.2004.01646.x

Jadhav, S. K., & Kumar, M. P. (2020). Biosorption of Lead from aqueous solution using a new fungal strain, Cunninghamella elegans TUFC20022. *Research Journal of Biotechnology, 15*, 35–42.

Jakhar, R., & Styszko, K. (2025). Recent trends in emerging greywater management in India: An overview. *Desalination and Water Treatment*. https://doi.org/10.1016/j.dwt.2025.101226

Jin, F. J., Hu, S., Wang, B. T., & Jin, L. (2021). Advances in genetic engineering technology and its application in the industrial fungus Aspergillus oryzae. *Frontiers in Microbiology*. https://doi.org/10.3389/fmicb.2021.644404

Juárez-Hernández, J., Castillo-Hernández, D., Pérez-Parada, C., Nava-Galicia, S., Cuervo-Parra, J. A., Surian-Cruz, E., et al. (2021). Isolation of fungi from a textile industry effluent and the screening of their potential to degrade industrial dyes. *Journal of Fungi*. https://doi.org/10.3390/jof7100805

Kallawar, G. A., & Bhanvase, B. A. (2024). A review on existing and emerging approaches for textile wastewater treatments: Challenges and future perspectives. *Environmental Science and Pollution Research*. https://doi.org/10.1007/s11356-023-31175-3

Kathing, C., & Saini, G. (2022). A review of various treatment methods for the removal of dyes from textile effluent. *Recent Progress in Materials*. https://doi.org/10.21926/rpm.2204028

Khan, S., & Malik, A. (2018). Toxicity evaluation of textile effluents and role of native soil bacterium in biodegradation of a textile dye. *Environmental Science and Pollution Research*. https://doi.org/10.1007/s11356-017-0783-7

Kishor, R., Purchase, D., Ferreira, L. F. R., Mulla, S. I., Bilal, M., & Bharagava, R. N. (2020). Environmental and health hazards of textile industry wastewater pollutants and its treatment approaches. *Handbook of Environmental Materials Management*. https://doi.org/10.1007/978-3-319-58538-3_230-1

Lanfranconi, I., Ceretta, M. B., Bertola, N., Wolski, E. A., & Durruty, I. (2022). Textile dyeing wastewater treatment by Penicillium chrysogenum: Design of a sustainable process. *Water Science and Technology*. https://doi.org/10.2166/wst.2022.204

Latif, W., Ciniglia, C., Iovinella, M., Shafiq, M., & Papa, S. (2023). Role of white rot fungi in industrial wastewater treatment: A review. *Applied Sciences*. https://doi.org/10.3390/app13148318

Li, D., Tang, Y., Lin, J., & Cai, W. (2017). Methods for genetic transformation of filamentous fungi. *Microbial Cell Factories*. https://doi.org/10.1186/s12934-017-0785-7

Magnoli, K., Carranza, C., Aluffi, M., Magnoli, C., & Barberis, C. (2023). Fungal biodegradation of chlorinated herbicides: An overview with an emphasis on 2,4-D in Argentina. *Biodegradation*. https://doi.org/10.1007/s10532-023-10022-9

Malik, S., Bora, J., Nag, S., Sinha, S., Mondal, S., Rustagi, S., & Almutary, A. G. (2023). Fungal-based remediation in the treatment of anthropogenic activities and pharmaceutical-pollutant-contaminated wastewater. *Water*. https://doi.org/10.3390/w15122262

Mandolini, E., Probst, M., & Peintner, U. (2021). Methods for studying bacterial–fungal interactions in the microenvironments of soil. *Applied Sciences*. https://doi.org/10.3390/app11199182

Mawad, A. M., Abd Hesham, E. L., Yousef, N. M., Shoreit, A. A., Gathergood, N., & Gupta, V. K. (2020). Role of bacterial-fungal consortium for enhancement in the degradation of industrial dyes. *Current Genomics*. https://doi.org/10.2174/1389202921999200505082901

Nødvig, C. S., Nielsen, J. B., Kogle, M. E., & Mortensen, U. H. (2015). A CRISPR-Cas9 system for genetic engineering of filamentous fungi. *PLoS One*. https://doi.org/10.1371/journal.pone.0133085

Ogunwumi, O. T., Adeniyi, A. F., Angus, M. C., & Oche, O. S. (2025). Biological wastewater treatment. In *Wastewater treatment and sludge management systems-the gutter-to-good approaches: The gutter-to-good approaches*. IntechOpen. https://doi.org/10.5772/intechopen.1004638

Okal, E. J., Heng, G., Magige, E. A., Khan, S., Wu, S., Ge, Z., & Xu, J. (2023). Insights into the mechanisms involved in the fungal degradation of plastics. *Ecotoxicology and Environmental Safety*. https://doi.org/10.1016/j.ecoenv.2023.115202

Oliveira, L. R., Gonçalves, A. R., Quintela, E. D., Colognese, L., Cortes, M. V. D. C. B., & de Filippi, M. C. C. (2024). Genetic engineering of filamentous fungi: Prospects for obtaining fourth-generation biological products. *Applied Microbiology*. https://doi.org/10.3390/applmicrobiol4020055

Pattnaik, P., Dangayach, G. S., & Bhardwaj, A. K. (2018). A review on the sustainability of textile industries wastewater with and without treatment methodologies. *Reviews on Environmental Health*. https://doi.org/10.1515/reveh-2018-0013

Pundir, A., Thakur, M. S., Prakash, S., Kumari, N., Sharma, N., Parameswari, E., & Kumar, M. (2024). Fungi as versatile biocatalytic tool for treatment of textile wastewater effluents. *Environmental Sciences Europe*. https://doi.org/10.1186/s12302-024-01007-3

Rajhans, G., Barik, A., Sen, S. K., & Raut, S. (2021). Degradation of dyes by fungi: An insight into mycoremediation. *Biotechnologia*. https://doi.org/10.5114/bta.2021.111109

Salazar-Cerezo, S., de Vries, R. P., & Garrigues, S. (2023). Strategies for the development of industrial fungal producing strains. *Journal of Fungi*. https://doi.org/10.3390/jof9080834

Selim, M. T., Salem, S. S., Mohamed, A. A., El-Gamal, M. S., Awad, M. F., & Fouda, A. (2021). Biological treatment of real textile effluent using aspergillus flavus and fusarium oxysporium and their consortium along with the evaluation of their Phytotoxicity. *Journal of Fungi*. https://doi.org/10.3390/jof7030193

Srikanth, M., Sandeep, T. S. R. S., Sucharitha, K., & Godi, S. (2022). Biodegradation of plastic polymers by fungi: A brief review. *Bioresources and Bioprocessing*. https://doi.org/10.1186/s40643-022-00532-4

Sudha, M., Saranya, A., Selvakumar, G., & Sivakumar, N. (2014). Microbial degradation of azo dyes: A review. *International Journal of Current Microbiology and Applied Sciences, 2*, 670–690.

Thirumalaivasan, N., Gnanasekaran, L., Kumar, S., Durvasulu, R., Sundaram, T., Rajendran, S., & Kanagaraj, K. (2024). Utilization of fungal and bacterial bioremediation techniques for the treatment of toxic waste and biowaste. *Frontiers in Materials*. https://doi.org/10.3389/fmats.2024.1416445

Vaksmaa, A., Guerrero-Cruz, S., Ghosh, P., Zeghal, E., Hernando-Morales, V., & Niemann, H. (2023). Role of fungi in bioremediation of emerging pollutants. *Frontiers in Marine Science*. https://doi.org/10.3389/fmars.2023.1070905

Vasconcelos, M. W., Goncalves, S., de Oliveira, E. C., Rubert, S., & de Castilhos Ghisi, N. (2022). Textile effluent toxicity trend: A scientometric review. *Journal of Cleaner Production*. https://doi.org/10.1016/j.jclepro.2022.132756

Yang, Z. Z., Guo, Y. N., Liu, A. L., Zhang, J., Wang, M. Y., Liu, Y., & Cao, J. G. (2019). Pollutant characteristics from wastewater of poplar pre-conditioning refiner chemical alkaline peroxide mechanical pulping pretreated with phanerochaete chrysosporium. *BioResources*. https://doi.org/10.15376/biores.14.2.4792-4805

Yin, H., Qiu, P., Qian, Y., Kong, Z., Zheng, X., Tang, Z., & Guo, H. (2019). Textile wastewater treatment for water reuse: A case study. *PRO*. https://doi.org/10.3390/pr7010034

Yu, S., Chen, Z., Li, M., Qiu, S., Lv, Z., & Ge, S. (2024). Principles, challenges, and optimization of indigenous microalgae-bacteria consortium for sustainable swine wastewater treatment. *Bioresource Technology*. https://doi.org/10.1016/j.biortech.2024.131055

Zhao, H., Lovett, B., & Fang, W. (2016). Genetically engineering entomopathogenic fungi. *Advances in Genetics*. https://doi.org/10.1016/bs.adgen.2015.11.001

Zhou, Y., Meng, J., Zhang, M., Chen, S., He, B., Zhao, H., & Wang, T. (2019). Which type of pollutants need to be controlled with priority in wastewater treatment plants: Traditional or emerging pollutants? *Environment International*. https://doi.org/10.1016/j.envint.2019.104982

Chapter 8
Fungal-Based Constructed Wetlands for Wastewater Treatment for Fungi in Wastewater Treatment

Sukanya Sonowal, Kabyashree Buragohain, Aparoop Das, and Ratul Nath

Abstract Water, often termed the "elixir of life," faces rising demand driven by economic development, urban expansion, and industrialization. The growing challenges of water scarcity and pollution have spurred researchers to develop sustainable approaches to wastewater treatment. Among these, fungi have gained recognition as effective agents in bioremediation, adept at eliminating dyes, heavy metals, and toxins through enzyme activity and collaborative interactions. Fungi contribute to wastewater purification through mycoremediation, which involves breaking down organic pollutants, and biosorption, where heavy metals are adsorbed, often enhanced by symbiotic relationships with other microorganisms. Recent advancements in this field include the use of genetically engineered fungal strains, fungal consortia, and immobilized fungi within bioreactors and mycofilters for large-scale treatment processes. These fungi-based methods are environmentally friendly, generating minimal harmful byproducts while adapting effectively to varied environmental conditions. This chapter highlights the growing role of fungi in wastewater management, emphasizing how ongoing research is paving the way for cleaner water and healthier ecosystems. By leveraging fungi's unique capabilities, innovative solutions to the global challenge of wastewater treatment are emerging.

Keywords Fungi · Waste-water remediation · Mycoremediation · Wastewater treatment · Constructed wastelands

S. Sonowal · K. Buragohain · R. Nath (✉)
Department of Life Sciences, Faculty of Biological Sciences, Dibrugarh University, Dibrugarh, Assam, India
e-mail: rs_sukanyasonowal@dibru.ac.in; ratulnath@dibru.ac.in

A. Das
Department of Pharmaceutical Sciences, Faculty of Science and Engineering, Dibrugarh University, Dibrugarh, Assam, India
e-mail: aparoopdas@dibru.ac.in

C. M. Hussain et al. (eds.), *Fungi in Waste Water Treatment*, Springer Water,
https://doi.org/10.1007/978-3-032-12693-1_8

8.1 Introduction

Wastewater treatment is a critical process designed to remove contaminants from water that has been used in various sectors, including domestic, industrial, and agricultural applications. The primary objective is to convert this wastewater into effluent that can be safely discharged into the environment or reused for irrigation and other purposes. The treatment process typically involves several stages: preliminary treatment to remove large solids, primary treatment for sedimentation, secondary biological treatment to degrade organic matter, and tertiary treatment for further purification and nutrient removal (Asano & Tchobanoglous, 1987; Vymazal, 2011).

Effective wastewater management is vital for public health, environmental sustainability, and resource conservation. With the global population increasing and urbanization expanding, the volume of wastewater generated continues to rise. This presents significant challenges, such as the presence of emerging contaminants, nutrient overloads (e.g., nitrogen and phosphorus), and the need for cost-effective technologies capable of handling diverse wastewater compositions (Bahri, 1988). The integration of innovative solutions is essential to ensure that wastewater treatment systems can meet regulatory standards while minimizing ecological impacts.

Current methods of wastewater treatment methods can be categorized into three main types: physical, chemical and biological. Physical method employs physical processes such as screening, sedimentation, and filtration to remove solids from wastewater without chemical additives. Techniques like aeration are also included to enhance the removal of dissolved gases (Bouwer & Chaney, 1974). Chemical treatment involves the application of chemical agents to eliminate impurities. Common practices include disinfection using chlorine or ozone and processes like coagulation and flocculation to aggregate suspended particles for easier removal (Pescod & Arar, 1988). The biological approach utilizes microorganisms to metabolize organic matter in wastewater. Biological treatment can be further divided into aerobic processes requiring oxygen and anaerobic processes that do not (Mara & Cairncross, 1989). Recent studies have highlighted the effectiveness of constructed wetlands (CWs) as a sustainable alternative for biological treatment due to their ability to integrate natural processes with engineered systems (Zhong et al., 2024).

Microorganisms are integral to biological wastewater treatment systems. They metabolize organic pollutants, converting them into less harmful substances while facilitating nutrient cycling within ecosystems. Bacteria are typically the dominant microorganisms in these processes; however, fungi are increasingly recognized for their unique capabilities in degrading complex organic compounds and heavy metals (Rose & Clark, 1986; Shuval et al., 1986). Recent research has shown that microbial communities in CWs contribute significantly to pollutant removal and overall system performance (Dimuro et al., 2023).

The microbial involvement in wastewater treatment encompasses a diverse array of organisms that work synergistically to break down pollutants. Bacterial communities primarily initiate the degradation of organic matter, while fungi contribute significantly through their enzymatic activities (Jewell et al., 1983). These microorganisms form complex biofilms that enhance their efficiency in pollutant removal, making them integral to both aerobic and anaerobic treatment systems (Polprasert et al., 1983). Studies have demonstrated that CWs not only improve microbial diversity but also enhance water quality by regulating nutrient concentrations and preventing eutrophication (Stottmeister et al., 2003; Vymazal, 2007).

Microorganisms typically do not exist in isolation but form interconnected networks within communities that regulate ecosystem structure and functionality (He et al., 2017). These microbes exhibit positive or negative co-occurrence patterns, where interactions can involve competition or complementarity in substrate decomposition and nutrient recycling. Interestingly, microbes occupying similar niches and utilizing the same substrates may also co-occur due to environmental filtering processes (de Menezes et al., 2017). Network analysis has emerged as a powerful tool for studying microbial interactions and has been successfully employed to investigate co-occurrence patterns in complex environments such as soils, pollen, and oceans (Barberán et al., 2012; Manirajan et al., 2018; Ruan et al., 2006). Applying such analyses could uncover novel fungal communities capable of cooperative pollutant degradation and enhance understanding of co-pathogenesis mechanisms. However, there is limited research available on the application of co-occurrence network analysis for fungal communities in wastewater treatment plants (Zhang et al., 2018).

Fungi are emerging as crucial players in wastewater treatment due to their ability to degrade recalcitrant compounds that are often resistant to bacterial degradation. Their unique metabolic pathways allow them to break down complex organic materials, including lignin and other phenolic compounds prevalent in industrial effluents (Hespanhol & Prost, 1994). Additionally, certain fungal species exhibit remarkable tolerance to heavy metals and other toxic substances, making them ideal candidates for bioremediation efforts (Glazer & Nikaido, 2007). Recent studies have highlighted the effectiveness of filamentous fungi in hybrid constructed wetlands for removing pollutants from treated river water (Mustafa et al., 2024). The production of specific enzymes such as lignin peroxidase and manganese peroxidase further enhances their efficacy in degrading pollutants (Asano & Tchobanoglous, 1987). Consequently, integrating fungi into constructed wetlands presents a promising approach to improving wastewater treatment outcomes while promoting ecological sustainability.

Fungi are widespread in the environment, exhibiting significant diversity, biomass levels comparable to bacteria, and a greater proportion than archaea and viruses (Maza-Márquez et al., 2016; Wei et al., 2018). Despite their distinct physiological and metabolic characteristics, fungi interact with other microorganisms across various environments, performing crucial ecosystem functions such as organic matter decomposition (Maza-Márquez et al., 2018). In particular, they are key members of microbial communities in wastewater treatment plants (WWTPs) (Zhang et al., 2018), contributing to the degradation of environmental organic chemicals, including proteins, complex carbohydrates, lipids, aromatic hydrocarbons, pharmaceutical compounds, heavy metals, and endocrine-disrupting chemicals. These processes are facilitated by an array of intra- and extracellular enzymes and sorption mechanisms (Harms et al., 2011; Naghdi et al., 2018; Siddiquee et al., 2015).

Moreover, fungi are linked to denitrification processes (Guest & Smith, 2002) and the stabilization of activated sludge cell aggregates (Weber et al., 2009). However, excessive fungal biomass can lead to operational challenges in WWTPs, such as membrane biofouling, foaming, and bulking (Luo et al., 2017; More et al., 2010).

8.2 Fungal Diversity in Wastewater Treatment

Fungi are essential players in wastewater treatment, contributing significantly to the breakdown of organic matter and the removal of various pollutants. The main types of fungi involved include filamentous fungi, yeasts and moulds. Filamentous fungi, such as *Aspergillus*, *Penicillium*, and *Fusarium*, are characterized by their long, thread-like structures (hyphae). They are particularly effective in removing pharmaceutical compounds and other contaminants from wastewater, achieving removal efficiencies ranging from 78% to 100% for various pollutants (Dalecka et al., 2023). Their ability to form biofilms enhances nutrient uptake and pollutant degradation, making them valuable in bioremediation processes. Yeasts, including genera such as *Candida*, *Hansenula*, and *Saccharomyces*, are unicellular fungi that thrive in high organic load environments. They have shown potential for reducing chemical oxygen demand (COD) and heavy metal concentrations in wastewater. For instance, certain yeast strains can remove up to 96% of lead and 40% of cadmium ions from contaminated waters (Nicula et al., 2023). Their ability to grow under semi-anaerobic conditions allows them to be effective in diverse wastewater treatment scenarios. Moulds, such as those from the genera *Trichoderma* and *Rhizopus*, contribute to the degradation of complex organic materials. They produce enzymes that facilitate the breakdown of pollutants, including dyes and phenolic compounds, which are commonly found in industrial effluents (Pundir et al., 2024).

Natural fungal strains and genetically engineered fungal strains are the two types of fungal strains utilized for wastewater remediation. Natural Fungal strains are isolated from environments such as wastewater treatment plants or contaminated sites. They are adapted to local conditions and exhibit resilience against various pollutants. For example, strains of Aspergillus have been shown to effectively biosorb heavy metals like iron and cobalt from wastewater (El-Gendy et al., 2023). Advances in genetic engineering have led to the development of modified fungal strains with enhanced capabilities for pollutant degradation. These engineered fungi can express specific enzymes that improve their efficiency in breaking down complex compounds found in wastewater (Ossio et al., 2023). For instance, white rot fungi have been genetically modified to enhance their lignin-degrading capabilities, making them suitable for treating lignin-rich industrial effluents.

Several characteristics make fungi particularly effective for wastewater treatment. The primary characteristics are tolerance to contaminants, enzyme production and mycoremediation capabilities. Many fungi possess remarkable tolerance to heavy metals and organic pollutants. For instance, certain fungal species can thrive in environments with high concentrations of chloride ions, demonstrating their potential for bioremediation in saline wastewater contexts (Zhou et al., 2023). They are also capable of producing a variety of enzymes that facilitate the degradation of complex organic substances. Lignin Peroxidase and Manganese Peroxidase are two key enzymes in this regard. Lignin Peroxidase is crucial for breaking down lignin, a complex polymer found in plant material that is often present in industrial effluents. Manganese Peroxidase, on the other hand, aids in degrading phenolic compounds and other aromatic pollutants commonly found in wastewater (Bilal et al., 2016). Mycoremediation refers to the use of fungi to degrade or sequester environmental contaminants. Fungi can effectively remove pollutants from soil and water through their metabolic processes, making them valuable tools for bioremediation efforts (Dalecka et al., 2023; Pundir et al., 2024).

Recent studies have emphasized the critical role of fungal communities in wastewater treatment systems, highlighting their diversity and functional contributions. For instance, filamentous fungi have been shown to effectively reduce pharmaceutical compounds within short time frames, demonstrating their potential in contaminant removal (Dalecka et al., 2023). Similarly, research on yeast strains has revealed their dual capability in removing heavy metals and reducing organic compounds from synthetic wastewater, further underscoring their suitability for practical applications in biotreatment technologies (Nicula et al., 2023). These findings highlight the need to deepen our understanding of fungal diversity and community dynamics to enhance the efficiency and effectiveness of wastewater treatment processes.

8.3 Mechanisms of Fungal Action in Wastewater Treatment

Wastewater pollution from industrial pollutants has become a significant environmental concern. This issue primarily stems from the release of untreated or insufficiently treated chemical substances during various industrial processes. Industries such as paper and pulp, agrochemicals, textiles, and petrochemicals generate vast quantities of wastewater containing toxic and persistent pollutants, including organic compounds, heavy metals, aromatic dyes, and other resistant chemicals. The discharge of these contaminants into natural water bodies and ecosystems without proper treatment poses serious risks to the environment and public health (Kuramae et al., 2021; Latif et al., 2023). Traditional wastewater treatment systems often struggle to remove certain pollutants due to their resistance to degradation effectively. Consequently, the persistence of these compounds in surface water and their potential environmental impact has raised concerns globally (Mohammed et al., 2024). These biological wastewater treatment systems produce significant amounts of low-value sludge, primarily bacterial biomass, which requires costly treatment before disposal. On the other hand, fungi are commonly used in industry to produce various valuable biochemicals. Fungi are microorganisms that decompose a significant portion of organic compounds in the environment. Since the 1980s, they have been employed in bioremediation techniques for treating water and soil (Rodríguez-Rodríguez et al., 2014). Fungal cell walls, comprising about 30% of their dry weight, are primarily made up of polysaccharides, which account for roughly 80% of this weight. This composition provides fungi with numerous cell wall components that effectively bind metals, making them highly efficient biosorbents (Mir-Tutusaus et al., 2018). Besides a small amount of glycoprotein, fungal cell walls are primarily composed of substantial quantities of chitin, chitosan, glucan, and mannan. These polymers often feature metal-binding ligands, such as carboxyl (R–COOH or R–CO_2H), hydroxyl (–OH), amine (NH_2), and phosphate (PO_4^-, PO_3^-) functional groups. Strains of various fungi, such as *Aspergillus niveus, Aspergillus fumigatus*, and *Aspergillus niger*, show promise for bioremediation applications. However, certain strains face limitations due to the extended time required for spore production. Notably, members of the Ascomycetes group are recognized for their effectiveness in reducing chemical oxygen demand (COD) in industrial wastewater, including effluents from distilleries (Saeed et al., 2022). Recently, employing white-rot fungi for the biological treatment of wastewater contaminated with pharmaceuticals and personal care products (PPCPs) has shown considerable promise (Mir-Tutusaus et al., 2018). Combining wastewater treatment with resource recovery could provide a cost-effective and sustainable approach to waste management. In this context, fungal-based wastewater treatment offers a promising alternative, using inexpensive organic substrates to generate high-value fungal byproducts while simultaneously treating wastewater. Constructed wetlands (CWs) mimic natural processes by using plants and soil to treat wastewater in a controlled environment (Parde et al., 2021). They are a highly effective technology for eliminating organic micropollutants through synergistic processes that involve interactions between plants and microorganisms, including hydrolysis, volatilization, sorption, biodegradation, and photolysis (Kaur et al., 2020; Mesacasa et al., 2024). Fungi offer additional benefits over bacteria in biological wastewater treatment, beyond producing valuable byproducts like amylase, chitin, and lactic acids. One key advantage is their production of extracellular enzymes, which break stubborn compounds such as phenolic compounds, dyes, and polyaromatic hydrocarbons (PAHs) through non-specific oxidation reactions (Sankaran et al., 2010). The Mechanisms Involved in Fungal Wastewater Treatment are-

8.3.1 Biosorption

Biosorption is a natural physicochemical process in which certain biomasses passively capture and bind contaminants to their cellular structures. Although frequently proposed as an alternative or complementary biotechnology, its practical application has ironically seen limited success (Volesky & Holan, 1995). The idea of utilizing fungi as biosorbents originated from studies on the toxic effects of heavy metals during commercial mushroom production. Since then, extensive research has been conducted in this field. Certain fungal species, including *Pythium* sp., *Dictyuchus sterile*, and *Scytalidium lignicola*, have been shown to adsorb heavy metals such as zinc, lead, and cadmium onto their mycelium from freshwater shrimp culture ponds (Duddridge & Wainwright, 1980). Numerous studies have since explored the effectiveness of various fungal species in removing heavy metals from wastewater. Currently, fungi are among the most extensively researched microorganisms in the field of biological pollutant remediation. These heterotrophic organisms are capable of transforming metal contaminants through chemical modifications or metabolic processes. Their ability to form extensive mycelial networks and utilize pollutants as growth substrates makes them well-suited for bioremediation efforts (Harms et al., 2011; Shakya et al., 2015). The mechanisms by which fungal pellets interact with acid-anionic dyes are highly complex, primarily involving processes of biodegradation and biosorption (Li et al., 2019). The primary mechanisms driving biosorption are complexation and electrostatic interactions. The process by which dyes are biosorbed depends largely on their diverse chemical structures. Fungal cell walls, which consist of polysaccharides (like chitin and chitosan), proteins, lipids, and melanin, feature various functional groups such as amines, carboxyl, and phosphate groups—that can effectively bind to dye molecules (Aksu et al., 2008).

8.3.2 Enzymes

Fungi have the capability to produce broad-spectrum oxidative enzymes, enabling them to break down pollutants with highly complex structures (Espinosa-Ortiz et al., 2016), including laccase, esterase, manganese peroxidase, lipase, and unspecific peroxygenases (UPOs). Fungi, as vital decomposers in nature, possess a rich array of enzymes that facilitate the breakdown of various complex substrates. It is estimated that fungi are responsible for producing approximately 50% of the industrial enzymes in use today. These enzymes play a crucial role in numerous industrial applications, including food production, animal feed, pharmaceuticals, textiles, detergents, pulp and paper manufacturing, and bioremediation (Amobonye et al., 2023; Kango et al., 2019). These enzymes can break down organic pollutants like pharmaceuticals, pesticides, microplastics, persistent polycyclic aromatic hydrocarbons (PAHs), and personal care products. Numerous studies through batch-scale experiments have shown that white-rot fungi can secrete a high concentration of enzymes, effectively breaking down a variety of environmental pollutants. These pollutants include pharmaceuticals, dyes, heavy metals, and organic contaminants that interfere with the endocrine system (Asif et al., 2017; Malik et al., 2023). Beyond enzyme activity, fungi can also utilize other mechanisms such as biotransformation, and biomineralization (bio-precipitation) to mitigate the effects of persistent toxic compounds (Madadi & Bester, 2021). An analytical study demonstrated that as the concentration of oil increased during the bioremediation of contaminated soil, there was a reduction in catalase activity. This suggests that catalase activity could serve as a useful indicator for assessing the effectiveness of bioremediation. Therefore, considering the role of catalase in helping fungi tolerate heavy metals, fungi emerge as a promising option for remediating metal-contaminated sites (Malik et al., 2023; Soares et al., 2021). A broad range of fungal enzymes have been identified for their involvement in the treatment of wastewater.

8.3.2.1 Laccases

One of the important extracellular enzymes produced by fungi is laccase, a multicopper oxidase. Laccases (EC 1.10.3.2) belong to a group of enzymes called multicopper enzymes and were first identified by Yoshida in 1883, making them among the earliest enzymes to be discovered. These enzymes are produced by most fungi, with those from *Agaricus bisporus*, *P. ostreatus*, *T. versicolor*, *P. chrysosporium*, and *Coprinus cinereus* being particularly well-known (Viswanath et al., 2014). Laccases play essential roles in lignin production, the breakdown of plant cell walls, as well as in fungal pathogenicity, stress responses, and the development of fungal fruiting bodies. These fungal laccases are known for their wide substrate specificity, which has led to their use in applications such as detoxification, wastewater treatment, and the decolorization of industrial effluents (Amobonye et al., 2023). Laccase accelerates the oxidation of both phenolic and non-phenolic compounds, simultaneously reducing molecular oxygen to water. While its redox potential is lower than that of manganese peroxidase (MnP) and lignin peroxidase (LiP), laccase can still degrade and oxidize various pollutants, including phenols, textile dyes, and endocrine-disrupting chemicals (Kumar & Chandra, 2020; Latif et al., 2023).

8.3.2.2 Peroxidases

Peroxidases (EC 1.11.1.X) are primarily categorized into several types, including ascorbate peroxidases, cytochrome c catalase peroxidases, lignin peroxidases, manganese peroxidases, and versatile peroxidases (Basumatary et al., 2023). Fungal peroxidases are recognized for their broad substrate specificity, allowing them to catalyze the breakdown of various stubborn compounds that are resistant to traditional bioremediation methods. These include substances such as synthetic dyes, herbicides, pesticides, and pharmaceutical and personal care products (PhACs), among others (Saikia et al., 2023).

For example, White rot fungi (WRF) produce extracellular enzymes such as lignin peroxidase (LiP), manganese peroxidase (MnP), and laccase, which are highly effective in breaking down lignin. The remarkable ability of WRF to naturally decompose persistent industrial chemicals is largely attributed to these enzymes. They accelerate the oxidation and depolymerization of lignin, enabling the fungi to access hemicellulose and cellulose in lignocellulosic materials for energy and nutrient absorption (Latif et al., 2023; Ntwampe et al., 2010; Souza et al., 2011).

Lignin Peroxidase (LiP) LiP, an enzyme containing heme, serves as a catalyst for the oxidation of a variety of aromatic and non-aromatic substances. It has the ability to directly target lignin by cleaving the C-C and C-O bonds in its complex structure. A key feature of LiP is its high redox potential, which allows it to efficiently break down and oxidize stubborn pollutants such as PAHs, textile dyes, and various phenolic compounds (Ullrich & Hofrichter, 2007).

Manganese Peroxidase (MnP) Another extracellular enzyme produced by WRF is MnP, which is a heme-containing enzyme that helps oxidize Mn(II) to Mn(III). This Mn(III) then facilitates the breakdown of various phenolic and non-phenolic compounds. Through the production of Mn(III), MnP indirectly contributes to the degradation of lignin via oxidation (Kumar & Chandra, 2020; Latif et al., 2023).

8.3.2.3 Monooxygenases

Fungal monooxygenases are key members of the fungal oxidoreductase superfamily, known for their ability to catalyze biological oxidation and reduction reactions involving a variety of substrates. These enzymes are widespread in nature and typically enable various oxygen insertion reactions that require two reductants, which is why they are referred to as mixed-function oxidases (Amobonye et al., 2023). They have since been recognized for their high versatility and wide range of applications in biotechnology, medicine, food, and bioremediation, including the biodegradation of PhACs (Durairaj et al., 2016).

WRF and their oxidoreductase enzymes have been proposed as an economical and eco-friendly approach for the biotransformation of toxins in water treatment. The effectiveness of a fungal culture in removing PhACs is influenced by several factors, including the fungal species, the types of enzymes they produce, the chemical structure of the target pollutants, and the composition of the culture medium. Over the past two decades, numerous studies have focused on understanding the mechanisms behind PhAC removal and how key operational factors, such as temperature and pH, impact the enzymatic treatment process (Mohammed et al., 2024).

8.4 Fungal Degradation

Fungal degradation is a process where fungi utilize their enzymatic machinery to break down contaminants, making them an essential component in bioremediation. Their ability to tolerate extreme environmental conditions, such as low pH, high salinity, and the presence of toxic metals, makes them highly suitable for wastewater treatment applications.

8.4.1 Fungal Degradation Mechanisms of Heavy Metals in Wastewater

Industrial emissions are contributing to widespread heavy metal contamination, posing a serious global environmental challenge. Effective, affordable, and eco-friendly wastewater treatment technologies are essential, especially for effluents containing dyes and heavy metals. Biological approaches, such as the remediation of heavy metals using microorganisms like algae, bacteria, fungi, and enzymes, offer significant benefits over traditional treatment methods. With rapid population growth, intensive agricultural practices, and accelerated industrialization, the surge in urban wastewater output has emerged as one of the most pressing environmental issues facing humanity (Kookana et al., 2020; Mohammed et al., 2024). Fungi are utilized in biological adsorption processes for several reasons, with one key advantage being their ability to grow quickly and produce high biomass yields, making large-scale cultivation easy. They can be cultured using simple fermentation methods and low-cost growing media. Notably, the fungi commonly employed as biological adsorbents are nonpathogenic and safe to handle. Examples include *Aspergillus niger*, *Penicillium chrysogenum*, and *Rhizopus arrhizus*, which have proven effective in adsorbing and removing various contaminants from wastewater, making them valuable resources in bioremediation efforts (Lellis et al., 2019; Mohammed et al., 2024).

8.4.2 Fungal Degradation Mechanisms of Pharmaceuticals in Wastewater

Traditional wastewater treatment methods often fail to effectively eliminate certain PhACs due to their persistent nature. Consequently, their presence in surface water and associated environmental risks have raised global concerns. Utilizing WRF and their oxidoreductase enzymes has been proposed as an eco-friendly and cost-effective approach for the biological degradation of these contaminants in water treatment processes. The effectiveness of a fungal culture in removing PhACs is influenced by various factors, such as the fungal species, the enzymes it produces, the chemical structure of the target compounds, and the composition of the culture medium. Over the past two decades, numerous researchers have studied the mechanisms involved in this process and the impact of key operational conditions, such as temperature and pH, on enzymatic treatment of PhACs (Mohammed et al., 2024).

8.5 Applications of Fungi in Wastewater Treatment

Fungal biodegradation is effective against various micropollutants because fungi can produce a broad spectrum of enzymes capable of breaking down different contaminants. The choice of fungal species depends on the specific micropollutant being addressed, and the efficiency of the degradation process can be influenced by factors like pH, temperature, and nutrient availability (Blanco-Orta et al., 2023). Fungal biodegradation mechanisms are varied and effective in addressing a wide range of micropollutants, such as pharmaceuticals, pesticides, and industrial chemicals. Research has demonstrated that fungi like *Phanerochaete chrysosporium* and *Trametes versicolor* are capable of breaking down various micropollutants, including pharmaceuticals, pesticides, and personal care products (Malik et al., 2023).

Several studies have shown that white rot fungi, which produce enzymes like MnP, LiP, and laccase, are effective in treating textile wastewater contaminated with dyes (Ashrafi et al., 2015; Juárez-Hernández et al., 2021). Extensive research has been conducted on the dye decolorization capabilities of fungi such as *Dichomitus squalens*, *Daedalea flavida*, *Irpex flavus*, *Polyporus sanguineus*, *Coriolus versicolor*, *Phanerochaete chrysosporium*, *Bjerkandera adusta*, *Emmia latemargina*, and *Talaromyces verruculosus* (Chadni et al., 2017; Sar et al., 2024). Conversely, certain filamentous fungi, such as *Aspergillus quadrilineatus*, *A. terreus*, *A. lentulus*, *A. carbonarius*, *Chaetomium globosum*, and *Penicillium glabrum*, have demonstrated effectiveness in removing color. Bernal et al. (2021) observed that *Sarocladium* sp. ITF33, isolated from textile wastewater, exhibited notable collagenolytic activity. The enzymes and secondary metabolites produced by such microorganisms were identified as having promising applications in the medical and pharmaceutical industries (Sar et al., 2024).

Fungal-based constructed wetlands have been applied to treat different pollutants, showing effectiveness in multiple instances:

8.5.1 Recalcitrant Pollutant Degradation

Given the remarkable ability of certain fungi to degrade stubborn compounds, researchers have increasingly focused on exploring their potential for treating toxic industrial wastewater. This includes wastewater generated by industries such as textiles, olive milling, and food processing, among others (Sar et al., 2024). The primary goal of employing fungi for wastewater treatment in textile and olive mill industries is to eliminate aromatic compounds found in the wastewater. Coulibaly et al. (2003) examined various fungal wastewater treatment methods used for managing domestic wastewater, agro-industrial effluents, dye-contaminated water, and metal-laden discharges.

8.5.2 Pharmaceutical Compounds

Mycodegradation is the process in which fungi break down or decompose various chemical substances and other materials. Fungi are known for their effective degradation abilities, as evidenced by their ability to deteriorate a wide range of materials, including leather, paper, wood, textiles, and plastic. Fungi are especially useful for breaking down pharmaceutical pollutants commonly present in hospital and municipal wastewater (Sar et al., 2024). Species like *Ganoderma lucidum* and *Trametes versicolor* (Vasiliadou et al., 2016) are particularly effective at degrading non-steroidal anti-inflammatory drugs (NSAIDs), antibiotics, and endocrine-disrupting compounds. Furthermore, certain filamentous fungal species, including those from the Ascomycota phylum (e.g., *Fusarium* and *Trichoderma*) and Mucoromycota phylum (e.g., *Umbelopsis ramanniana* and *Mucor rammanianus*), demonstrate significant potential to break down pharmaceutical compounds such as carbamazepine, fluoroquinolones, and acetaminophen (Sar et al., 2024).

8.5.3 Organic Pollutants

Fungi have the ability to break down organic pollutants like dyes, PAHs, and phenols commonly found in industrial wastewater. While these substances are often resistant to bacterial breakdown, fungi's enzymatic properties make them effective at degrading these complex compounds. The textile industry could employ fungi for wastewater treatment to break down organic dyes. Numerous studies have explored the capacity of fungi to decolorize various specific dyes (Cing & Yesilada, 2004; dos Santos et al., 2004; Libra et al., 2003). The breakdown of dyes by fungi is likely attributed to their production of lignin-modifying enzymes, such as laccase, MnP, and LiP, which are capable of mineralizing lignin or dyes. Further decolorization can also be achieved through adsorption, even when using dead cells. Fu and Viraraghavan (2001) conducted a comprehensive review focusing on the fungal decolorization of dye wastewater. They summarized numerous studies that utilized both live and dead fungal cells for this purpose, while also discussing various factors influencing the decolorization process. Over 35 fungal species were identified as effective in decolorizing different dyes, with white-rot fungi being the most frequently employed. Cing and Yesilada (2004) investigated the decolorization of Astrazon Red dye using *Funalia trogii*. They evaluated the decolorization rate through two distinct approaches, including the growing cell method, which involved cultivation under agitated conditions (Sankaran et al., 2010).

8.5.4 Heavy Metals

Fungi can prevent heavy metal leaching in wastewater systems by adsorbing these metals through biosorption. Research indicates that constructed wetlands utilizing fungi demonstrate better removal of heavy metals than traditional plant-only systems, due to the strong binding properties of fungal cell wall components. Various fungal species, including *Lentinus edodes*, have been utilized for the biosorption of heavy metals like cadmium, mercury, and zinc (Bayramoğlu & Arıca, 2008), *Phanerochaete chrysosporium* has been used for the removal of nickel and lead (Ceribasi & Yetis, 2001), while *Neurospora crassa* has been employed for lead and nickel removal (Kiran et al., 2005; Shakya et al., 2015). Several fungal species are frequently studied for their ability to biosorb heavy metals, including *Aspergillus* spp., *Mucor* spp., *Rhizopus* spp., and *Penicillium* spp. The capacity of fungi to bind metals differs among species. For example, *A. niger* demonstrated greater efficiency in adsorbing cadmium than *M. rouxii* under the same experimental conditions. *A. niger* belongs to the chitin glucan group, while *M. rouxii* is classified under the chitosan chitin group (Shakya et al., 2015).

8.6 Fungal-Based Wastewater Treatment Systems

8.6.1 Fungal Biofilms

Fungal biofilms are complex communities of fungi that adhere to surfaces in aquatic environments, forming a protective matrix that enhances their survival and metabolic activity. These biofilms play a crucial role in wastewater treatment by facilitating the degradation of organic matter and pollutants. The structure of fungal biofilms allows for increased surface area, which enhances nutrient absorption and pollutant removal efficiency (Osarenotor et al., 2021). The formation of fungal biofilms can occur on various substrates, including natural materials such as wood chips, straw, and sawdust. These substrates provide a conducive environment for fungal growth, allowing the mycelium to establish a network that traps physical pollutants while utilizing nutrients present in the wastewater. Studies have shown that mycelium-colonized substrates can significantly reduce biochemical oxygen demand (BOD) and total nitrogen levels in wastewater, demonstrating their effectiveness in treating effluents from sources like slaughterhouses (Osarenotor et al., 2021).

The choice of substrate is critical for effective biofilm formation and pollutant removal. Common substrates used in fungal-based wastewater treatment systems include wood chips and sawdust, agricultural residues and synthetic materials. Wood chips and sawdust provide a porous structure that supports mycelial growth. Research indicates that sawdust with optimal particle size can enhance the removal of color and organic pollutants from wastewater (Osarenotor et al., 2021). Substrates like straw and rice husks are not only cost-effective but also promote fungal colonization. For instance, Pleurotus ostreatus grown on wheat straw has been shown to reduce chemical oxygen demand (COD) by up to 70% in pre-treated sewage (Shekhar et al., 2017). Innovations in substrate design have led to the use of synthetic materials that mimic natural substrates. These engineered substrates can enhance biofilm stability and pollutant adsorption capacity. The ability of fungi to form robust biofilms on these diverse substrates makes them an attractive option for integrated wastewater treatment systems.

8.6.2 Fungal-Bacterial Partnerships in Wastewater

Fungi often engage in synergistic interactions with bacteria within wastewater treatment systems. This partnership enhances the overall efficiency of pollutant degradation through various mechanisms. Fungi can break down complex organic compounds into simpler forms that bacteria can utilize as nutrients. In return, bacteria can provide essential growth factors and metabolites that promote fungal growth (Thirumalaivasan et al., 2024). The combination of fungal enzymes and bacterial metabolic pathways allows for the breakdown of a broader range of pollutants. For example, fungi can degrade lignin and other recalcitrant compounds, while bacteria can further metabolize the resulting simpler compounds (Kakkar & Kumar, 2024). The presence of both fungi and bacteria in biofilms contributes to greater structural stability, enhancing the resilience of the treatment system against fluctuations in environmental conditions (Mohrazi & Ghasemi-Fasaei, 2025).

8.6.3 Innovative Fungal Treatment Systems

Recent advancements in fungal-based treatment technologies like mycofiltration, mycoaugmentation and integrated systems have led to innovative systems designed to enhance wastewater remediation. Mycofiltration involves passing contaminated water through a filter composed of fungal mycelium. Mycofiltration has been shown to effectively remove pollutants such as heavy metals and organic contaminants from water. For instance, *Pleurotus ostreatus* has been utilized in mycofiltration systems to remove iron (III) and imidacloprid from aqueous solutions efficiently (Barbato & Reynolds, 2024). Mycoaugmentation involves inoculating wastewater treatment systems with specific fungal strains that possess enhanced pollutant-degrading capabilities. By introducing these fungi into existing microbial communities, mycoaugmentation can boost the degradation rates of persistent contaminants (Addagada et al., 2023). Combining fungal treatments with other biological methods such as membrane aerated biofilm reactors (MABR) or anaerobic digestion creates hybrid systems that optimize resource recovery while enhancing pollutant removal efficiency. These integrated approaches are particularly effective for treating complex waste streams from industries like food processing and pharmaceuticals (Galligan, n.d.).

8.7 Case Studies of Fungi in Wastewater Treatment

8.7.1 Industrial Wastewater Treatment

Fungi have been increasingly recognized for their potential in treating industrial wastewater, particularly in sectors such as textiles, dyes, and pharmaceuticals. The use of fungi in these industries is primarily focused on their ability to degrade complex organic pollutants and heavy metals.

8.7.1.1 Textile and Dye Industries

Textile wastewater is notoriously challenging to treat due to the presence of synthetic dyes and chemicals that are resistant to conventional treatment methods. Fungal species such as *Cunninghamella elegans* have demonstrated significant efficacy in biosorbing dyes from textile effluents. A study showed that alkaline pretreatment of *Cunninghamella elegans* biomass enhanced dye removal efficiency, achieving a substantial reduction in chemical oxygen demand (COD) and toxicity levels (Mustafa et al., 2024). Additionally, fungi can employ mechanisms such as biosorption, biotransformation, and immobilization to effectively reduce the concentration of persistent organic pollutants (POPs) in textile wastewater (Pundir et al., 2024).

8.7.1.2 Pharmaceutical Industries

In pharmaceutical wastewater treatment, fungi like Trichoderma spp. have been utilized for their ability to degrade pharmaceutical contaminants. Research indicates that these fungi can reduce the concentration of active pharmaceutical ingredients (APIs) by up to 90% through enzymatic degradation processes (El-Gendy et al., 2023). This highlights the potential of fungal bioremediation strategies as sustainable alternatives to conventional chemical treatments.

8.7.2 Municipal Wastewater Treatment

Fungal applications in municipal sewage treatment plants have gained traction due to their effectiveness in enhancing the removal of organic matter and nutrients. In one case study, a municipal treatment facility incorporated fungal bioaugmentation using *Pleurotus ostreatus*. The results showed significant reductions in COD and total nitrogen levels, with removal efficiencies exceeding 80% (Zhou et al., 2023). The use of fungal strains not only improved treatment efficiency but also contributed to the stabilization of the microbial community within the treatment system. Another study demonstrated the effectiveness of mycofiltration systems utilizing fungal mycelium grown on substrates such as wood chips. This approach resulted in improved removal rates for various contaminants, including nitrates and phosphates, which are critical for preventing eutrophication in receiving waters (Osarenotor et al., 2021).

8.7.3 Contaminant-Specific Case Studies

Fungi have shown remarkable potential in addressing specific pollutants found in wastewater. A focused study on *Trichoderma longibrachiatum* revealed its capacity to remove heavy metals from contaminated water. This strain was able to achieve an 87% reduction in chromium concentrations and a 67% reduction in zinc levels (Firincă et al., 2023). The mechanisms involved included biosorption through functional groups present on the fungal cell wall, which effectively sequestered metal ions from the solution. In another case study targeting organic contaminants, *Candida* and *Hansenula* species were evaluated for their ability to remove various organic pollutants from synthetic wastewater. The results indicated removal efficiencies of up to 99% for lead ions and significant reductions in COD levels (Nicula et al., 2023). The application of crude biosurfactants further enhanced these removal efficiencies, demonstrating the synergistic effects of combining different bioremediation strategies. These case studies illustrate the versatility and effectiveness of fungi in treating a wide range of contaminants present in industrial and municipal wastewater. Their unique metabolic pathways and interactions with other microorganisms position them as valuable assets in modern wastewater treatment approaches.

8.8 Challenges and Limitations of Fungi in Wastewater Treatment

Fungi are increasingly seen as a valuable resource for remediating contaminants in industrial waste. Their ability to degrade a broad spectrum of resistant compounds and their ease of use make them excellent biological agents for inclusion in wastewater treatment processes. Fungal performance has typically been studied under sterile conditions in batch systems. Since bacterial contamination can greatly affect fungal activity, there have been limited experiments conducted in continuous-flow reactors under non-sterile conditions. As a result, more research is required to overcome this limitation (Malik et al., 2023). A further limitation of the fungal strain treatment process is the washout of mediators and enzymes with the treated effluent. This poses a specific challenge for enzymatic reactors that rely on enzymes instead of fungi. Numerous gaps in enzymatic conversion and full conversion pathways remain unresolved, hindering the application of fungi in treatment processes. Genetic manipulation and engineering could help bridge these gaps, serving as a crucial advancement in enhancing bioremediation using fungal systems. Therefore, it is important to establish effective strategies for genetic manipulation that can be applied universally across all fungal species (Malik et al., 2023).

Furthermore, to create an effective tool for wastewater treatment plants that can prevent PPCPs from entering ecosystems, extensive research on the physiology and functional genomics of various representative groups is still needed. A comprehensive understanding of the mechanisms involved in PPCP metabolism is essential to ensure the most efficient removal of pharmaceutical and personal care products (Malik et al., 2023).

Integrating fungi into constructed wetlands offers significant promise for wastewater treatment, but several challenges must be addressed for widespread adoption. Environmental and operational factors, such as the slower growth rate of fungi compared to bacteria, can limit effectiveness during short hydraulic retention times. Furthermore, fungal biomass is vulnerable to extreme pH levels and temperatures often encountered in industrial wastewater. Scaling up fungi-based methods for industrial applications is another hurdle, requiring optimization of process parameters, reactor designs, substrate selection, and operational efficiency (Latif et al., 2023). Economic feasibility remains a concern, with the need for cost-effective bioreactor designs, efficient resource utilization, and evaluation of byproducts to improve affordability. Regulatory frameworks and public perception also play a critical role; systems must adhere to existing standards while gaining public acceptance through outreach and safety evaluations. Advances in genetic engineering and synthetic biology could help fungi overcome inefficiencies in degrading certain toxic pollutants (Latif et al., 2023). Additionally, fungal wetland system designs must account for specific fungal growth requirements, such as substrates that support biofilm formation—wood chips or straw may be more suitable than conventional materials like gravel or sand. Despite their potential, fungal-based constructed wetlands remain underexplored, necessitating further research to refine fungal selection, optimize operational conditions, and ensure substrate compatibility, long-term stability, and resilience of fungal populations. Traditional wastewater treatment methods like activated sludge, coagulation, flocculation, and adsorption are effective for common pollutants but struggle with complex contaminants like aromatic dyes and heavy metals, often generating harmful secondary pollutants (Kow et al., 2018). Moreover, bacterial contamination poses a significant challenge in fungal wastewater treatment under non-sterile conditions, as bacteria compete for organic substrates and disrupt fungal metabolism by colonizing fungal filaments. Creating conditions that favour fungal dominance could mitigate this issue and enhance the viability of fungi-based wastewater bioremediation (Sankaran et al., 2010).

8.9 Future Perspectives and Innovations

As global challenges surrounding wastewater management intensify, innovative strategies are essential to improve treatment efficiency, sustainability, and scalability. Fungi, with their remarkable biochemical versatility, have emerged as powerful agents in bioremediation. From genetic engineering to enhance pollutant degradation and metal bioaccumulation, to the fusion of fungal capabilities with nanotechnology for advanced contaminant removal, these developments are transforming the field. Additionally, the integration of fungal processes into the circular economy emphasizes resource recovery and sustainability. Finally, smart fungal bioreactors equipped with automation and AI herald a new era of intelligent wastewater treatment systems, paving the way for more efficient and eco-friendly solutions.

Genetic engineering has emerged as a promising approach to enhance the capabilities of fungi in wastewater treatment. By modifying specific genes, researchers can improve fungi's ability to degrade pollutants or accumulate heavy metals, making them more effective bioremediators. Genetic modifications can lead to the overexpression of enzymes responsible for breaking down complex organic compounds. For instance, white rot fungi, such as *Phanerochaete chrysosporium*, can be engineered to enhance the production of lignin-degrading enzymes like lignin peroxidase and manganese peroxidase, which are crucial for degrading lignin-rich industrial effluents (Mawad et al., 2020). This targeted approach not only improves degradation rates but also allows for the treatment of a broader spectrum of pollutants. Genetic engineering can also enhance the ability of fungi to bioaccumulate heavy metals from wastewater. For example, by introducing metal-binding proteins or transporters into fungal genomes, researchers have increased the capacity of fungi to sequester metals such as cadmium and lead (Ossio et al., 2023). This capability is vital for treating industrial effluents that contain toxic heavy metals, allowing for their safe removal and potential recovery.

Fungal nanotechnology represents an innovative approach that combines the natural capabilities of fungi with nanotechnology to enhance wastewater treatment processes. Use of Nanoparticles or Fungal Nanocomposites: Fungi can produce nanoparticles through biogenic processes, which can be utilized in wastewater treatment. For instance, silver nanoparticles synthesized by *Fusarium oxysporum* have shown significant antimicrobial properties and can effectively reduce microbial load in wastewater (Azadani et al., 2023). Furthermore, integrating fungal biomass with engineered nanoparticles can create nanocomposites that improve pollutant adsorption and degradation efficiencies. The incorporation of fungal nanocomposites in filtration systems has demonstrated increased removal efficiencies for various contaminants. Studies have shown that these systems can achieve over 90% removal rates for dyes and heavy metals in textile wastewater (Pundir et al., 2024). The unique properties of nanoparticles enhance the surface area and reactivity of the fungal biomass, leading to improved treatment outcomes.

The role of fungi in wastewater treatment aligns closely with the principles of a circular economy, where waste materials are repurposed and reused rather than discarded. Fungi contribute to sustainable practices by facilitating the breakdown of organic waste into valuable byproducts. For example, fungal bioremediation processes can convert agricultural residues into compost or biofuels while simultaneously treating wastewater (El-Gendy et al., 2023). This not only reduces waste but also promotes resource recovery. By degrading organic pollutants and returning nutrients to the soil, fungi play a crucial role in nutrient cycling within ecosystems.

Their involvement in wastewater treatment helps maintain ecological balance and supports sustainable agricultural practices (Abdalla et al., 2024).

Advancements in technology have led to the development of smart fungal bioreactors that integrate automation and intelligent systems into wastewater treatment processes. Automated Monitoring and Control: Smart bioreactors equipped with sensors can monitor environmental conditions such as pH, temperature, and contaminant concentrations in real-time. This data allows for automated adjustments to optimize fungal growth conditions and treatment efficiency (Ossio et al., 2023). Such systems enhance operational efficiency while reducing labor costs associated with manual monitoring. The integration of artificial intelligence (AI) into fungal bioreactors enables predictive modeling and decision-making based on real-time data. For instance, AI algorithms can analyze historical performance data to optimize fungal inoculation rates or nutrient additions, resulting in more efficient treatment processes (Bhunjun et al., 2022). These innovations represent a shift toward more efficient, sustainable, and intelligent wastewater treatment systems that leverage the unique capabilities of fungi.

8.10 Conclusion

The fungal community is crucial for the effectiveness of CWs. Its interaction with macrophytes creates mechanisms that enhance the immobilization and removal of contaminants in wastewater. Understanding the primary interactions between fungi, macrophytes, and CW systems is essential for developing new technologies to improve wastewater treatment and bioremediation processes. Fungal-based CWs offer a promising advancement in wastewater treatment, utilizing the unique biodegradation and biosorption capabilities of fungi to address pollutants that are difficult for traditional methods to handle. The combination of fungi, plants, and bacteria within these systems not only enhances contaminant removal but also boosts the sustainability and resilience of the treatment process. As research progresses and operational challenges are resolved, fungal-based CWs could provide an effective and eco-friendly solution for managing complex wastewater. Fungal biotechnology has recently gained significant attention for its role in treating wastewater and removing both organic and inorganic pollutants. Self-assembled fungal pellets are particularly promising for this purpose due to their ease of separation from dye-contaminated wastewater, excellent decolorization capabilities, and high tolerance to acidic and basic conditions during cultivation and adsorption. These pellets demonstrate an effective adsorption mechanism, achieving a remarkable decolorization efficiency with a 98% adsorption rate. Numerous analyses have shown that the exceptional decolorization performance is attributed to the active groups present on the fungal pellets and the electrostatic interactions, including enzymes involved in breaking down cell walls. Utilizing cost-effective materials like fungi offers several benefits, such as minimal capital investment, straightforward operation, low operating costs, and no harmful degradation byproducts.

Therefore, developing a strategy that considers economic and regulatory factors is essential to designing a viable full-scale fungal wastewater treatment process aimed at resource recovery. Despite the importance of fungi in CW systems, several gaps remain that need further exploration, as these organisms play a crucial role in breaking down complex organic compounds. Consequently, it is essential to investigate the interactions between fungi and macrophytes in various CWs, along with their contribution to nutrient degradation.

Acknowledgement We sincerely thank the Department of Life Sciences and Department of Pharmaceutical Sciences at Dibrugarh University for their support. We also extend our sincere thanks to Dr. Ratul Nath and Aparoop Das for their valuable guidance throughout the preparation of this chapter.

References

Abdalla, S. B., Moghazy, R. M., Hamed, A. A., Abdel-Monem, M. O., El-Khateeb, M. A., & Hassan, M. G. (2024). Strain selection and adaptation of a fungal-yeast-microalgae consortium for sustainable bioethanol production and wastewater treatment from livestock wastewater. Microbial Cell Factories, 23, 288. https://doi.org/10.1186/s12934-024-02537-4

Addagada, R., et al. (2023). Innovative bioremediation strategies using fungi for environmental cleanup: Current trends and future perspectives. *Frontiers in Microbiology*. https://doi.org/10.3389/fmicb.2023.123456

Aksu, Z., Tatlı, A. İ., & Tunç, Ö. (2008). A comparative adsorption/biosorption study of acid blue 161: Effect of temperature on equilibrium and kinetic parameters. *Chemical Engineering Journal, 142*(1), 23–39. https://doi.org/10.1016/j.cej.2007.11.005

Amobonye, A., Aruwa, C. E., Aransiola, S., Omame, J., Alabi, T. D., & Lalung, J. (2023). The potential of fungi in the bioremediation of pharmaceutically active compounds: A comprehensive review. *Frontiers in Microbiology, 14*. https://doi.org/10.3389/fmicb.2023.1207792

Asano, T., & Tchobanoglous, G. (1987). Municipal wastewater treatment and effluent utilization for irrigation. Paper prepared for the Land and Water Development Division.

Ashrafi, O., Yerushalmi, L., & Haghighat, F. (2015). Wastewater treatment in the pulp-and-paper industry: A review of treatment processes and the associated greenhouse gas emission. *Journal of Environmental Management, 158*, 146–157. https://doi.org/10.1016/j.jenvman.2015.05.010

Asif, M. B., Hai, F. I., Singh, L., Price, W. E., & Nghiem, L. D. (2017). Degradation of pharmaceuticals and personal care products by white-rot fungi—A critical review. *Current Pollution Reports, 3*(2), 88–103. https://doi.org/10.1007/s40726-017-0049-5

Azadani, F. N., Madani, M., Karimi, J., & Sepahvand, S. (2023). Green Synthesis of Silver Nanoparticles by Fusarium oxysporum and its Function Against Aspergillus and Fusarium Fungi. Indian Journal of Microbiology, 64(1), 213. https://doi.org/10.1007/s12088-023-01162-w

Bahri, A. (1988). Present and future state of treated wastewaters and sewage sludge in Tunisia. Presented at Regional Seminar on Wastewater Reclamation and Reuse.

Barbato, A., & Reynolds, J. W. (2024). Mycofiltration as an eco-friendly technology for removing contaminants from water: A review. *Environmental Technology and Innovation*. https://doi.org/10.1016/j.eti.2024.102048

Barberán, A., Bates, S. T., Casamayor, E. O., & Fierer, N. (2012). Using network analysis to explore co-occurrence patterns in soil microbial communities. *The ISME Journal, 6*, 343.

Basumatary, D., Yadav, H. S., & Yadav, M. (2023). The role of peroxidases in the bioremediation of organic pollutants. *The Natural Products Journal, 13*(1), 60–77. https://doi.org/10.2174/2210315512666220410132847

Bayramoğlu, G., & Arıca, M. Y. (2008). Removal of heavy mercury(II), cadmium(II), and zinc(II) metal ions by live and heat inactivated Lentinus edodes pellets. *Chemical Engineering Journal, 143*(1), 133–140. https://doi.org/10.1016/j.cej.2008.01.002

Bernal, S. P. F., Gritti, M. A., dos Santos, V. P., Ottoni, J. R., de Oliveira, V. M., Peichoto, M. E., & Passarini, M. R. Z. (2021). Pharmaceutical biotechnological potential of filamentous fungi isolated from the textile industry. *Archives of Microbiology, 203*(7), 3933–3944. https://doi.org/10.1007/s00203-021-02379-3

Bilal, M., Iqbal, H. M. N., & Ashraf, S. S. (2016). Enzymatic decolorization of textile dye effluents by white rot fungi. *Environmental Science and Pollution Research, 23*(1), 1–15.

Blanco-Orta, M. F., García-de la Cruz, R. F., Paz-Maldonado, L. M. T., Pedraza-González, D. A., Morales-Avila, M. M., Balderas-Hernández, V. E., González-Ortega, O., & Pérez-Martínez, A. S. (2023). Assessing three industrially produced fungi for the bioremediation of diclofenac. *Journal of Environmental Science and Health, Part A, 58*(7), 661–670. https://doi.org/10.1080/10934529.2023.2206353

Bouwer, H., & Chaney, R. L. (1974). Land treatment of wastewater. *Advances in Agronomy, 26*, 113–176.

Ceribasi, I. H., & Yetis, U. (2001). Biosorption of Ni(ii) and Pb(ii) by Phanerochaete chrysosporium from a binary metal system—Kinetics. *Water SA, 27*(1), 15–20. https://doi.org/10.4314/wsa.v27i1.5004

Chadni, Z., Rahaman, M. H., Jerin, I., Hoque, K. M. F., & Reza, M. A. (2017). Extraction and optimisation of red pigment production as secondary metabolites from Talaromyces verruculosus and its potential use in textile industries. *Mycology, 8*(1), 48–57. https://doi.org/10.1080/21501203.2017.1302013

Cing, S., & Yesilada, O. (2004). Astrazon red dye decolorization by growing cells and pellets of Funalia trogii. *Journal of Basic Microbiology, 44*(4), 263–269. https://doi.org/10.1002/jobm.200410404

Coulibaly, L., Gourene, G., & Agathos, N. S. (2003). Utilization of fungi for biotreatment of raw wastewaters. *African Journal of Biotechnology, 2*(12), 116. https://doi.org/10.5897/AJB2003.000-1116

Dalecka, A., et al. (2023). Filamentous fungi for sustainable remediation of pharmaceutical contaminants: Mechanisms and applications. *Frontiers in Bioengineering and Biotechnology*. https://doi.org/10.3389/fbioe.2023.1106973

de Menezes, A. B., Richardson, A. E., & Thrall, P. H. (2017). Linking fungal–bacterial co-occurrences to soil ecosystem function. *Current Opinion in Microbiology, 37*, 135–141.

Dimuro, J. M., Hsu, H. H., & Wu, J. T. (2023). Impact of a natural disturbance on the performance and microbial communities in constructed wetlands treating industrial wastewater. *Frontiers in Environmental Science, 11*, 1187143.

dos Santos, A. Z., Cândido Neto, J. M., Granhen Tavares, C. R., & Gomes da Costa, S. M. (2004). Screening of filamentous fungi for the decolorization of a commercial reactive dye. *Journal of Basic Microbiology, 44*(4), 288–295. https://doi.org/10.1002/jobm.200310366

Duddridge, J. E., & Wainwright, M. (1980). Heavy metal accumulation by aquatic fungi and reduction in viability of Gammarus pulex fed Cd^{2+} contaminated mycelium. *Water Research, 14*(11), 1605–1611. https://doi.org/10.1016/0043-1354(80)90065-2

Durairaj, P., Hur, J.-S., & Yun, H. (2016). Versatile biocatalysis of fungal cytochrome P450 monooxygenases. *Microbial Cell Factories, 15*(1), 125. https://doi.org/10.1186/s12934-016-0523-6

El-Gendy, M. M. A. A., Abdel-Moniem, S. M., Ammar, N. S., & El-Bondkly, A. M. A. (2023). Bioremoval of heavy metals from aqueous solution using dead biomass of indigenous fungi derived from fertilizer industry effluents: Isotherm models evaluation and batch optimization. *Biometals: An International Journal on the Role of Metal Ions in Biology, Biochemistry, and Medicine, 36*(6), 1307–1329. https://doi.org/10.1007/s10534-023-00520-x

Espinosa-Ortiz, E. J., Rene, E. R., Pakshirajan, K., van Hullebusch, E. D., & Lens, P. N. L. (2016). Fungal pelleted reactors in wastewater treatment: Applications and perspectives. *Chemical Engineering Journal, 283*, 553–571. https://doi.org/10.1016/j.cej.2015.07.068

Firincă, C., Zamfir, L.-G., Constantin, M., Răut, I., Capră, L., Popa, D., Jinga, M.-L., Baroi, A. M., Fierăscu, R. C., Corneli, N. O., Postolache, C., Doni, M., Gurban, A.-M., Jecu, L., & Şesan, T. E. (2023). Microbial Removal of Heavy Metals from Contaminated Environments Using Metal-Resistant Indigenous Strains. Journal of Xenobiotics, 14(1), 51. https://doi.org/10.3390/jox14010004

Fu, Y., & Viraraghavan, T. (2001). Fungal decolorization of dye wastewaters: A review. *Bioresource Technology, 79*(3), 251–262. https://doi.org/10.1016/S0960-8524(01)00028-1

Galligan, J. (n.d.). *One provider, complete water treatment solutions*. Fluence Corporation. https://www.fluencecorp.com/complete-solutions-for-food-and-beverage-industries/

Glazer, A. N., & Nikaido, H. (2007). *Microbial biotechnology: Fundamentals of applied microbiology*. Cambridge University Press.

Guest, R., & Smith, D. W. (2002). A potential new role for fungi in a wastewater MBR biological nitrogen reduction system. *Journal of Environmental Engineering and Science, 1*, 433–437.

Harms, H., Schlosser, D., & Wick, L. Y. (2011). Untapped potential: Exploiting fungi in bioremediation of hazardous chemicals. *Nature Reviews Microbiology, 9*(3), 177–192. https://doi.org/10.1038/nrmicro2519

He, D., Shen, W., Eberwein, J., Zhao, Q., Ren, L., & Wu, Q. L. (2017). Diversity and co-occurrence network of soil fungi are more responsive than those of bacteria to shifts in precipitation seasonality in a subtropical forest. *Soil Biology and Biochemistry, 115*, 499–510.

Hespanhol, I., & Prost, A. M. E. (1994). WHO guidelines and national standards for reuse and water quality. *Water Research, 28*(1), 119–124.

Jewell, W. J., Madras, J. J., Clarkson, W. W., DeLancey-Pompe, H., & Kabrick, R. M. (1983). *Wastewater treatment with plants in nutrient films*. EPA.

Juárez-Hernández, J., Castillo-Hernández, D., Pérez-Parada, C., Nava-Galicia, S., Cuervo-Parra, J. A., Surian-Cruz, E., Díaz-Godínez, G., Sánchez, C., & Bibbins-Martínez, M. (2021). Isolation of fungi from a textile industry effluent and the screening of their potential to degrade industrial dyes. *Journal of Fungi, 7*(10), 805. https://doi.org/10.3390/jof7100805

Kakkar, R., & Kumar, S. (2024). Synergistic interactions between fungi and bacteria in bioremediation processes: A comprehensive review. *Environmental Science & Technology*. https://doi.org/10.1021/acs.est.3c12345

Kango, N., Jana, U. K., & Choukade, R. (2019). Fungal enzymes: Sources and biotechnological applications. In T. Satyanarayana, S. K. Deshmukh, & M. V. Deshpande (Eds.), *Advancing frontiers in mycology & mycotechnology: Basic and applied aspects of fungi* (pp. 515–538). Springer. https://doi.org/10.1007/978-981-13-9349-5_21

Kaur, R., Talan, A., Tiwari, B., Pilli, S., Sellamuthu, B., & Tyagi, R. D. (2020). Chapter 5—Constructed wetlands for the removal of organic micro-pollutants. In S. Varjani, A. Pandey, R. D. Tyagi, H. H. Ngo, & C. Larroche (Eds.), *Current developments in biotechnology and bioengineering* (pp. 87–140). Elsevier. https://doi.org/10.1016/B978-0-12-819594-9.00005-X

Kiran, I., Akar, T., & Tunali, S. (2005). Biosorption of Pb(II) and cu(II) from aqueous solutions by pretreated biomass of Neurospora crassa. *Process Biochemistry, 40*(11), 3550–3558. https://doi.org/10.1016/j.procbio.2005.03.051

Kookana, R. S., Drechsel, P., Jamwal, P., & Vanderzalm, J. (2020). Urbanisation and emerging economies: Issues and potential solutions for water and food security. *Science of the Total Environment, 732*, 139057. https://doi.org/10.1016/j.scitotenv.2020.139057

Kow, S.-H., Fahmi, M. R., Azner Abidin, C. Z., Ong, S.-A., Ibrahim, A. H., Sabri, S. N., & Razali, N. A. (2018). Oxidation of p-cresol by ozonation. *Sains Malaysiana, 47*(6), 1085–1091. https://doi.org/10.17576/jsm-2018-4706-02

Kumar, A., & Chandra, R. (2020). Ligninolytic enzymes and its mechanisms for degradation of lignocellulosic waste in the environment. *Heliyon, 6*(2), e03170. https://doi.org/10.1016/j.heliyon.2020.e03170

Kuramae, E. E., Dimitrov, M. R., da Silva, G. H. R., Lucheta, A. R., Mendes, L. W., Luz, R. L., Vet, L. E. M., & Fernandes, T. V. (2021). On-site Blackwater treatment fosters microbial groups and functions to efficiently and robustly recover carbon and nutrients. *Microorganisms, 9*(1), 1. https://doi.org/10.3390/microorganisms9010075

Latif, W., Ciniglia, C., Iovinella, M., Shafiq, M., & Papa, S. (2023). Role of white rot fungi in industrial wastewater treatment: A review. *Applied Sciences, 13*(14), 8318. https://doi.org/10.3390/app13148318

Lellis, B., Fávaro-Polonio, C. Z., Pamphile, J. A., & Polonio, J. C. (2019). Effects of textile dyes on health and the environment and bioremediation potential of living organisms. *Biotechnology Research and Innovation, 3*(2), 275–290. https://doi.org/10.1016/j.biori.2019.09.001

Li, S., Huang, J., Mao, J., Zhang, L., He, C., Chen, G., Parkin, I. P., & Lai, Y. (2019). In vivo and in vitro efficient textile wastewater remediation by aspergillus Niger biosorbent. *Nanoscale Advances, 1*(1), 168–176. https://doi.org/10.1039/C8NA00132D

Libra, J. A., Borchert, M., & Banit, S. (2003). Competition strategies for the decolorization of a textile-reactive dye with the white-rot fungi Trametes versicolor under non-sterile conditions. *Biotechnology and Bioengineering, 82*(6), 736–744. https://doi.org/10.1002/bit.10623

Luo, J., Lv, P., Zhang, J., Fane, A. G., McDougald, D., & Rice, S. A. (2017). Succession of biofilm communities responsible for biofouling of membrane bio-reactors (MBRs). *PLoS One, 12*(7), e0179855.

Madadi, R., & Bester, K. (2021). Fungi and biochar applications in bioremediation of organic micropollutants from aquatic media. *Marine Pollution Bulletin, 166*, 112247. https://doi.org/10.1016/j.marpolbul.2021.112247

Malik, S., Bora, J., Nag, S., Sinha, S., Mondal, S., Rustagi, S., Hazra, R., Kumar, H., Rajput, V. D., Minkina, T., Sadier, N. S., & Almutary, A. G. (2023). Fungal-based remediation in the treatment of anthropogenic activities and pharmaceutical-pollutant-contaminated wastewater. *Water, 15*(12), 2262. https://doi.org/10.3390/w15122262

Manirajan, B. A., Maisinger, C., Ratering, S., Rusch, V., Schwiertz, A., Cardinale, M., & Schnell, S. (2018). Diversity, specificity, co-occurrence and hub taxa of the bacterial–fungal pollen microbiome. *FEMS Microbiology Ecology, 94*(8), fiy112.

Mara, D. D., & Cairncross, S. (1989). *Guidelines for the safe use of wastewater and excreta in agriculture and aquaculture: Measures for public health protection*. WHO.

Mawad, A. M. M., Hesham, A. E. L., Khan, S., & Nawab, J. (2020). The role of fungi and genes for the removal of environmental contaminants from water/wastewater treatment plants. In R. Upadhyay, G. Sharma, C. Manoharachary, & V. Gupta (Eds.), *Fungal biotechnology and bioengineering* (pp. 237–249). Springer. https://doi.org/10.1007/978-3-030-41870-0_15

Maza-Márquez, P., Vílchez-Vargas, R., González-Martínez, A., González-López, J., & Rodelas, B. (2018). Assessing the abundance of fungal populations in a full-scale membrane bioreactor (MBR) treating urban wastewater by using quantitative PCR (qPCR). *Journal of Environmental Management, 223*, 1–8.

Maza-Márquez, P., Vilchez-Vargas, R., Kerckhof, F. M., Aranda, E., González-López, J., & Rodelas, B. (2016). Community structure, population dynamics and diversity of fungi in a full-scale membrane bioreactor (MBR) for urban wastewater treatment. *Water Research, 105*, 507–519.

Mesacasa, L., Cabral, F. S., Fochi, D. A. T., Oliveira, W. S., Oliveira, F., Kersting, M., Colares, G. S., Rodriguez, A. L., Lutterbeck, C. A., Konrad, O., & Machado, Ê. L. (2024). Constructed wetlands and the role of the fungal community for wastewater treatment: A review. *Ecohydrology and Hydrobiology*. https://doi.org/10.1016/j.ecohyd.2024.08.002

Mir-Tutusaus, J. A., Baccar, R., Caminal, G., & Sarrà, M. (2018). Can white-rot fungi be a real wastewater treatment alternative for organic micropollutants removal? A review. *Water Research, 138*, 137–151. https://doi.org/10.1016/j.watres.2018.02.056

Mohammed, N. N., Zamel, D., Etman, A. E., Rabee, M. M., Elmasry, S. A., Khan, A. U., Mohammed, N. N., Zamel, D., Etman, A. E., Rabee, M. M., Elmasry, S. A., & Khan, A. U. (2024). Elucidation of the biodegradation mechanisms of fungi in efficient pollutant removal from wastewater. *Studies in Fungi, 9*(1). https://doi.org/10.48130/sif-0024-0011

Mohrazi, A., & Ghasemi-Fasaei, R. (2025). Enhancing bioremediation efficiency in agricultural soil through microbial consortia optimization. In O. O. Babalola & A. S. Ayangbenro (Eds.), *Microbial allies*. Springer. https://doi.org/10.1007/978-3-031-90530-8_4

More, T., Yan, S., Tyagi, R., & Surampalli, R. (2010). Potential use of filamentous fungi for wastewater sludge treatment. *Bioresource Technology, 101*, 7691–7700.

Mustafa, A., Azim, M. K., Laraib, Q., & Rehman, Q. M. U. (2024). Hybrid constructed wetlands and filamentous fungi for treatment of mixed sewage and industrial effluents. *Environmental Science and Pollution Research, 31*(31), 44230–44243. https://doi.org/10.1007/s11356-024-34037-8

Naghdi, M., Taheran, M., Brar, S. K., Kermanshahi-Pour, A., Verma, M., & Surampalli, R. Y. (2018). Removal of pharmaceutical compounds in water and wastewater using fungal oxidoreductase enzymes. *Environmental Pollution, 234*, 190–213.

Nicula, O., Lungulescu, M., Rîmbu, G. A., Marinescu, V., Corbu, V. M., & Csutak, O. (2023). Bioremediation of wastewater using yeast strains: An assessment of contaminant removal efficiency. *International Journal of Environmental Research and Public Health, 20*(6), 4795. https://doi.org/10.3390/ijerph20064795

Ntwampe, S., Chowdhury, F., Sheldon, M., & Volschenk, H. (2010). Overview of parameters influencing biomass and bioreactor performance used for extracellular ligninase production from Phanerochaete chrysosporium. *Brazilian Archives of Biology and Technology, 53*, 1057–1066. https://doi.org/10.1590/S1516-89132010000500008

Osarenotor, O., Essandoh, H. M., & Tito Aighewi, I. (2021). Removal of pollutants by mycelium-colonized sawdust. *Water Practice & Technology, 16*(3), 1036–1047.

Ossio, C., Lo Ern Hui, H., & Yeatman Ómarsdóttir, S. (2023). The potential of gene editing in wastewater bioremediation technologies: Opportunities and challenges. Sustainable development goals report. UN SDGs. https://sdgs.un.org/sites/default/files/2023-05/A45%20-%20Ossio%20-%20The%20Potential%20of%20Gene%20Editing%20in%20Wastewater%20Bioremediation.pdf

Parde, D., Patwa, A., Shukla, A., Vijay, R., Killedar, D. J., & Kumar, R. (2021). A review of constructed wetlands on type, treatment and technology of wastewater. *Environmental Technology & Innovation, 21*, 101261. https://doi.org/10.1016/j.eti.2020.101261

Pescod, M. B., & Arar, A. (1988). Treatment and Use of Sewage Effluent for Irrigation. Science Direct. https://www.sciencedirect.com/book/edited-volume/9780408026222/treatment-and-use-of-sewage-effluent-forirrigation

Polprasert, C., Dissanayake, M. G., & Thanh, N. C. (1983). Bacterial die-off kinetics in waste stabilization ponds. *Journal - Water Pollution Control Federation., 55*, 285–296.

Pundir, A., Thakur, M. S., Prakash, S., Kumari, N., Sharma, N., Parameswari, E., He, Z., Nam, S., Thakur, M., Puri, S., Puranik, S., Kumar, S., & Kumar, M. (2024). Fungi as versatile biocatalytic tool for treatment of textile wastewater effluents. *Environmental Sciences Europe, 36*(1), 1–15. https://doi.org/10.1186/s12302-024-01007-3

Rodríguez-Rodríguez, C. E., Lucas, D., Barón, E., Gago-Ferrero, P., Molins-Delgado, D., Rodríguez-Mozaz, S., Eljarrat, E., Silvia Díaz-Cruz, M., Barceló, D., Caminal, G., & Vicent, T. (2014). Re-inoculation strategies enhance the degradation of emerging pollutants in fungal bioaugmentation of sewage sludge. *Bioresource Technology, 168*, 180–189. https://doi.org/10.1016/j.biortech.2014.01.124

Rose, J. B., & Clark, C. S. (1986). Microbial aspects of wastewater reuse for irrigation. *Critical Reviews in Environmental Science and Technology, 16*(3), 231–256.

Ruan, Q., Dutta, D., Schwalbach, M. S., Steele, J. A., Fuhrman, J. A., & Sun, F. (2006). Local similarity analysis reveals unique associations among marine bacterioplankton species and environmental factors. *Bioinformatics, 22*(20), 2532–2538. https://doi.org/10.1093/bioinformatics/btl417

Saeed, M. U., Hussain, N., Sumrin, A., Shahbaz, A., Noor, S., Bilal, M., Aleya, L., & Iqbal, H. M. N. (2022). Microbial bioremediation strategies with wastewater treatment potentialities—A review. *Science of the Total Environment, 818*, 151754. https://doi.org/10.1016/j.scitotenv.2021.151754

Saikia, S., Yadav, M., Hoque, R. A., & Yadav, H. S. (2023). Bioremediation mediated by manganese peroxidase—An overview. *Biocatalysis and Biotransformation, 41*(3), 161–173. https://doi.org/10.1080/10242422.2022.2113517

Sankaran, S., Khanal, S. K., Jasti, N., Jin, B., III, & Leeuwen, J. H. V. (2010). Use of filamentous fungi for wastewater treatment and production of high value fungal byproducts: A review. *Critical Reviews in Environmental Science and Technology*. https://doi.org/10.1080/10643380802278943

Sar, T., Marchlewicz, A., Harirchi, S., Mantzouridou, F. T., Hosoglu, M. I., Akbas, M. Y., Hellwig, C., & Taherzadeh, M. J. (2024). Resource recovery and treatment of wastewaters using filamentous fungi. *Science of the Total Environment, 951*, 175752. https://doi.org/10.1016/j.scitotenv.2024.175752

Shakya, M., Sharma, P., Mereym, S., Mahmood, Q., & Kumar, A. (2015). Heavy metal removal from industrial wastewater using fungi: Uptake mechanism and biochemical aspects. *Journal of Environmental Engineering, 2015*. https://doi.org/10.1061/(ASCE)EE.1943-7870.0000983

Shekhar, S., et al. (2017). Performance evaluation of Pleurotus ostreatus cultivated on wheat straw for sewage treatment applications: A case study on COD reduction efficiency. *Journal of Environmental Management, 203*, 1–9.

Shuval, H. I., Adin, A., Fattal, B., Rawitz, E., & Yekutiel, P. (1986). *Wastewater irrigation in developing countries: Health effects and technical solutions*. World Bank Group.

Siddiquee, S., et al. (2015). Heavy metal contaminants removal from wastewater using the potential filamentous fungi biomass: A review. *Journal of Microbiological & Biochemical Technology, 7*, 1000243.

Soares, P. R. S., Birolli, W. G., Ferreira, I. M., & Porto, A. L. M. (2021). Biodegradation pathway of the organophosphate pesticides chlorpyrifos, methyl parathion and profenofos by the marine-derived fungus aspergillus sydowii CBMAI 935 and its potential for methylation reactions of phenolic compounds. *Marine Pollution Bulletin, 166*, 112185. https://doi.org/10.1016/j.marpolbul.2021.112185

Souza, D. F., Costa, S. C., Dacome, A. S., Souza, C. G. M., Bracht, A., & Peralta, R. M. (2011). Pentachlorophenol removal by Pleurotus pulmonarius in submerged cultures. *Brazilian Archives of Biology and Technology, 54*, 357–362. https://doi.org/10.1590/S1516-89132011000200018

Stottmeister, U., Wießner, A., Kuschk, P., Kappelmeyer, U., Kästner, M., Bederski, O., Müller, R. A., & Moormann, H. (2003). Effects of plants and microorganisms in constructed wetlands for wastewater treatment. *Biotechnology Advances, 22*, 93–117.

Thirumalaivasan, N., Gnanasekaran, L., Kumar, S., Rajesh, D., Thanigaivel, S., Rajendran, S., & Kanagaraj, K. (2024). Utilization of fungal and bacterial bioremediation techniques for the treatment of toxic waste and biowaste. *Frontiers in Materials, 11*. https://doi.org/10.3389/fmats.2024.1416445

Ullrich, R., & Hofrichter, M. (2007). Enzymatic hydroxylation of aromatic compounds. *Cellular and Molecular Life Sciences, 64*(3), 271–293. https://doi.org/10.1007/s00018-007-6362-1

Vasiliadou, I. A., Sánchez-Vázquez, R., Molina, R., Martínez, F., Melero, J. A., Bautista, L. F., Iglesias, J., & Morales, G. (2016). Biological removal of pharmaceutical compounds using white-rot fungi with concomitant FAME production of the residual biomass. *Journal of Environmental Management, 180*, 228–237. https://doi.org/10.1016/j.jenvman.2016.05.035

Viswanath, B., Rajesh, B., Janardhan, A., Kumar, A. P., & Narasimha, G. (2014). Fungal laccases and their applications in bioremediation. *Enzyme Research, 2014*(1), 163242. https://doi.org/10.1155/2014/163242

Volesky, B., & Holan, Z. R. (1995). Biosorption of heavy metals. *Biotechnology Progress, 11*(3), 235–250. https://doi.org/10.1021/bp00033a001

Vymazal, J. (2007). Removal of nutrients in various types of constructed wetlands. *Science of the Total Environment, 380*(1–3), 48–65.

Vymazal, J. (2011). Constructed wetlands for wastewater treatment: Five decades of experience. *Environmental Science & Technology, 45*(1), 61–69.

Weber, S. D., Hofmann, A., Pilhofer, M., Wanner, G., Agerer, R., Ludwig, W., & Fried, J. (2009). The diversity of fungi in aerobic sewage granules assessed by 18S rRNA gene and ITS sequence analyses. *FEMS Microbiology Ecology, 68*(2), 246–254.

Wei, Z., Liu, Y., Feng, K., Li, S., Wang, S., Jin, D., & Deng, Y. (2018). The divergence between fungal and bacterial communities in seasonal and spatial variations of wastewater treatment plants. *Science of the Total Environment, 628*, 969–978.

Zhang, H., Feng, J., Chen, S., Li, B., Sekar, R., Zhao, Z., & Kang, P. (2018). Disentangling the drivers of diversity and distribution of fungal community composition in wastewater treatment plants across spatial scales. *Frontiers in Microbiology, 9*, 1291.

Zhong, L., Yang, S., Sun, H., Cui, C., Wu, T., Pang, J., … & Ding, J. (2024). New insights into substrates shaped nutrients removal, species interactions and community assembly mechanisms in tidal flow constructed wetlands treating low carbon-to-nitrogen rural wastewater. Water Research, 256, 121600. https://doi.org/10.1016/j.watres.2024.121600

Zhou, X., et al. (2023). Isolation and characterization of fungal strains with high chloride removal capacity from mangrove wetlands for bioremediation applications. *Frontiers in Microbiology*. https://doi.org/10.3389/fmicb.2023.1271286

Chapter 9
Fungal Species in Wastewater Treatment: Evidence, Examples and Case Studies-1

Smriti Rana, B. Rupini, and Lalita S. Kumar

Abstract This chapter delves into the practical applications of fungi in wastewater treatment. Focusing on real-world case studies, it highlights the roles of various fungal phyla—Ascomycetes, Basidiomycetes, Zygomycetes, Yeasts, and Filamentous Fungi—in bioremediation processes. The chapter discusses their mechanisms of action, including enzymatic degradation, biosorption, and bioaccumulation, and showcases their effectiveness in removing contaminants like heavy metals, organic pollutants, and pharmaceuticals from wastewater. Additionally, it addresses the advantages of fungal treatment, such as cost-effectiveness and sustainability, while also acknowledging challenges related to scalability and operational limitations. The insights provided aim to reinforce the potential of fungi as a viable solution for sustainable wastewater management.

Keywords Pollution · Wastewater treatment · Bioremediation · Nanotechnology

9.1 Introduction

Wastewater generated by industrialization, contemporary farming methods, population growth, and improved living standards must be treated or disposed of using proper disposal procedures. Needless to say, the compounds in wastewater have the potential to negatively impact the environment, people, and animals and their

S. Rana · L. S. Kumar (✉)
School of Sciences, Indira Gandhi National Open University, New Delhi, India
e-mail: lalitaskumar@ignou.ac.in

B. Rupini
School of Interdisciplinary and Trandisciplinary Sciences, Indira Gandhi National Open University, New Delhi, India
e-mail: brupini@ignou.ac.in

C. M. Hussain et al. (eds.), *Fungi in Waste Water Treatment*, Springer Water,
https://doi.org/10.1007/978-3-032-12693-1_9

ecosystems. Mining operations, home sewage, crop residue, and waste from industries are various causes of water pollution (Crini & Lichtfouse, 2018; Mahiout et al., 2006; Rana et al., 2024). According to Morin-Crini and Crini, these pollutants can be divided into two categories. Point-source pollution, which comes from a single, identifiable source, and non-point-source pollution, which comes from diffuse sources like agricultural fields or urban runoff (Eaux Industrielles Contaminées, 2017). Industries in particular contribute significantly to water pollution by discharging a range of pollutants into water bodies, such as heavy metals (such as lead, mercury, and chromium), organic compounds (such as benzene and bisphenol A), and nutrients (such as nitrates and phosphates) (Anjaneyulu et al., 2005; Wastewater Treatment, 2020). In aquatic environments, these contaminants frequently cause toxicity, bioaccumulation, and eutrophication, among other ecological disruptions (Forgacs et al., 2004). Certain industrial sectors, like metal finishing, agro-food processing, and textiles, are notorious for their severe chemical contamination and excessive water consumption (Rana et al., 2004).

The European Union enacted the Water Framework Directive (WFD) in 2000 to address these issues, imposing strict limits on pollutant discharge and guidance to maintain water quality (Eaux Industrielles Contaminées, 2017) Compounds are divided into two sections under the WFD: the "Black List" for persistent, hazardous pollutants and the "Grey List" for compounds that pose a serious harm to the environment (Khalaf, 2021). Industrial waste waters are complicated to purify because of the variety and severity of the pollutants. Flotation, precipitation, oxidation, adsorption, biodegradation, and membrane filtration are some of the physical, chemical, and biological techniques that have been developed to address this problem (Crini, 2009; Henze, 2001). However, no single treatment technique can adequately handle every kind of industrial wastewater. In order to balance cost, efficiency, and environmental effect, enterprises frequently use a variety of techniques to produce the best possible water quality (Sharma & Sanghi, 2012).

This chapter synthesizes current knowledge on the methods used for water quality assessment, the treatment of waste water by various fungal species and the challenges associated with wastewater treatment, especially from industrial sources.

9.1.1 Methods for Measuring Waste Water Quality

To evaluate water quality, several methods are employed to detect and quantify pollutants. Analytical techniques such as Inductively Coupled Plasma (ICP) are commonly used to measure concentrations of metals and other inorganic substances. Additionally, chemical indicators like chemical oxygen demand (COD), biochemical oxygen demand (BOD), and total organic carbon (TOC) help assess the organic content of water and its potential impact on aquatic life. Biological indicators, such as fish and aquatic microorganisms, are also used to assess the toxicity and ecological health of water bodies (Eaux Industrielles Contaminées, 2017).

9.1.2 Nature of Effluents and Industrial Contribution

There are several sources of water contamination, including homes, businesses, mines, and irrigation, one of the largest being the industries. According to Crini and Badot (2007), there are four main types of wastewaters.

1. Rainwater (runoff from impermeable surfaces)
2. Wastewater from homes
3. Agricultural wastewater
4. Industrial wastewater
5. Cooling water, washing effluent (of variable composition), and manufacturing or process water (biodegradable and/or potentially hazardous) comprise the final group (Crini, 2009).

Numerous toxic contaminants found in wastewater from different industrial sectors pose a threat to aquatic beings, human life, and agriculture. Heavy metals such as chromium (Cr), zinc (Zn), lead (Pb), copper (Cu), iron (Fe), cadmium (Cd), nickel (Ni), arsenic (As), and mercury (Hg) are examples of such contaminants (Lokhande et al., 2011; Wang & Yang, 2016). The industries that produce paint and dyes, textiles, pharmaceuticals, paper, and fine chemicals are responsible for the majority of the heavy metal pollution. Among the main contaminants found in industrial effluents are phenol and phenolic compounds (Wang & Yang, 2016). The bulk medication manufacture, phenol-formaldehyde resin, and oil refineries are the main sources of their discharge. Wastewater contains a variety of poorly biodegradable refractory pollutants produced by the petrochemical industries, including petroleum hydrocarbons, sulfides, aniline, naphthalenic acid, organochlorines, olefins, nitrobenzene, alkanes, and chloroalkanes (Hai et al., 2007). Petrochemical wastes have an extremely complicated chemical makeup, and biological methods of treating them are slow and ineffective.

Process water often contains biodegradable compounds but may also include potentially toxic chemicals, depending on the nature of the industrial processes involved (Wesenberg, 2003).

Certain industrial sectors, such as the pulp and paper industry, textile manufacturing, and metal finishing, are known for producing particularly harmful effluents. Effluents from paper mills, for example, are often nutrient-rich, leading to eutrophication, a process that causes excessive algae growth in water bodies and disrupts aquatic ecosystems (Lacorte, 2003; Pokhrel & Viraraghavan, 2004; Heisler et al., 2008).

9.1.3 Wastewater Treatment

The effluents contain different kinds of pollutants depending on where these come from; the issues that arise during wastewater treatment are generally quite complicated. Therefore, there are various effluent kinds to handle, each with unique properties that calls for particular treatment procedures. For instance, effluents from food processing industries may be high in organic matter but are generally biodegradable, while effluents from mining and metal industries often contain toxic metals that require specialized treatment (Pokhrel & Viraraghavan, 2004; Wastewater Treatment, 2020). Accordingly, the wastewater properties will determine the best approach. The feasibility, efficiency, practicability, reliability, environmental impact, sludge production, operating difficulty, pre-treatment needs, and the production of potentially harmful byproducts are the main factors of each treatment in addition to cost (Henze, 2001; Mahiout et al., 2006), (Pokhrel & Viraraghavan, 2004; Cooney, 1999; Barakat, 2011; Berefeld et al., 1982; Chen, 2004; Chuah et al., 2005; Mohan & Pittman, 2007; Parsons, 2015; Rathoure & Dhatwalia, 2015; Sonune & Ghate, 2004; Wojnárovits & Takács, 2008), a recent development in nanofiltration membrane technology has been depicted in Fig. 9.1. Typically, a purification procedure consists of the five phases listed below and explained thereafter.

1. Physical and mechanical initial treatment or Pre-treatment;
2. Physical and Chemical / Chemical primary treatment;
3. Chemical and Biological secondary treatment or purification;
4. Tertiary or final treatment (physical and chemical); and
5. Managing the resulting sludge (supervised tipping or recycling or incineration).

1. Pre-treatment: The first step initially involves the physical separation of the solid particles and suspended substances. Finally, it undergoes mechanical treatment like floatation process to inhibit the pollutants from interfering further treatment procedures.
2. Primary treatment (physicochemical and chemical primary treatment): Chemical precipitation occurs during this stage of wastewater treatment, where the wastewater is treated for sedimentation and coagulation to absorb the pollutants and separate the products. Large chemical consumption, poor metal ion removal,

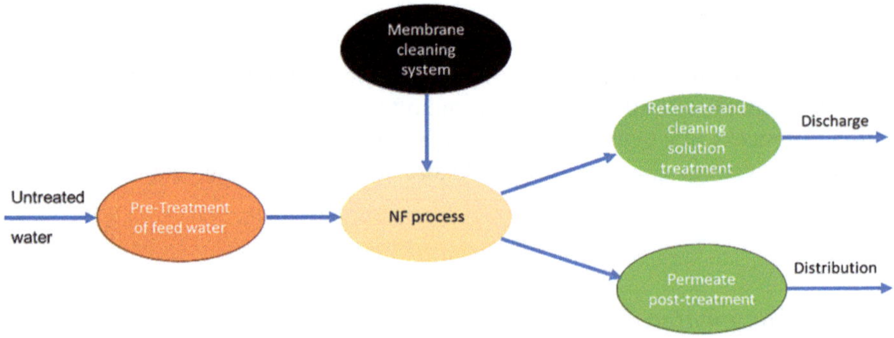

Fig. 9.1 Primary steps are involved in the nanofiltration process. Adapted from (Abdel-Fatah, 2018)

particularly for low concentrations of arsenic, significant sludge generation, pH-dependent selectivity, and a high initial capital cost are the challenges.

3. Chemical and biological secondary treatment or purification: The secondary treatment phase involves treatment methods like chemical oxidation, biological methods and adsorption. Chemical oxidation is carried out in the presence of various oxidizing agents such as Ozone, Hydrogen peroxide, Hypochlorite and $KMnO_4$. In Biological treatment methods bioreactors, Biological Activated Sludge (BAS), Microbiological treatments and Enzymatic decomposition is used. The main challenges involved in these treatment methods involve short half-life of ozone, no effect on salinity (Ozone treatment) emission of VOCs and aromatic amines (Hypochlorite treatment). In case of biological methods, low biodegradability of dyes, poor decolourization, slow process, sludge bulking and foaming require management and maintenance of the microorganisms and/or physicochemical pre-treatment (inefficient on non-degradable compounds or when toxic compounds are present). In adsorption method the main challenge involves high cost, requirement of various adsorbents and elimination of adsorbents after use which requires unsustainable procedures.
4. Tertiary or final treatment (physical and chemical): The tertiary level of treatment is the advanced stage of treatment. Thermal adsorption, catalytic oxidation, and membrane filtering methods are employed in this last phase of wastewater treatment. The primary obstacles are radioactive waste materials and matrix deterioration over some time. In continuous operation, anode passivation and sludge accumulation on the electrodes can prevent the electrolytic process. It requires post-treatment to eliminate elevated iron, and aluminum ion concentrations, high energy needs. Membrane filtration systems' designs can vary to a good extent, may have high operating and maintenance expenses, frequent issues with the evaporation of effluents containing free cyanide, and quick membrane clogging (fouling with high concentrations).
5. Managing the sludge: The generated sludge is treated by incineration, recycling and landfilling methods. Despite being inexpensive, the process of landfilling produces greenhouse gases and may contaminate leachate. Compared to landfilling, incineration is more costly and, if improperly managed, can pollute the air (Abdel-Fatah, 2018).

9.2 Fungal Biodiversity in Waste Water Treatment

Conventional wastewater treatment techniques, including coagulation, flocculation, adsorption, and the activated sludge process, have long been successful in eliminating common environmental contaminants (di Cicco, Iovinella, et al., 2021; di Cicco, Masiello, et al., 2021; Rosa et al., 2019). Nevertheless, these techniques have trouble breaking down complex and enduring contaminants, such as heavy metals and aromatic dyes, which are frequently present in industrial effluents (Kow et al.,

2018). Besides these drawbacks, the techniques involve high costs and energy consumption. Furthermore, throughout the degrading process, these traditional methods may create dangerous secondary pollutants, increasing the risks to the environment. Consequently, there has been a surge in interest in investigating novel approaches, especially those that incorporate biological agents. Microorganisms present a possible answer because of their innate capacity to break down complex contaminants. By efficiently treating difficult contaminants and reducing the generation of harmful byproducts, these more recent bioremediation techniques seek to overcome the drawbacks of conventional systems and provide a more effective and sustainable solution as depicted in Fig. 9.2 (Iovinella et al., 2020, 2022; Kow et al., 2019; Manfredi et al., 2023; Palmieri et al., 2022; Sirakov et al., 2021). The advantage of using the microorganisms as a strategy in treatment of waste water is due to the involvement of low costs, sustainability and the eco-friendly methods. Amongst the microorganisms, bacteria are known to play major role, however, there are various classes and strains of fungi that have been employed for mitigating the problem. This chapter deals with some of the case studies that have used fungi in waste water treatment.

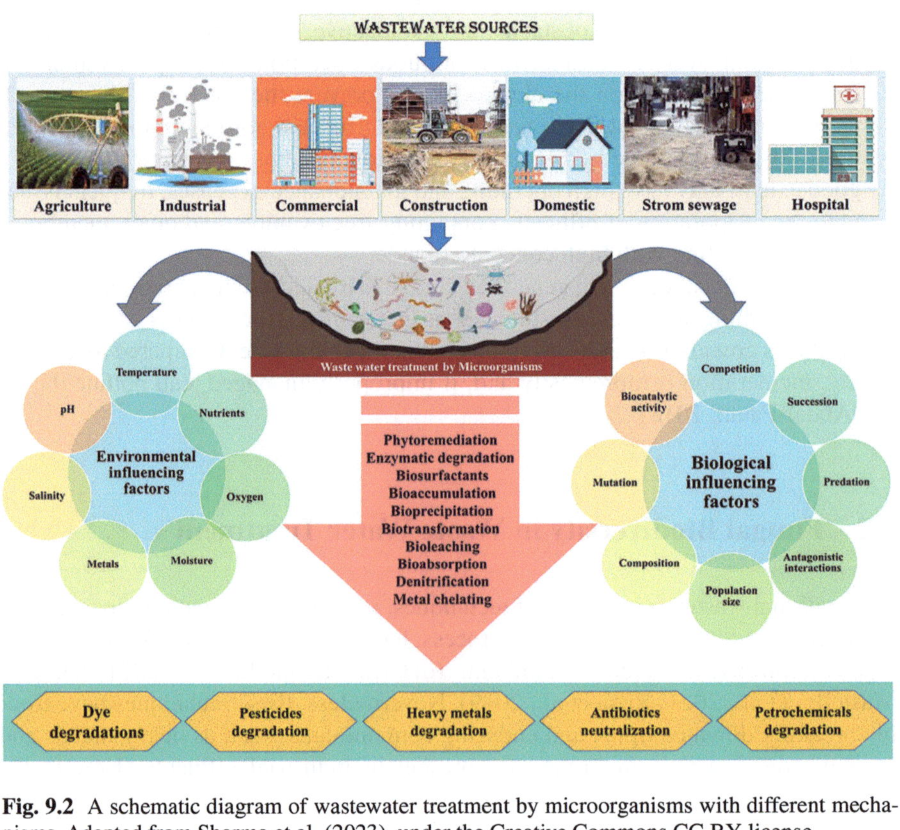

Fig. 9.2 A schematic diagram of wastewater treatment by microorganisms with different mechanisms. Adapted from Sharma et al. (2023), under the Creative Commons CC BY license

9.2.1 *Ascomycetes*

Ascomycetes are among the most versatile fungal groups, known for their enzymatic capabilities in breaking down diverse organic pollutants. Their ecological role spans nutrient cycling and degradation of recalcitrant compounds in industrial effluents making them an attractive bio-alternative for wastewater treatment. They are ecologically significant due to their ability to decompose a wide array of organic materials, including lignin, cellulose, and complex carbohydrates, making them essential in nutrient cycling. Their metabolic versatility allows them to thrive in various environments, including soil, decaying organic matter, and even wastewater (Ferrari et al., 2021).

The phylum Ascomycota, known as sac fungi, constitutes a diverse and monophyletic group of eukaryotic microorganisms, including both unicellular and multicellular species. These fungi are distinguished by their production of non-motile sexual spores, termed ascospores, within specialized sac-like structures called asci, often housed in fruiting bodies (ascocarps). As the largest fungal phylum, Ascomycota encompasses over 93,000 identified species across three subphyla:

- Pezizomycotina: Includes multicellular filamentous fungi with diverse fruiting body morphologies, such as truffles and cup fungi.
- Saccharomycotina: Comprises true yeasts like Saccharomyces cerevisiae, which predominantly reproduce asexually via budding.
- Taphrinomycotina: Encompasses both yeasts and filamentous fungi, some of which exhibit no known sexual reproduction, but molecular studies reveal genetic links to ascus-forming taxa (Liu, 2024; Volk, 2013).

The bioremediation potential of ascomycetes has been successfully demonstrated in controlled environments using synthetic wastewater enriched with sterile glucose-based spike media. Nevertheless, the application of these fungi to real wastewater treatment remains insufficiently understood. For the technology to progress to industrial-scale implementation, it is essential to examine the resilience and adaptability of fungal communities in the intricate and fluctuating conditions typical of real wastewater systems (Hanna et al., 2022).

9.2.1.1 Bioremediation of Selenium-Contaminated Industrial and Municipal Wastewaters-A Case Study

Filamentous Ascomycete fungi have demonstrated considerable potential for the bioremediation of selenium (Se)-contaminated water, particularly in reducing the toxic forms of selenate (Se(VI)) and selenite (Se(IV)) to elemental Se(0) or volatile Se(-II). In a study assessing their Se removal capacity, two strains of Se-reducing Ascomycetes were tested under bench-scale conditions using growth media containing Se(IV) or Se(VI) as well as Se-containing municipal (approximately 25 μg/L Se) and industrial (approximately 2000 μg/L Se) wastewaters. The results showed that both fungi efficiently removed Se(IV) from solution, with the highest Se removal occurring in municipal wastewater, where a significant portion of Se was reduced and removed to solid phases.

Under optimal conditions, the fungi removed approximately 75% of the initial Se(IV) concentration within 7 days, reaching levels that comply with the U.S. Environmental Protection Agency (USEPA) recommendations for Se in standing and flowing water. The removal rate for Se(IV) was notably higher compared to Se(VI), suggesting that these fungi are more effective at reducing and removing Se(IV) in aqueous solutions. In contrast, Se removal from industrial wastewater, which contained high levels of other contaminants, was less efficient, with fungi removing a smaller proportion of Se from the solution, although a higher percentage of Se was sequestered in solid forms like nanoparticles. Additional experiments demonstrated that the inclusion of carbohydrate-based carbon sources enhanced Se removal in municipal wastewater but had limited impact on industrial wastewater, highlighting the influence of carbon source availability on fungal Se reduction efficiency. These findings suggest that fungal systems, particularly Ascomycetes, hold significant promise for the bioremediation of selenium in contaminated waters. However, further research into the specific metabolic pathways responsible for Se reduction is needed to optimize these systems, and future studies should explore strategies for utilizing the produced selenium nanoparticles in biotechnological applications, rather than disposing of them as hazardous waste (Sabuda et al., 2020).

The significance of isolating fungi from polluted sources lies in the possibility of uncovering new fungal strains that have naturally evolved or adapted enzymatic systems capable of dealing with high levels of contaminants. These enzymes could degrade or transform pollutants, making these fungi valuable for bioremediation efforts (Agosta & Klemens, 2008).

9.2.1.2 Bioremediation of Polycyclic Aromatic Hydrocarbons (PAHs): A Case Study

A recent case study highlighted the application of Ascomycetes in the bioremediation of polycyclic aromatic hydrocarbons in contaminated environments. This study aimed to investigate the ability of fungi present in a PAH-contaminated pond to act as remediation agents. Polycyclic aromatic hydrocarbons, a worldwide environmental contaminant of anthropogenic origin are released into aquatic systems mainly due to accidental fuel spills or leakages from corroded tanks at petrol stations or refineries. Their degradation rate is commonly slower in comparison to the concentration released, leading to an accumulation in the contaminated sites that can remain for decades (Patel et al., 2020; Singh et al., 2023). Eight of the ten isolated strains belonging to Ascomycota could remove anthracene after 21 days. Among these, *Penicillium oxalicum* was selected for further investigation. Time-course experiments revealed that *P. oxalicum* completely removed anthracene and dibenzothiophene within 4 days, and also degraded phenanthrene and dibenzofuran, though at slower rates. The study focused on the induction of oxygenase activity in the fungal mycelium when exposed to these aromatic compounds (Aranda et al., 2017).

9.2.2 Basidiomycetes

A vast and varied group of fungi belonging to the phylum Basidiomycota, the Basidiomycetes are important in both ecological and industrial contexts because of their remarkable capacity to break down lignocelluloses, a key component of plant cell walls. They are very useful in environmental sustainability and biotechnology because of this capability. In nature, bacteria and fungus are the primary decomposers of lignocellulose. Three primary categories of fungus are capable of breaking down lignocellulose: brown-rot and white-rot fungus as well as soft-rot fungus. White-rot fungi have the best chances for growth and application among these fungi because they can secrete extracellular ligninolytic enzymes that are the most effective at breaking down lignocellulose material.

Basidiomycetes' extensive enzymatic capabilities make them versatile tools in biotechnology. Their enzymes not only help degrade plant materials but also have potential uses in the production of value-added products such as bio-based chemicals, and biodegradable plastics, and even in improving the digestibility of animal feed.

By using natural, efficient biological systems to break down plant materials, Basidiomycetes offer more sustainable alternatives to traditional chemical processes, reducing the environmental impact of industrial activities. Their use in converting agricultural waste into valuable products can also contribute to reducing waste and promoting circular economy practices (Hussain et al., 2022).

9.2.2.1 Use of White-Rot Fungi in Municipal Wastewater: A Case Study

Because of their exceptional capacity to break down complex organic contaminants, white-rot fungi (WRF), in particular *Phanerochaete chrysosporium* and *Trametes versicolor*, have attracted more and more interest in the wastewater treatment sector. This case study looks at how white-rot fungi are used worldwide to treat municipal wastewater, emphasizing how they can improve conventional treatment techniques by decomposing contaminants that are otherwise challenging to eliminate with conventional methods (Khiyami et al., 2006).

Mechanism White-rot fungi may break down a variety of organic contaminants because they have a special set of ligninolytic enzymes, including laccase, manganese peroxidase, and lignin peroxidase. These enzymes target and break down chemicals that frequently remain in urban wastewater systems, such as pesticides, aromatic compounds, polycyclic aromatic hydrocarbons (PAHs), and even medicines. WRF is very useful in the treatment of municipal wastewater, which contains a wide variety of organic pollutants, due to its capacity to breakdown such a wide range of contaminants (Malik et al., 2023; Sayara & Sánchez, 2020).

Some global case studies are given in brief below.

1. Deepwater Horizon Oil leak (2010): Alcanivorax borkumensis and other oil-degrading bacteria demonstrated their abilities in the Deepwater Horizon oil leak, helping to restore the impacted water and marine life through a natural cleanup process.
2. The Arcata Marsh and Wildlife Sanctuary in California is a sustainable and organic filtration system for urban runoff. It is a prime example of how bacteria and plants work together to filter wastewater.
3. Garfield County Landfill, Oklahoma: By introducing particular bacteria to increase trash decomposition and reduce environmental effect, bioremediation procedures at the Garfield County Landfill improved overall cleanliness.
4. Plants in Southeast Asia that behave as hyperaccumulators include Pteris vittata, which absorbs and accumulates heavy metals like arsenic from disturbed soil, so offering a natural and efficient remediation method.
5. Innovative approaches are being used in Maharashtra, India, to address the wastewater management dilemma. The Bhandup Wastewater Treatment Facility is changing dramatically because to the application of effective microorganism technology. In addition to addressing water pollution, this project protects the abundant biodiversity in the area.

9.2.3 Zygomycetes

The majority of zygomycetes are saprophytic fungi, i.e., they thrive on decomposing organic waste. They are widely distributed and can be found in manure, soil, and decomposing plant matter. They can also be found in the digestive tracts of certain animals. *Rhizopus, Mucor,* and *Absidia* are among the genera that belong to this class of fungi. The development of a thick-walled zygospore during sexual reproduction is the characteristic that distinguishes zygomycetes. They are also well-known for growing quickly, which makes them effective decomposers of organic materials like lignocellulose, dead plant matter, and other complex organic substances.

In both natural and man-made settings, zygomycetes are crucial to the breakdown of complex organic materials, including cellulose and lignin. They accomplish this by secreting extracellular enzymes, including cellulases, ligninases, and proteases, which degrade complex macromolecules into simpler substances that other species can use or mineralize (Payne et al., 2015).

- Organic waste degradation: Zygomycetes play an important role in the breakdown of organic waste in sites such as landfills, sewage treatment plants, and composting facilities. They aid in the breakdown of materials into simpler compounds like sugars, amino acids, and fatty acids by colonising decaying organic waste. Microorganisms like bacteria can then further metabolize these molecules, further mineralizing the nitrogen and carbon.

- Bioconversion of agricultural wastes: Crop leftovers, animal dung, and byproducts of food processing are examples of agricultural wastes that frequently include high levels of nutrients and nitrogen. By decomposing lignocellulose and using the nitrogen it contains, zygomycetes contribute to the bioconversion of these wastes. In order to preserve soil fertility, they also aid in the cycling of vital elements like carbon and nitrogen.
- Bioremediation of organic pollutants: Certain zygomycetes are helpful in bioremediation applications because they can break down organic contaminants, such as petroleum hydrocarbons and polycyclic aromatic hydrocarbons (PAHs). They can survive and break down organic contaminants in contaminated soil and water because they can grow in nutrient-poor settings (Battaglia et al., 2011; Ribes et al., 2000).
- Organic nitrogen degradation: By secreting proteases, zygomycetes are able to break down organic nitrogen molecules like proteins, peptides, and amino acids. The funguses lower the nitrogen burden in the discharge by decomposing these organic nitrogen sources. Additionally, this process makes it easier for organic nitrogen to mineralize into inorganic forms that plants and other creatures can use (Khiyami et al., 2006). Both organic and inorganic nitrogen can be assimilated by zygomycetes. Fungal growth depends on the intake of nitrogen, either in the form of nitrate ion (NO_3^-) or ammonium ion (NH_4^+). We may simulate this by assuming that nitrogen is taken up and integrated into fungal biomass for understanding (Zayed et al., 2023).

9.2.3.1 Zygomycetes in Wastewater and Pollutant Bioremediation: Case Studies

In a compelling comparative study by Guiraud et al. (Guiraud et al., 2003) two strains of Absidia *fusca* Linnemann—a zygomycete—were evaluated for their ability to biodegrade a range of xenobiotics, including lignin-derived phenolics, chlorinated aromatics, and polycyclic aromatic hydrocarbons (PAHs), commonly found in contaminated soils, sediments, and water. The researchers compared a reference strain from a fungal culture collection (grown under non-inducing conditions) with a strain isolated from pesticide-contaminated soil in Annaba, Algeria. The contaminated-soil strain demonstrated markedly higher degradation efficiency across multiple compounds—particularly pentachlorophenol, phenol, catechol, guaiacol, and ferulic acid—while the collection strain exhibited only minimal activity.

These findings highlight not only the promise of A. fusca in xenobiotic bioremediation but also the adaptive advantage of environmental strains pre-exposed to pollutants, which seemingly harbor more effective and robust enzymatic machinery for broad-spectrum degradation.

Similarly, Pereira et al. (2009), *Rhizopus oryzae*—a zygomycete—was evaluated as a biological treatment agent for the detoxification of bleached kraft pulp mill effluent. This study compared *R. oryzae* treatment to photo-Fenton oxidation as tertiary steps after secondary wastewater processing. Remarkably, treatment with *R. oryzae* alone offered mitigation of toxicity when applied biologically. However, when combined with photo-Fenton treatment, the effluent toxicity increased significantly, as assessed using a battery of sensitive freshwater organisms (including bacteria like Vibrio fischeri, green algae such as Pseudokirchneriella subcapitata, and Daphnia magna).

9.2.4 Yeasts

Yeasts are versatile microorganisms that can thrive in a wide array of environments, including both aquatic and terrestrial habitats. They are found in freshwater and marine ecosystems, as well as in environments with fluctuating conditions such as wastewater treatment plants (Nagahama, n.d.).

They are unicellular fungi that typically exhibit ovoid, circular, or cylindrical shapes (Chavez et al., 2023). They can be broadly classified into two categories:

- Fermentative yeasts: These yeasts primarily ferment six-carbon sugars to produce alcohol and carbon dioxide. They are widely used in industries such as bread-making, brewing, and winemaking (Maicas, 2020).
- Oxidative yeasts: These are known for their strong oxidative capabilities and limited or no fermentation ability; examples include *Candida* and *Hansenula polymorpha*. These yeasts are particularly valuable in wastewater treatment and the petroleum processing industry due to their ability to metabolize various organic and complex compounds (Hartner & Glieder, 2006).

The size of yeasts varies significantly based on their species and environmental conditions, generally ranging from 3–4 μm in diameter, with some species capable of reaching sizes up to 40 μm (Walker et al., 2002). Yeasts thrive optimally at temperatures ranging from 20 to 28 °C and under mildly acidic conditions, with an ideal pH of 3.5–4. Commonly employed media for yeast cultivation and isolation include Wickerham's YM agar, malt agar, potato dextrose agar, yeast peptone dextrose agar, yeast mold agar, gyp agar, and Sabouraud's agar (Phale, 2018). They exhibit properties such as resistance to acidic conditions (Fletcher et al., 2017), osmotic pressure (Hohmann, 2002), elevated temperatures (Huang et al., 2018), and high metabolic efficiency (Zakhartsev et al., 2015). Furthermore, they possess a robust enzymatic system, enabling them to adapt to a wide range of specialized environments. These organisms are extensively distributed across soil, freshwater, and marine ecosystems, as well as on the surfaces and within the bodies of living organisms (Nagahama, n.d.).

9.2.4.1 Yeast-Based Treatment of Heavy Metal Ion-Contaminated Wastewater

Yeasts are particularly resilient, capable of adapting to diverse climates, and propagate through the dispersal of spores. These microorganisms help degrade organic pollutants and heavy metals (Ledesma-Amaro, 2022). Recently, bioremediation with microorganisms such as yeast has played a significant role in removing heavy metals from contaminated sources (Bahafid et al., 2017).

In a recent study by Nicula et al. (2023), five yeast strains (Kluyveromyces marxianus, Saccharomyces cerevisiae, and Pichia anomala) were tested for their ability to remove contaminants from synthetic wastewater contaminated with Pb^{2+} (43 mg/L) and Cd^{2+} (39 mg/L). The results showed removal efficiencies of up to 70% for COD, 97% for nitrate, 80% for nitrite, 93% for phosphate, and 70% for sulfate ions. However, ammonium ion concentrations increased, particularly in the presence of Pb^{2+}.The yeast strains effectively reduced Pb^{2+} by up to 96% and Cd^{2+} by up to 40%. The addition of a biosurfactant from Yarrowia lipolytica CMGB32 enhanced Pb^{2+} removal to 99% and Cd^{2+} to 56%, along with an 11-fold increase in biomass. These results, obtained under neutral pH and without aeration, suggest that yeast-based bioremediation is a cost-effective and sustainable approach for treating wastewater contaminated with heavy metals and organic pollutants.

9.2.4.2 Treatment of High Concentration Organic Wastewater by Yeast

Recently, Biochar-based materials combined with microbial components, like yeast, have indicated a promising avenue for the development of sustainable and efficient bioadsorbents aimed at the remediation of water contaminated with pollutants such as dyes. Staroń and Chwastowski (2024) investigated the development of a novel biocomposite containing Yarrowia lipolytica yeast cells integrated with magnetic properties for the sorption of organic pollutants like methylene blue (MB) from aqueous solutions. The study provides compelling evidence for the biocomposite's potential as an eco-friendly and cost-effective bioadsorbent, with applications in water purification. By examining parameters such as initial MB concentration (50–250 mg/dm^3), contact time (0.167–24 h), and temperature (25–40 °C), the researchers demonstrated that the adsorption process aligns well with the Freundlich isotherm model (R^2 = 0.9959; ARE = 3.41%) and pseudo-second-order kinetics (R^2 = 0.9989; ARE = 1.14%).

The biocomposite exhibited a sorption capacity ranging from 32.10 mg/g at 25 °C to 23.64 mg/g at 40 °C, confirming that the process is exothermic and temperature-sensitive. The material maintained its adsorption efficiency over five reuse cycles, with a desorption rate decreasing modestly from 90% to 84%, further underscoring its viability for multiple applications.

9.2.4.3 Biodegradation of Phenolic Compounds and Treatment of Olive Mill Wastewater by *Rhodotorula Mucilaginosa:* A Case Study

A recent study (Jarboui et al., 2012) investigated the potential of *Rhodotorula mucilaginosa* for the degradation of phenolic compounds in olive mill wastewater (OMW). This yeast was isolated from OMW and identified through 18S RNA sequencing. The study focused on assessing the yeast's ability to degrade key phenolic compounds and its overall performance in reducing organic pollutants in OMW. The findings are given below.

- Degradation of phenolic compounds: *Rhodotorula mucilaginosa* successfully degraded several phenolic compounds found in OMW. Complete degradation was achieved for protocatechuic acid, vanillic acid, and p-coumaric acid within 24 h, while tyrosol was degraded in 96 h. However, catechol and gallic acid showed partial inhibition of yeast growth.
- Pollutant removal efficiency
  - Chemical oxygen demand (COD): The yeast strain was effective in reducing COD by 38.38%, 47.69%, and 56.91% at initial concentrations of 26,700 mg/L, 14,400 mg/L, and 6500 mg/L, respectively.
  - Phenol reduction: Phenol concentrations were reduced by 5.84%, 27.89%, and 34.81% at the three initial COD levels.
  - Color reduction: Significant color reduction was observed with 7.84%, 46.31%, and 56.08% reductions at the corresponding COD concentrations.

9.2.4.4 Pollutant Removal Process

- The yeast utilized OMW as the sole carbon and energy source, demonstrating a strong growth response even at high COD concentrations. After an initial lag period, *Rhodotorula mucilaginosa* adapted and grew at varying rates depending on the initial COD concentration: 0.06 day^{-1} at 26,700 mg/L, 0.086 day^{-1} at 14,400 mg/L, and 0.20 day^{-1} at 6500 mg/L.
- The reduction in phenolic compounds and COD was attributed to the yeast's ability to metabolize the sugars and phenolic compounds in the OMW. The preference for assimilating simpler carbon sources, such as sugars, was observed first, with phenolic compounds being used later as secondary carbon sources.

The results demonstrate that *Rhodotorula mucilaginosa* can effectively reduce pollutants in high-strength organic wastewater, especially from olive oil production. This yeast not only purifies OMW but also contributes to the removal of organic dyes and phenolic compounds.

9.2.5 *Filamentous Fungi*

Filamentous fungi are a diverse group of fungi characterized by their thread-like structures called hyphae, which form the main body of the organism known as the mycelium (Bardana, 2003). These fungi play essential roles in ecosystems as decomposers, symbionts, and pathogens (Fischer & Glass, 2019).

Filamentous fungi play a significant role in bioaugmentation and microbial degradation due to their ability to degrade complex organic compounds and enhance wastewater treatment processes. These are particularly useful in the treatment of industrial wastewater, where high concentrations of organic and inorganic contaminants are often present. Their ability to adapt to diverse environments, such as varying pH levels, temperatures, and nutrient availability, makes them valuable in bioengineering applications. By enhancing microbial populations in bioreactors, filamentous fungi can improve the efficiency of pollutant removal and facilitate the breakdown of toxic substances that may be resistant to traditional methods.

9.2.5.1 Bioaugmentation with Filamentous Fungi in Fluidized Bed Bioreactors for Pharmaceutical Removal from Municipal Wastewater: A Case Study

This study evaluated the potential of filamentous fungi, specifically *Aspergillus luchuensis*, to remove pharmaceuticals from municipal wastewater using a fluidized bed pelleted bioreactor. The bioreactor was designed with biofilm-supporting carriers to promote fungal attachment and pollutant degradation. The system operated under non-sterile conditions for 28 days, with periodic additions of fungal biomass at concentrations of 10 g and 50 g per 1.25 L reactor (Dalecka et al., 2021). The key findings are given below.

1. Pharmaceutical degradation
   - *A. luchuensis* achieved removal efficiencies >90% for ibuprofen, ketoprofen, and metoprolol within 3 h at higher biomass concentrations (50 g).
   - Lower removal efficiencies (<30%) were observed for recalcitrant pharmaceuticals such as diclofenac, carbamazepine, and sulfamethoxazole.
2. Nutrient removal
   - The fungus reduced ammonium nitrogen (NH_4^+-N) concentrations by over 80% within 12 days. However, nitrate (NO_3^-–N) and phosphate (PO_4^{3-}–P) removal remained limited.

9.3 Mechanisms of Action

Mycoremediation is one of the potential approach uses fungi to break down or change toxic materials into less toxic or non-toxic forms. Combining the terms "myco" (fungus) and "remediation" (to clean, resolve, or rectify), "mycoremediation" uses fungi, especially mushrooms, to create low-tech biomass that can efficiently break down environmental and industrial toxins (Delsarte et al., 2020). Various fungal species are used as mycoremediators, such as *Rhizopus arrhizus*, *Pleurotus ostreatus*, *Phanerochaete chrysosporium*, *Phanerochaete sordida*, *Trametes hirsuta*, *Trametes versicolour*, *Lentinus tigrinus*, and *Lentinus edodes* (Chavda & Shah, 2017). Using the metabolic powers of fungi, fungal bioremediation is an environmentally friendly method that breaks down and purifies a variety of contaminants. The wide range of enzymes found in fungi, such as laccases, peroxidases, and hydrolases, allow complex organic molecules, heavy metals, and xenobiotics to be broken down into less toxic forms. Fungal bioremediation has demonstrated efficacy in the immobilisation and transformation of heavy metals through biosorption and bioaccumulation, as well as in the treatment of persistent organic pollutants such petroleum hydrocarbons, polycyclic aromatic hydrocarbons (PAHs), and polychlorinated biphenyls (PCBs). Here we have presented some environmentally significant fungal bioremediations.

The production of organic acids and the binding of metals to components of fungal cell walls are two of the main mechanisms involved in the chemical process of bioremediation employing fungus such as *Aspergillus Niger* and *Trichoderma harzianum* for heavy metal detoxification. A probable explanation of the chemical mechanisms is provided below:

Aspergillus niger and other fungi release organic acids, especially citric acid, into their surroundings. These acids are essential for breaking down heavy metals like lead (Pb) and cadmium (Cd) into less hazardous and more soluble forms. For instance, citric acid forms soluble metal-organic complexes by chelating metal ions such as Cd^{2+} and Pb^{2+}. By chelating the metals, their bioavailability and environmental toxicity are decreased by transforming them into forms that can be precipitated or immobilized. The metal ions and organic acids combine to produce soluble, stable complexes. This diminishes the metals' harmful effects by stopping them from interacting with other substances or environmental microbes (Zheng et al., 2024).

9.3.1 *Bioremediation of Hydrocarbons*

Hydrocarbons present in contaminated soils and streams are broken down by fungi like *Pleurotus ostreatus* and *Phanerochaete chrysosporium.* Laccase and peroxidase are two examples of the ligninolytic enzymes produced by these fungi that decompose complicated hydrocarbons into simpler ones. Within 30 days, *P.*

chrysosporium can break down up to 95% of the polycyclic aromatic hydrocarbon benzo[a]pyrene (Deng & Zhao, 2015; Wang & Xu, 2012).

This degradation capability is significant compared to bacterial strains, which achieve lower degradation rates under similar conditions (Favour Oghenero Efenudu, 2024). This high degradation efficiency is attributed to the fungi's ability to produce non-specific extracellular enzymes that can attack a broad range of chemical bonds in these complex molecules (Okal et al., 2023).

9.3.2 Bioremediation of Dyes

The enzymatic action of fungi such as *Pleurotus ostreatus* and *Trametes versicolour* are well known for their capacity to decolorise and break down synthetic dyes. Laccase and manganese peroxidase are two examples of enzymes that break down the intricate structures of dyes, rendering them non-toxic. *T. versicolour*, for example, can break down 90% of Reactive Black 5 dye in just 7 days (Pereira et al., 2009).

9.3.3 Chelation and Precipitation

The process by which metal ions (such as lead, cadmium, and aluminium) attach to organic acids (such as citric and oxalic acids) to create stable complexes is known as chelation. Because of their binding, these metals have a lower bioavailability, which makes them less dangerous because living things are less likely to absorb them. Metals like cadmium can be chelated by Aspergillus niger's citric acid, which lessens their toxicity and makes it easier for the environment to immobilise them. This facilitates their removal from polluted areas. Fungi that secrete oxalic acid help precipitate metal oxalates, which are insoluble and readily extracted from soils and water. Soluble metal ions are transformed into solid metal oxalates during the precipitation process, which can be separated using either filtration or sedimentation. *Aspergillus niger*, for instance, has been demonstrated to efficiently reduce lead by precipitating lead as lead oxalate levels in water that has been tainted (Nagahama, n.d.; Chavez et al., 2023).

9.3.4 Enzymatic Efficiency and Diversity

Numerous enzymes found in fungi are capable of breaking down a broad spectrum of contaminants. These include ligninolytic enzymes that may break down complex organic compounds, such as laccases, peroxidases, and oxidoreductases. For instance, lignin and a variety of xenobiotic substances, including as polycyclic aromatic hydrocarbons (PAHs) and chlorinated phenols, can be broken down by extracellular enzymes produced by white-rot fungi such as *Phanerochaete chrysosporium* (Shoun et al., 2012).

All the above things considered, Mycoremediation is an innovative and exciting way to tackle some of the most urgent environmental problems. However, more research and development are required to realize its full potential.

9.3.5 Mechanism of Pollutant Removal

The primary mechanism driving pharmaceutical removal was biosorption, rather than enzymatic activity. Biosorption occurs through the interaction of pharmaceutical compounds with the fungal cell wall, which contains functional groups like hydroxyl, carboxyl, and amine groups capable of binding contaminants. A schematic representation of a fungal fluidized bed bioreactor with a pelletized design, illustrating the experimental setup and fungal interaction with pollutants, is shown in Fig. 9.3. Key observations supporting this include the following.

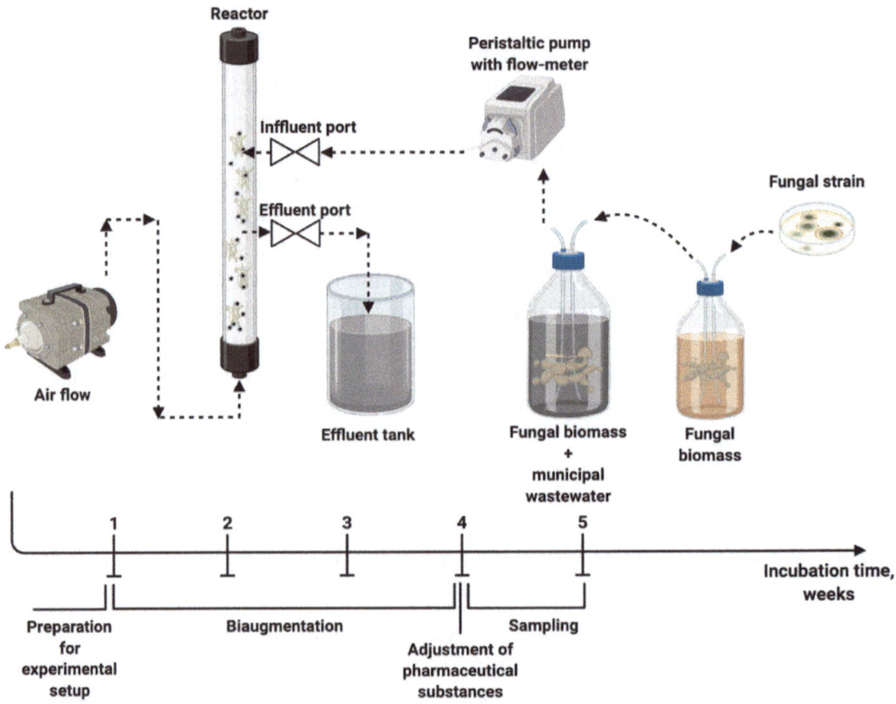

Fig. 9.3 A schematic representation of a fungal fluidized bed bioreactor with a pelletized design and experimental setup. Reprinted with permission Dalecka et al. (2021), Copyright 2021 Elsevier

- Absence of Laccase Activity: No significant laccase or other extracellular enzymatic activity was detected during the experiment, ruling out oxidative enzymatic degradation as the primary pathway (Dalecka et al., 2021).
- Surface Adsorption: The fungal cell walls acted as adsorbents, capturing pharmaceutical molecules through ionic interactions and hydrogen bonding. This process was influenced by the surface properties of the fungal biofilm and the chemical characteristics of the pharmaceuticals (Garcia-Rubio et al., 2020).
- Selective Adsorption: Compounds like ibuprofen and ketoprofen, which have higher hydrophobicity (Alsop et al., 2015; Chavda & Shah, 2017) were more effectively adsorbed compared to highly polar compounds like sulfamethoxazole and carbamazepine.

These findings highlight the potential of filamentous fungi to adsorb and immobilize pharmaceutical contaminants, making biosorption a key mechanism in fungal bioaugmentation for wastewater treatment. Future optimization could focus on enhancing cell wall modifications or integrating enzymatic pathways to address recalcitrant compounds.

9.4 Advantages of Fungi in Wastewater Treatment

Fungi offer numerous advantages in wastewater treatment due to their unique biological and metabolic properties. Their ability to adapt to diverse conditions, coupled with their cheap cost, and being eco-friendly, makes them a promising alternative to conventional chemical treatments (Akerman-Sanchez & Rojas-Jimenez, 2021). The factors influencing fungal biodegradation are given below.

9.4.1 Factors Influencing Fungal Biodegradation

The effectiveness of fungal biodegradation is significantly impacted by physicochemical parameters, including temperature, pH levels, nutrient composition, redox mediators, and the bioreactor configuration. Temperature significantly impacts fungal efficiency in wastewater treatment by affecting growth, enzymatic activity, and biofilm dynamics. Elevated temperatures can inhibit fungal development, reduce enzyme production, and alter biofilm structure, potentially decreasing pollutant degradation and increasing the risk of pathogen release. Maintaining optimal temperature conditions is essential for stable and effective fungal treatment processes (Calero Preciado et al., 2021; Yusuf et al., 2024). Similarly, pH significantly influences the efficiency of fungal biodegradation by modulating enzyme activity and the transport of pollutants across the cell membrane (Kane, 2016). Fungi generally exhibit higher biodegradation rates with better growth at acidic or neutral pH levels compared to bacteria, as their enzymatic systems are optimized in these conditions

(Rousk et al., 2009). Bioreactors are essential for optimizing fungal performance in wastewater treatment by maintaining controlled conditions such as pH and temperature, which are critical for fungal growth. Bioaugmentation, where specific microorganisms are added to enhance fungal activity, has been shown to improve the removal of pharmaceutical contaminants in municipal wastewater (Dalecka et al., 2021).

The key advantages of using fungi in wastewater treatment processes are briefly given in the following subsection.

9.4.2 Biodegradability and Efficiency

Fungi excel in degrading a wide range of complex pollutants, including recalcitrant organic compounds such as phenols, hydrocarbons, dyes, and pharmaceuticals. Their extracellular enzyme systems, such as ligninases, peroxidases, and oxidases, enable the breakdown of high-molecular-weight pollutants into simpler, less toxic compounds (Pereira et al., 2009). This high biodegradative capacity makes fungi particularly effective for treating industrial wastewater, where traditional methods often fail to address persistent contaminants. Additionally, fungi are capable of removing heavy metals through biosorption and bioaccumulation processes, further broadening their application in wastewater treatment (Filote et al., 2021; Ramya et al., 2021).

This method requires the activity of living fungal cells, as the degradation is facilitated by enzymes secreted by the fungi. Biodegradation has a wide range of applications, including the treatment and management of wastewater sludge, where fungal activity plays a crucial role in breaking down organic matter and reducing environmental pollution (Lu et al., 2016).

9.4.3 Cost-Effectiveness Compared to Conventional Methods

Fungal-based treatments are often more cost-effective than conventional methods ion exchange, chemical precipitation, membrane filtration, coagulation-flocculation, solvent extraction, biological methods, and (Sathinathan et al., 2023) chemical treatments typically require significant investment in reagents and infrastructure (Yin & Allen, 1999), as well as secondary treatment to mitigate chemical byproducts. In contrast, fungi can grow on low-cost substrates, such as agricultural residues or wastewater itself (Assress et al., 2019), reducing operational costs. Furthermore, fungal bioreactors, such as fluidized bed systems, are energy-efficient and capable of handling high pollutant loads without the need for intensive aeration, making them economically viable for large-scale applications (Mir-Tutusaus & Sarrà, 2020).

9.4.4 Environmental Benefits and Sustainability

Fungi offer a sustainable and eco-friendly alternative to traditional wastewater treatment methods. Unlike chemical treatments, which can produce secondary pollutants, fungal-based processes are inherently green and contribute to the circular economy (Subasri et al., 2020). For example, fungal biomass generated during wastewater treatment can be repurposed as biofertilizers, animal feed, or as a source of bioactive compounds for pharmaceutical and agricultural applications (Amara & El-Baky, 2023; Isaza-Pérez et al., 2020; Odoh et al., 2020). The integration of fungi with other microorganisms, such as bacteria and microalgae has gained increasing attention in wastewater treatment due to their synergistic interactions, which enhance pollutant degradation and resource recovery. Recently, Mi et al. (Mi et al., 2024) investigated the potential of a fungi-microalgae consortium (FMC), demonstrating its superior performance in addressing nutrient pollution and emerging contaminants. *Cordyceps militaris* (fungus) and *Navicula seminulum* (microalgae) exhibited a synergistic relationship that improved pollutant removal efficiency. The FMC effectively reduced nutrient pollutants, optimized bacterial community structures, and achieved superior removal of emerging contaminants, including heavy metals, antibiotics, and pathogenic Vibrios. For instance, the FMC outperformed single-species treatments by showing a stronger positive correlation between the enrichment of functional bacteria and pollutant removal performance.

This symbiotic system leverages fungi's enzymatic degradation capabilities and microalgae's photosynthetic activity, enhancing pollutant elimination and promoting mutualistic benefits. The findings highlight the feasibility of fungal integration with other microorganisms to establish sustainable and efficient water treatment technologies, particularly for wastewater management.

Mycoremediation is also a sustainable and environmentally conscious approach that is an efficient alternative to conventional methods like incineration or chemical treatments, often associated with high energy demands and pollutant emissions. This highly adaptable technique can address a wide spectrum of contaminants, including pesticides, hydrocarbons, heavy metals, and synthetic compounds (Navina et al., 2024).

9.5 Challenges and Limitations

The removal of contaminants using special fungi, including polyaromatic hydrocarbons, heavy metals, plastics, dyes, herbicides, pesticides, detergents, phthalates, cyanotoxins, and antibiotics. However, certain challenges and limitations are presented here under (Kadri et al., 2017; Khandaker et al., 2021; Mahamuni-Badiger et al., 2020; Mousavi et al., 2021; Negi & Das, 2023; Shivalkar et al., 2020).

S. No.	Challenge	Limitation
1.	In Mycoremediation, only a small number of fungal species have been found to be useful for bio-carrier mycelial pellets	This restricts the technology's ability to scale for broad application in pollutant remediation
2.	Certain pollutants are hard for fungi to break down because they are hydrophobic and have low bioavailability, especially polycyclic aromatic hydrocarbons (PAHs). The challenge of degrading hydrophobic pollutants without the inclusion of extra agents, including surfactants, which are required to boost their bioavailability	The challenge of degrading hydrophobic pollutants without the inclusion of extra agents, including surfactants, which are required to boost their bioavailability
3.	Despite being eco-friendly, mycoremediation is often slower than other techniques, including bacterial bioremediation, particularly for specific kinds of pollutants	Fungi have a long life cycle and a slower rate of growth, which might make remediation take longer and may not always be appropriate for urgent cleanup needs
4.	Fungi have difficulty with the limited bioavailability of hydrophobic contaminants, such as PAH pollution	The mycoremediation process is made more difficult and expensive by the requirement for surfactants to improve the uptake and degradation of hydrophobic substances
5.	The practical difficulties in mass culture and our limited knowledge of fungal species limit the scalability of mycoremediation for extensive environmental cleanups	Overcoming financial and logistical obstacles to enable large-scale, economically feasible, and effective fungal-based bioremediation
6.	Fungi is beneficial because of their strong growth, extensive hyphal network, production of adaptable extracellular enzymes, high floor area to volume ratio, resistance to complex pollutants, ability to adapt to changes in pH and temperature, and presence of metal-binding proteins	One of the things preventing PAH from being used in mycoremediation is its poor bioavailability, which results from its hydrophobic nature and delayed passage through the cell membrane

9.6 Conclusion

In view of the generation of waste waters from various sources and the search for improved techniques for waste water treatment, this chapter has detailed the uses of fungi as one of the upcoming techniques. It is because of the eco-friendly and economic viability that various fungi have been employed for this purpose. More precisely it covers the fungi belonging to the phyla Ascomycetes, Basidiomycetes, Zygomycetes, Yeasts, and Filamentous fungi as powerful bioremediation agents due to their enzymatic diversity, enabling them to degrade persistent organic pollutants, immobilize heavy metals, and mitigate ecological risks.

Thus, two or more case studies for each phylum have been covered illustrating their application in selenium reduction, dye degradation and hydrocarbon degradation. The case studies indicate fungi's adaptability and efficiency in diverse industrial and environmental contexts. The chapter also covers the critical challenges faced by the researchers despite their proliferating use with limitations such as scalability, pollutant specificity, and operational constraints. Advancements in bioreactor technologies, microbial consortia, and the optimization of fungal strains hold the key to addressing these limitations. The insights gathered in this chapter reinforce fungi's status as a sustainable, cost-effective alternative to conventional wastewater treatment methods, urging.

The promising applications of fungi in wastewater treatment highlight the need for continued advancements and strategic implementation to maximize their potential. The key recommendations include a biotechnological approach which is widely adopted due to its cost-effectiveness, enhanced efficiency, and the generation of non-toxic value-added products, making it a sustainable and environmentally favorable solution in various industrial applications with significant advantages over conventional processes (Ezeonu et al., 2012; Abdel-Fatah, 2018; Ezeorba et al., 2024). Scaling up fungal-based systems from laboratory to industrial levels remains a challenge due to environmental and operational constraints, such as maintaining optimal growth conditions (e.g., pH, temperature) and microbial competition. Pilot-scale studies using advanced bioreactor designs, such as fluidized beds or membrane bioreactors (Dalecka et al., 2021), should be prioritized to assess their feasibility in large-scale applications (Pereira et al., 2009).

The application of fungi in wastewater treatment holds immense potential for addressing environmental pollution. By fostering innovations in fungal biotechnology, scaling up operations, and conducting thorough field studies, fungal-based systems can become a cornerstone of sustainable wastewater management solutions.

Acknowledgement S.R. gratefully acknowledges the Research Unit, IGNOU, for the support provided through the Savitribai Phule Fellowship.

Conflict of Interest The authors have no competing interests to declare that are relevant to the content of this article.

References

Abdel-Fatah, M. A. (2018). Nanofiltration systems and applications in wastewater treatment: Review article. *Ain Shams Engineering Journal, 9*(4), 3077–3092. https://doi.org/10.1016/j.asej.2018.08.001

Agosta, S. J., & Klemens, J. A. (2008). Ecological fitting by phenotypically flexible genotypes: Implications for species associations, community assembly and evolution. *Ecology Letters, 11*(11), 1123–1134. https://doi.org/10.1111/j.1461-0248.2008.01237.x

Akerman-Sanchez, G., & Rojas-Jimenez, K. (2021). Fungi for the bioremediation of pharmaceutical-derived pollutants: A bioengineering approach to water treatment. *Environmental Advances, 4*, 100071. https://doi.org/10.1016/j.envadv.2021.100071

Alsop, R. J., Armstrong, C. L., Maqbool, A., Toppozini, L., Dies, H., & Rheinstädter, M. C. (2015). Cholesterol expels ibuprofen from the hydrophobic membrane core and stabilizes lamellar phases in lipid membranes containing ibuprofen. *Soft Matter, 11*(24), 4756–4767. https://doi.org/10.1039/c5sm00597c

Amara, A. A., & El-Baky, N. A. (2023). Fungi as a source of edible proteins and animal feed. *Journal of Fungi, 9*(1), 73. https://doi.org/10.3390/jof9010073

Anjaneyulu, Y., Sreedhara Chary, N., Raj, S. S., & D. (2005). Decolourization of industrial effluents—Available methods and emerging technologies—A review. *Reviews in Environmental Science and Bio/Technology, 4*(4), 245–273. https://doi.org/10.1007/s11157-005-1246-z

Aranda, E., Godoy, P., Reina, R., Badia-Fabregat, M., Rosell, M., Marco-Urrea, E., & García-Romera, I. (2017). Isolation of Ascomycota fungi with capability to transform PAHs: Insights into the biodegradation mechanisms of Penicillium oxalicum. *International Biodeterioration & Biodegradation, 122*, 141–150. https://doi.org/10.1016/j.ibiod.2017.05.015

Assress, H. A., Selvarajan, R., Nyoni, H., Ntushelo, K., Mamba, B. B., & Msagati, T. A. M. (2019). Diversity, co-occurrence and implications of fungal communities in wastewater treatment plants. *Scientific Reports, 9*(1). https://doi.org/10.1038/s41598-019-50624-z

Bahafid, W., Joutey, N. T., Asri, M., Sayel, H., Tirry, N., & Ghachtouli, N. E. (2017). Yeast biomass: An alternative for bioremediation of heavy metals. *Yeast - Industrial Applications*. https://doi.org/10.5772/intechopen.70559

Barakat, M. A. (2011). New trends in removing heavy metals from industrial wastewater. *Arabian Journal of Chemistry, 4*(4), 361–377. https://doi.org/10.1016/j.arabjc.2010.07.019

Bardana, E. J. (2003). Indoor air quality and health. *Immunology and Allergy Clinics of North America, 23*(2), 291–309. https://doi.org/10.1016/s0889-8561(02)00081-4

Battaglia, E., Benoit, I., van den Brink, J., Wiebenga, A., Coutinho, P. M., Henrissat, B., & de Vries, R. P. (2011). Carbohydrate-active enzymes from the zygomycete fungus Rhizopus oryzae: A highly specialized approach to carbohydrate degradation depicted at genome level. *BMC Genomics, 12*(1). https://doi.org/10.1186/1471-2164-12-38

Berefeld, L. D., Judkins, J. F., & Weand, B. L. (1982). *Process chemistry for water and wastewater treatment*. Prentice-Hall.

Calero Preciado, C., Boxall, J., Soria-Carrasco, V., Martínez, S., & Douterelo, I. (2021). Implications of climate change: How does increased water temperature influence biofilm and water quality of chlorinated drinking water distribution systems? *Frontiers in Microbiology, 12*. https://doi.org/10.3389/fmicb.2021.658927

Chavda, V. P., & Shah, D. (2017). Self-emulsifying delivery systems: One step ahead in improving solubility of poorly soluble drugs. *Nanostructures for Cancer Therapy*, 653–718. https://doi.org/10.1016/b978-0-323-46144-3.00025-8

Chavez, C. M., Groenewald, M., Hulfachor, A. B., Kpurubu, G., Huerta, R., Hittinger, C. T., & Rokas, A. (2023). The cell morphological diversity of *Saccharomycotina* yeasts. *FEMS Yeast Research, 24*. https://doi.org/10.1093/femsyr/foad055

Chen, G. (2004). Electrochemical technologies in wastewater treatment. *Separation and Purification Technology, 38*(1), 11–41. https://doi.org/10.1016/j.seppur.2003.10.006

Chuah, T. G., Jumasiah, A., Azni, I., Katayon, S., & Thomas Choong, S. Y. (2005). Rice husk as a potentially low-cost biosorbent for heavy metal and dye removal: An overview. *Desalination, 175*(3), 305–316. https://doi.org/10.1016/j.desal.2004.10.014

Cooney, D. O. (1999). *Adsorption design for wastewater treatment*. Lewis Publishers.

Crini, G., & Lichtfouse, E. (2018). Wastewater treatment: An overview. *Environmental Chemistry for a Sustainable World, 18*, 1–21. https://doi.org/10.1007/978-3-319-92111-2_1

Crini, G. (2009). *Traitement et épuration des eaux industrielles polluées : procédés membranaires, bioadsorption et oxydation chimique*. Presses Universitaires De Franche-Comté.

Dalecka, B., Strods, M., Cacivkins, P., Ziverte, E., Rajarao, G. K., & Juhna, T. (2021). Removal of pharmaceutical compounds from municipal wastewater by bioaugmentation with fungi: An emerging strategy using fluidized bed pelleted bioreactor. *Environmental Advances, 5*, 100086. https://doi.org/10.1016/j.envadv.2021.100086

Delsarte, I., Cohen, G. J. V., Momtbrun, M., Höhener, P., & Atteia, O. (2020). Soil carbon dioxide fluxes to atmosphere: The role of rainfall to control CO2 transport. *Applied Geochemistry, 127*, 104854–104854. https://doi.org/10.1016/j.apgeochem.2020.104854

Deng, Y., & Zhao, R. (2015). Advanced oxidation processes (AOPs) in wastewater Treatment. *Current Pollution Reports, 1*(3), 167–176. https://doi.org/10.1007/s40726-015-0015-z

di Cicco, M. R., Iovinella, M., Palmieri, M., Lubritto, C., & Ciniglia, C. (2021). Extremophilic microalgae Galdieria gen. For urban wastewater treatment: Current state, the case of "power" system, and future prospects. *Plants, 10*(11), 2343. https://doi.org/10.3390/plants10112343

di Cicco, M. R., Masiello, A., Spagnuolo, A., Vetromile, C., Borea, L., Giannella, G., Iovinella, M., & Lubritto, C. (2021). Real-time monitoring and static data analysis to assess energetic and environmental performances in the wastewater sector: A case study. *Energies, 14*(21), 6948. https://doi.org/10.3390/en14216948

Eaux Industrielles Contaminées. (2017). Presses universitaires de Franche-Comté eBooks. https://doi.org/10.4000/books.pufc.10857.

Ezeonu, C. S., Tagbo, R., Anike, E. N., Oje, O. A., & Onwurah, I. N. E. (2012). Biotechnological tools for environmental sustainability: Prospects and challenges for environments in Nigeria—A standard review. *Biotechnology Research International, 2012*, 1–26. https://doi.org/10.1155/2012/450802

Ezeorba, T. P. C., Okeke, E. S., Mayel, M. H., Nwuche, C. O., & Ezike, T. C. (2024). Recent advances in biotechnological valorization of agro-food wastes (AFW): Optimizing integrated approaches for sustainable biorefinery and circular bioeconomy. *Bioresource Technology Reports, 26*, 101823. https://doi.org/10.1016/j.biteb.2024.101823

Favour Oghenero Efenudu. (2024). Bioremediation of PAH'S using Pleurotus ostreatus. *American Journal of Environment Studies, 7*(1), 1–15. https://doi.org/10.47672/ajes.1721

Ferrari, R., Gautier, V., & Silar, P. (2021). Lignin degradation by ascomycetes. *Wood Degradation and Ligninolytic Fungi, 99*, 77–113. https://doi.org/10.1016/bs.abr.2021.05.006

Filote, C., Roşca, M., Hlihor, R. M., Cozma, P., Simion, I. M., Apostol, M., & Gavrilescu, M. (2021). Sustainable application of biosorption and bioaccumulation of persistent pollutants in wastewater treatment: Current practice. *Process, 9*(10), 1696. https://doi.org/10.3390/pr9101696

Fischer, M. S., & Glass, N. L. (2019). Communicate and fuse: How filamentous fungi establish and maintain an interconnected mycelial network. *Frontiers in Microbiology, 10*. https://doi.org/10.3389/fmicb.2019.00619

Fletcher, E., Feizi, A., Bisschops, M. M. M., Hallström, B. M., Khoomrung, S., Siewers, V., & Nielsen, J. (2017). Evolutionary engineering reveals divergent paths when yeast is adapted to different acidic environments. *Metabolic Engineering, 39*, 19–28. https://doi.org/10.1016/j.ymben.2016.10.010

Forgacs, E., Cserháti, T., & Oros, G. (2004). Removal of synthetic dyes from wastewaters: A review. *Environment International, 30*(7), 953–971. https://doi.org/10.1016/j.envint.2004.02.001

Garcia-Rubio, R., de Oliveira, H. C., Rivera, J., & Trevijano-Contador, N. (2020). The fungal cell wall: Candida, Cryptococcus, and aspergillus species. *Frontiers in Microbiology, 10*, 2993. https://doi.org/10.3389/fmicb.2019.02993

Guiraud, P., Villemain, D., Kadri, M., Bordjiba, O., & Steiman, R. (2003). Biodegradation capability of Absidia fusca Linnemann towards environmental pollutants. *Chemosphere, 52*(4), 663–671. https://doi.org/10.1016/s0045-6535(03)00229-7

Hai, F. I., Yamamoto, K., & Fukushi, K. (2007). Hybrid treatment systems for dye wastewater. *Critical Reviews in Environmental Science and Technology, 37*, 315–377. https://doi.org/10.1080/10643380601174723

Hanna, W., Abd, W., Zahuri, A. A., Ibrahim, M. F., Show, P.-L., Ilham, Z., Jamaludin, A. A., Patah, A., Usuldin, A., & Rowan, N. J. (2022). Role of ascomycete and basidiomycete fungi in meeting established and emerging sustainability opportunities: A review. *Bioengineered, 13*(7–12), 14903–14935. https://doi.org/10.1080/21655979.2023.2184785

Hartner, F. S., & Glieder, A. (2006). Regulation of methanol utilisation pathway genes in yeasts. *Microbial Cell Factories, 5*(1), 39. https://doi.org/10.1186/1475-2859-5-39

Heisler, J., Glibert, P. M., Burkholder, J. M., Anderson, D. M., Cochlan, W., Dennison, W. C., Dortch, Q., Gobler, C. J., Heil, C. A., Humphries, E., Lewitus, A., Magnien, R., Marshall, H. G., Sellner, K., Stockwell, D. A., Stoecker, D. K., & Suddleson, M. (2008). Eutrophication and harmful algal blooms: A scientific consensus. *Harmful Algae, 8*(1), 3–13. https://doi.org/10.1016/j.hal.2008.08.006

Henze, M. (Ed.). (2001). *Wastewater treatment—Biological and chemical processes*. Springer.
Hohmann, S. (2002). Osmotic adaptation in yeast–control of the yeast osmolyte system. *International Review of Cytology, 215*, 149–187. https://doi.org/10.1016/s0074-7696(02)15008-x
Huang, C.-J., Lu, M.-Y., Chang, Y.-W., & Li, W.-H. (2018). Experimental evolution of yeast for high-temperature tolerance. *Molecular Biology and Evolution, 35*(8). https://doi.org/10.1093/molbev/msy077
Hussain, A., Rehman, F., Rafeeq, H., Waqas, M., Asghar, A., Afsheen, N., Rahdar, A., Bilal, M., & Iqbal, H. M. N. (2022). In-situ, ex-situ, and nano-remediation strategies to treat polluted soil, water, and air—A review. *Chemosphere, 289*, 133252. https://doi.org/10.1016/j.chemosphere.2021.133252
Iovinella, M., Carbone, D. A., Cioppa, D., Davis, S. J., Innangi, M., Esposito, S., & Ciniglia, C. (2020). Prevalent pH controls the capacity of Galdieria maxima to use ammonia and nitrate as a nitrogen source. *Plants, 9*(2), 232. https://doi.org/10.3390/plants9020232
Iovinella, M., Lombardo, F., Ciniglia, C., Palmieri, M., Rosa, M., Trifuoggi, M., Race, M., Manfredi, C., Lubritto, C., Fabbricino, M., Stefano, M. D., & Davis, S. J. (2022). Bioremoval of yttrium (III), cerium (III), europium (III), and terbium (III) from single and quaternary aqueous solutions using the extremophile Galdieria sulphuraria (Galdieriaceae, Rhodophyta). *Plants, 11*(10), 1376–1376. https://doi.org/10.3390/plants11101376
Isaza-Pérez, F., Ramírez-Carmona, M., Rendón-Castrillón, L., & Ocampo-López, C. (2020). Potential of residual fungal biomass: A review. *Environmental Science and Pollution Research International, 27*(12), 13019–13031. https://doi.org/10.1007/s11356-020-08193-6
Jarboui, R., Baati, H., Fetoui, F., Gargouri, A., Gharsallah, N., & Ammar, E. (2012). Yeast performance in wastewater treatment: Case study of*Rhodotorula mucilaginosa. Environmental Technology, 33*(8), 951–960. https://doi.org/10.1080/09593330.2011.603753
Kadri, T., Rouissi, T., Kaur Brar, S., Cledon, M., Sarma, S., & Verma, M. (2017). Biodegradation of polycyclic aromatic hydrocarbons (PAHs) by fungal enzymes: A review. *Journal of Environmental Sciences, 51*, 52–74. https://doi.org/10.1016/j.jes.2016.08.023
Kane, P. M. (2016). Proton transport and pH control in fungi. *Advances in Experimental Medicine and Biology, 892*, 33–68. https://doi.org/10.1007/978-3-319-25304-6_3
Khalaf, M. N. (2021). *Green polymers and environmental pollution control*. Apple Academic Press.
Khandaker, S., Das, S., Hossain, M. T., Islam, A., Miah, M. R., & Awual, M. R. (2021). Sustainable approach for wastewater treatment using microbial fuel cells and green energy generation—A comprehensive review. *Journal of Molecular Liquids, 344*, 117795. https://doi.org/10.1016/j.molliq.2021.117795
Khiyami, M. A., Pometto, A. L., & Kennedy, W. J. (2006). Ligninolytic enzyme production by *Phanerochaete chrysosporium* in plastic composite support biofilm stirred tank bioreactors. *Journal of Agricultural and Food Chemistry, 54*(5), 1693–1698. https://doi.org/10.1021/jf051424l
Kow, S.-H., Fahmi, M. R., Azner Abidin, C. Z., Ong, S.-A., Ibrahim, A. H., Sabri, S. N., & Razali, N. A. (2018). Oxidation of p-cresol by ozonation. *Sains Malaysiana, 47*(6), 1085–1091. https://doi.org/10.17576/jsm-2018-4706-02
Kow, S. H., Ridwan, F., & Ibrahim, A. H. (2019). Phenol removal from wastewaters by electrochemical biosensor. *American International Journal of Biology and Life Sciences*, 24–32. https://doi.org/10.46545/aijbls.v1i2.53
Lacorte, S. (2003). Organic compounds in paper-mill process waters and effluents. *TrAC Trends in Analytical Chemistry, 22*(10), 725–737. https://doi.org/10.1016/s0165-9936(03)01009-4
Ledesma-Amaro, R. (2022). Yeasts inhabiting extreme environments and their biotechnological applications. *Microorganisms, 10*(4). https://doi.org/10.3390/microorganisms10040794
Liu, D. (2024). Classification of medically important fungi. In *Molecular medical microbiology* (pp. 2763–2777). Elsevier. https://doi.org/10.1016/b978-0-12-818619-0.00034-4
Lokhande, R. S., Singare, P. U., & Pimple, D. S. (2011). Toxicity study of heavy metals pollutants in waste water effluent samples collected from Taloja industrial Estate of Mumbai, India. *Resources and Environment, 1*(1), 13–19.

Lu, T., Zhang, Q.-L., & Yao, S.-J. (2016). Application of biosorption and biodegradation functions of fungi in wastewater and sludge Treatment. In *Fungal applications in sustainable environmental biotechnology* (pp. 65–90). Springer. https://doi.org/10.1007/978-3-319-42852-9_4

Mahamuni-Badiger, P. P., Patil, P. M., Badiger, M. V., Patel, P. R., Thorat-Gadgil, B. S., Pandit, A., & Bohara, R. A. (2020). Biofilm formation to inhibition: Role of zinc oxide-based nanoparticles. *Materials Science and Engineering: C, 108*, 110319. https://doi.org/10.1016/j.msec.2019.110319

Mahiout, A., Damann, R., Pera, J., A. Luonsi, & Aslan, F. (2006). *Industrial liquid effluents in the pulp and paper industry.*

Maicas, S. (2020). The role of yeasts in fermentation processes. *Microorganisms, 8*(8), 1142. https://doi.org/10.3390/microorganisms8081142

Malik, S., Bora, J., Nag, S., Sinha, S., Mondal, S., Rustagi, S., Hazra, R., Kumar, H., Rajput, V. D., Minkina, T., Sadier, N. S., & Almutary, A. G. (2023). Fungal-based remediation in the treatment of anthropogenic activities and pharmaceutical-pollutant-contaminated wastewater. *Water, 15*(12), 2262. https://doi.org/10.3390/w15122262

Manfredi, C., Amoruso, A. J., Ciniglia, C., Iovinella, M., Palmieri, M., Lubritto, C., El Hassanin, A., Davis, S. J., & Trifuoggi, M. (2023). Selective biosorption of lanthanides onto Galdieria sulphuraria. *Chemosphere, 317*, 137818. https://doi.org/10.1016/j.chemosphere.2023.137818

Mi, R., Wang, X., Dong, Y., Li, S., Zhao, Z., Guan, X., Jiang, J., Gao, S., Fu, Z., & Zhou, Z. (2024). Sustainable treatment of aquaculture water employing fungi-microalgae consortium: Nutrients removal enhancement, bacterial communities optimization, emerging contaminants elimination, and mechanism analysis. *The Science of the Total Environment, 929*, 172600–172600. https://doi.org/10.1016/j.scitotenv.2024.172600

Mir-Tutusaus, J. A., & Sarrà, M. (2020). Fungal reactors: A solution for the removal of pharmaceuticals in urban and hospital wastewater. In *Removal and degradation of pharmaceutically active compounds in wastewater treatment* (pp. 145–162). Springer. https://doi.org/10.1007/698_2020_660

Mohan, D., & Pittman, C. U. (2007). Arsenic removal from water/wastewater using adsorbents—A critical review. *Journal of Hazardous Materials, 142*(1-2), 1–53. https://doi.org/10.1016/j.jhazmat.2007.01.006

Mousavi, S. M., Hashemi, S. A., Iman Moezzi, S. M., Ravan, N., Gholami, A., Lai, C. W., Chiang, W.-H., Omidifar, N., Yousefi, K., & Behbudi, G. (2021). Recent advances in enzymes for the bioremediation of pollutants. *Biochemistry Research International, 2021*, 1–12. https://doi.org/10.1155/2021/5599204

Nagahama, T. (n.d.). Yeast biodiversity in freshwater, marine and deep-sea environments. In *Biodiversity and ecophysiology of yeasts* (pp. 241–262). Springer. https://doi.org/10.1007/3-540-30985-3_12

Navina, B. K., Velmurugan, N. K., Senthil Kumar, P., Rangasamy, G., Palanivelu, J., Thamarai, P., Vickram, A. S., Saravanan, A., & Shakoor, A. (2024). Fungal bioremediation approaches for the removal of toxic pollutants: Mechanistic understanding for biorefinery applications. *Chemosphere, 350*, 141123. https://doi.org/10.1016/j.chemosphere.2024.141123

Negi, B. B., & Das, C. (2023). Mycoremediation of wastewater, challenges, and current status: A review. *Bioresource Technology Reports, 22*, 101409. https://doi.org/10.1016/j.biteb.2023.101409

Nicula, N.-O., Lungulescu, E.-M., Rîmbu, G. A., Marinescu, V., Corbu, V. M., & Csutak, O. (2023). Bioremediation of wastewater using yeast strains: An assessment of contaminant removal efficiency. *International Journal of Environmental Research and Public Health, 20*(6), 4795. https://doi.org/10.3390/ijerph20064795

Odoh, C. K., Eze, C. N., Obi, C. J., Anyah, F., Egbe, K., Unah, U., Akpi, U. K., & Adobu, U. S. (2020). Fungal biofertilizers for sustainable agricultural productivity. In *Agriculturally important fungi for sustainable agriculture* (pp. 199–225). Springer. https://doi.org/10.1007/978-3-030-45971-0_9

Okal, E. J., Heng, G., Magige, E. A., Khan, S., Wu, S., Ge, Z., Zhang, T., Mortimer, P. E., & Xu, J. (2023). Insights into the mechanisms involved in the fungal degradation of plastics. *Ecotoxicology and Environmental Safety, 262*, 115202. https://doi.org/10.1016/j.ecoenv.2023.115202

Palmieri, M., Iovinella, M., Davis, S. J., di Cicco, M. R., Lubritto, C., Race, M., Papa, S., Fabbricino, M., & Ciniglia, C. (2022). Galdieria sulphuraria ACUF427 freeze-dried biomass as novel biosorbent for rare earth elements. *Microorganisms, 10*(11), 2138. https://doi.org/10.3390/microorganisms10112138

Parsons, S. (2015). *Advanced oxidation processes for water and wastewater treatment. Water intelligence online*. IWA Publishing. https://doi.org/10.2166/9781780403076

Patel, A. B., Shaikh, S., Jain, K. R., Desai, C., & Madamwar, D. (2020). Polycyclic aromatic hydrocarbons: Sources, toxicity, and remediation approaches. *Frontiers in Microbiology, 11*(11). https://doi.org/10.3389/fmicb.2020.562813

Payne, C. M., Knott, B. C., Mayes, H. B., Hansson, H., Himmel, M. E., Sandgren, M., Ståhlberg, J., & Beckham, G. T. (2015). Fungal cellulases. *Chemical Reviews, 115*(3), 1308–1448. https://doi.org/10.1021/cr500351c

Pereira, R., Antunes, S. C., Gonçalves, A. M. M., Marques, S. M., Gonçalves, F., Ferreira, F., Freitas, A. C., Rocha-Santos, T. A. P., Diniz, M. S., Castro, L., Peres, I., & Duarte, A. C. (2009). The effectiveness of a biological treatment with Rhizopus oryzae and of a photo-Fenton oxidation in the mitigation of toxicity of a bleached Kraft pulp mill effluent. *Water Research, 43*(9), 2471–2480. https://doi.org/10.1016/j.watres.2009.03.013

Phale, S. (2018). Yeast: Characteristics and economic significance. *Journal of Bioprocessing & Biotechniques, 08*(05). https://doi.org/10.4172/2155-9821.1000337

Pokhrel, D., & Viraraghavan, T. (2004). Treatment of pulp and paper mill wastewater—A review. *Science of the Total Environment, 333*(1-3), 37–58. https://doi.org/10.1016/j.scitotenv.2004.05.017

Ramya, D., Michellin, J., & Joseph Thatheyus, A. (2021). Biosorption of heavy metals using fungal biosorbents—A review. In *Fungi bio-prospects in sustainable agriculture, environment and nano-technology* (pp. 331–352). Elsevier. https://doi.org/10.1016/b978-0-12-821925-6.00015-0

Rana, S., Kumar, Y., Kaushik, R., & Kumar, L. S. (2024). Recent developments in metal-organic frameworks for water purification: A Mini Review. *Chemistry Select, 9*(35). https://doi.org/10.1002/slct.202403309

Rana, T., Gupta, S., Kumar, D., Sharma, S., Rana, M., Rathore, V. S., & Pereira, B. M. J. (2004). Toxic effects of pulp and paper-mill effluents on male reproductive organs and some systemic parameters in rats. *Environmental Toxicology and Pharmacology, 18*(1), 1–7. https://doi.org/10.1016/j.etap.2004.04.005

Rathoure, A. K., & Dhatwalia, V. K. (2015). Toxicity and waste management using bioremediation. In *Advances in environmental engineering and green technologies book series*. IGI Global. https://doi.org/10.4018/978-1-4666-9734-8

Ribes, J. A., Vanover-Sams, C. L., & Baker, D. J. (2000). Zygomycetes in human disease. *Clinical Microbiology Reviews, 13*(2), 236–301. https://doi.org/10.1128/CMR.13.2.236

Rosa, M., Spagnuolo, A., Masiello, A., Vetromile, C., Lubritto, C., Nappa, M., & Corbo, G. (2019). Energy monitoring of a wastewater Treatment Plant in Salerno, Campania region (southern Italy). In *Frontiers in water-energy-nexus—nature-based solutions, advanced technologies and best practices for environmental sustainability* (pp. 107–111). Elsevier. https://doi.org/10.1007/978-3-030-13068-8_26

Rousk, J., Brookes, P. C., & Baath, E. (2009). Contrasting soil pH effects on fungal and bacterial growth suggest functional redundancy in carbon mineralization. *Applied and Environmental Microbiology, 75*(6), 1589–1596. https://doi.org/10.1128/aem.02775-08

Sabuda, M. C., Rosenfeld, C. E., DeJournett, T. D., Schroeder, K., Wuolo-Journey, K., & Santelli, C. M. (2020). Fungal bioremediation of selenium-contaminated industrial and municipal wastewaters. *Frontiers in Microbiology, 11*. https://doi.org/10.3389/fmicb.2020.02105

Sathinathan, P., Parab, H. M., Yusoff, R., Ibrahim, S., Vello, V., & Ngoh, G. C. (2023). Photobioreactor design and parameters essential for algal cultivation using industrial wastewater: A review. *Renewable and Sustainable Energy Reviews, 173*, 113096. https://doi.org/10.1016/j.rser.2022.113096

Sayara, T., & Sánchez, A. (2020). Bioremediation of PAH-contaminated soils: Process enhancement through composting/compost. *Applied Sciences, 10*(11), 3684. https://doi.org/10.3390/app10113684

Sharma, M., Agarwal, S., Malik, R., Kumar, G., Pal, D. B., Mandal, M., Sarkar, A., Bantun, F., Haque, S., Singh, P., Srivastava, N., & Gupta, V. K. (2023). Recent advances in microbial engineering approaches for wastewater treatment: A review. *Bioengineered, 14*(1). https://doi.org/10.1080/21655979.2023.2184518

Sharma, S. K., & Sanghi, R. (2012). *Advances in water Treatment and pollution prevention*. Springer. https://doi.org/10.1007/978-94-007-4204-8

Shivalkar, S., Singh, V., Sahoo, A. K., Samanta, S. K., & Gautam, P. K. (2020). *Bioremediation: A potential ecological tool for waste management* (pp. 1–21). https://doi.org/10.1016/B978-0-12-820318-7.00001-0

Shoun, H., Fushinobu, S., Jiang, L., Kim, S.-W., & Wakagi, T. (2012). Fungal denitrification and nitric oxide reductase cytochrome P450nor. *Philosophical Transactions of the Royal Society, B: Biological Sciences, 367*(1593), 1186–1194. https://doi.org/10.1098/rstb.2011.0335

Singh, V. B., Negi, R., Jacob, M., Gayathri, A., Rokade, A., Sarma, H., Kalita, J., Tasfia, S. T., Bharti, R. K., Wakid, A., Suthar, S., Kolipakam, V., & Qureshi, Q. (2023). Polycyclic aromatic hydrocarbons (PAHs) in aquatic ecosystem exposed to the 2020 Baghjan oil spill in upper Assam, India: Short-term toxicity and ecological risk assessment. *PLoS One, 18*(11), e0293601. https://doi.org/10.1371/journal.pone.0293601

Sirakov, M., Palmieri, M., Iovinella, M., Davis, S. J., Petriccione, M., di Cicco, M. R., De Stefano, M., & Ciniglia, C. (2021). Cyanidiophyceae (Rhodophyta) tolerance to precious metals: Metabolic response to palladium and gold. *Plants, 10*(11), 2367. https://doi.org/10.3390/plants10112367

Sonune, A., & Ghate, R. (2004). Developments in wastewater treatment methods. *Desalination, 167*, 55–63. https://doi.org/10.1016/j.desal.2004.06.113

Staroń, P., & Chwastowski, J. (2024). Yeast-based magnetic biocomposite for efficient sorption of organic pollutants. *Applied Sciences, 14*(2), 655–655. https://doi.org/10.3390/app14020655

Subasri, M., Gomathi, V., & Mary, J. K. (2020). Fungi: An approach to treat wastewater. *International Journal of Chemical Studies, 8*(6), 163–166. https://doi.org/10.22271/chemi.2020.v8.i6c.11067

Volk, T. J. (2013). Fungi. In *Encyclopedia of biodiversity* (pp. 624–640). Elsevier. https://doi.org/10.1016/b978-0-12-384719-5.00062-9

Walker, K., Skelton, H., & Smith, K. (2002). Cutaneous lesions showing giant yeast forms of Blastomyces dermatitidis. *Journal of Cutaneous Pathology, 29*(10), 616–618. https://doi.org/10.1034/j.1600-0560.2002.291009.x

Wang, J. L., & Xu, L. J. (2012). Advanced oxidation processes for wastewater treatment: Formation of hydroxyl radical and application. *Critical Reviews in Environmental Science and Technology, 42*(3), 251–325. https://doi.org/10.1080/10643389.2010.507698

Wang, Q., & Yang, Z. (2016). Industrial water pollution, water environment treatment, and health risks in China. *Environmental Pollution, 218*, 358–365. https://doi.org/10.1016/j.envpol.2016.07.011

Wastewater Treatment. (2020). *CRC Press eBooks*. Inform. https://doi.org/10.1201/9781003075752

Wesenberg, D. (2003). White-rot fungi and their enzymes for the treatment of industrial dye effluents. *Biotechnology Advances, 22*(1–2), 161–187. https://doi.org/10.1016/j.biotechadv.2003.08.011

Wojnárovits, L., & Takács, E. (2008). Irradiation treatment of azo dye containing wastewater: An overview. *Radiation Physics and Chemistry, 77*(3), 225–244. https://doi.org/10.1016/j.radphyschem.2007.05.003

Yin, Y., & Allen, H. (1999). Technology evaluation report in situ chemical treatment. https://www.clu-in.org/download/toolkit/inchem.pdf

Yusuf, F., Muhammad, J. B., Usman, S., Jagaba, A. H., Yusuf, M. R., & Yakasai, H. M. (2024). Biodegradation of dimethyl yellow by a locally isolated fungus from dye contaminated wastewater. *Case Studies in Chemical and Environmental Engineering, 9*, 100650–100650. https://doi.org/10.1016/j.cscee.2024.100650

Zakhartsev, M., Yang, X., Reuss, M., & Pörtner, H. O. (2015). Metabolic efficiency in yeast Saccharomyces cerevisiae in relation to temperature dependent growth and biomass yield. *Journal of Thermal Biology, 52*, 117–129. https://doi.org/10.1016/j.jtherbio.2015.05.008

Zayed, O., Hewedy, O. A., Abdelmoteleb, A., Ali, M., Youssef, M. S., Roumia, A. F., Seymour, D., & Yuan, Z. (2023). Nitrogen journey in plants: From uptake to metabolism, stress response, and microbe interaction. *Biomolecules, 13*(10), 1443–1443. https://doi.org/10.3390/biom13101443

Zheng, X., Lin, H., Du, D., Li, G., Alam, O., Cheng, Z., Liu, X., Jiang, S., & Li, J. (2024). Remediation of heavy metals polluted soil environment: A critical review on biological approaches. *Ecotoxicology and Environmental Safety, 284*, 116883–116883. https://doi.org/10.1016/j.ecoenv.2024.116883

Chapter 10
Integration of Fungal and Algal Treatment Systems for Fungi in Waste Water Treatment

Shreya, Bhawna, Peeyush Phogat, Ranjana Jha, and Sukhvir Singh

Abstract The chapter explores the integration of fungal and algal treatment systems for wastewater management, addressing the challenges associated with fungal contaminants. Conventional wastewater treatments often struggle with fungal contaminants due to their resilience and ability to form biofilms. Fungal contaminants pose health risks and hinder effective treatment, making their removal a critical concern. The chapter proposes combining fungal and algal systems as a solution, leveraging the strengths of both. Fungi, known for breaking down complex organic pollutants, can degrade materials that traditional methods cannot handle, while algae excel in nutrient removal and oxygen production. This integration allows for enhanced pollutant degradation and nutrient management, offering a dual approach to improving wastewater treatment. The chapter also discusses various integration strategies, such as co-cultivation and sequential processes, emphasizing the importance of optimizing conditions like pH, temperature, and nutrient availability to enhance the system's performance. Case studies illustrate successful applications of integrated systems in diverse wastewater contexts, demonstrating improvements in treatment efficiency, reduced sludge production, and economic benefits through resource recovery. The chapter concludes that integrated fungal and algal treatment systems represent a promising approach to wastewater management, offering improved efficiency and sustainability. It highlights the need for continued research and innovation to address challenges related to scaling and system optimization,

Shreya (✉) · P. Phogat
Research Lab for Energy Systems, Department of Physics, Netaji Subhas University of Technology, New Delhi, India

Council of Scientific and Industrial Research-National Institute of Science Communication and Policy Research, New Delhi, India

Bhawna · R. Jha
Research Lab for Energy Systems, Department of Physics, Netaji Subhas University of Technology, New Delhi, India

S. Singh
Department of Nanotechnology, Delhi Skill Entrepreneurship University, New Delhi, India

C. M. Hussain et al. (eds.), *Fungi in Waste Water Treatment*, Springer Water,
https://doi.org/10.1007/978-3-032-12693-1_10

aiming to advance the adoption of these technologies for better environmental outcomes.

Keywords Fungal treatment · Algal treatment · Wastewater management · Co-cultivation · Sequential process

10.1 Introduction

Wastewater treatment is a critical component of modern environmental management and public health (Angelakis et al., 2022). As industrialization, urbanization, and population growth continue to escalate, the volume and complexity of wastewater generated have increased substantially (Sheriff & Kachalla, n.d.). Wastewater encompasses a wide range of contaminants, including organic matter, nutrients, heavy metals, and pathogens, which pose significant risks to aquatic ecosystems and human health if not properly managed. The primary goal of wastewater treatment is to remove or neutralize these contaminants to ensure the safety of discharged effluents and to reclaim valuable resources.

Traditional wastewater treatment methods typically involve physical, chemical, and biological processes to address various pollutants (Ahmed et al., n.d.). Physical treatments, such as sedimentation and filtration, remove particulate matter and large debris. Chemical treatments involve the use of chemicals to neutralize or precipitate contaminants. Biological treatments harness the metabolic activities of microorganisms to degrade organic matter and nutrients. Despite their effectiveness, these conventional methods often face challenges in dealing with complex and persistent pollutants, particularly fungal contaminants (Ahmed et al., n.d.).

Fungal contaminants in wastewater are of particular concern due to their resilience and potential health impacts. Fungi, including yeasts, molds, and filamentous fungi, can proliferate in wastewater environments and contribute to the formation of biofilms, which complicate treatment processes and lead to operational issues (Sankaran et al., 2010). These microorganisms are known for their ability to produce a wide array of enzymes that can degrade various organic substances. However, their persistence and adaptability often make them difficult to remove using traditional treatment methods.

The challenges associated with fungal contaminants include their resistance to chemical treatments and their tendency to form complex, multi-layered biofilms that are not easily disrupted by physical or chemical means. Additionally, fungal spores and metabolites can pose health risks, including allergic reactions and infections, particularly in sensitive populations (Egbuta & Mwanza, 1995). As a result, there is a growing need for innovative treatment approaches that can effectively manage and remove fungal contaminants from wastewater.

This chapter aims to explore the integration of fungal and algal treatment systems as a novel approach to addressing the challenges posed by fungal contaminants in wastewater. By combining the unique capabilities of fungi and algae, this integrated approach seeks to enhance the efficiency and effectiveness of wastewater treatment processes.

The chapter is structured to provide a comprehensive overview of the current state of wastewater treatment, with a particular focus on the limitations of existing methods in managing fungal contaminants. It begins with a review of conventional treatment methods, including their advantages and drawbacks, followed by a detailed examination of fungal and algal treatment systems. The chapter then delves into the concept of integration, discussing various strategies for combining fungal and algal technologies and their potential benefits.

Key factors influencing the performance of integrated systems, such as environmental conditions, nutrient management, and microbial interactions, are also explored. Recent advances and case studies are presented to illustrate the practical applications and successes of these integrated systems. Finally, the chapter discusses the economic and environmental benefits of adopting integrated fungal-algal systems and outlines future research directions to further advance this field.

By providing a thorough analysis of these topics, the chapter aims to contribute valuable insights for researchers, practitioners, and policymakers interested in developing more effective and sustainable wastewater treatment technologies. The integration of fungal and algal systems represents a promising frontier in wastewater management, offering the potential for enhanced performance and greater environmental benefits.

10.2 Overview of Conventional Wastewater Treatment Methods

Conventional wastewater treatment methods play a crucial role in managing and mitigating the environmental impact of wastewater generated by various industrial, agricultural, and domestic activities. These methods are designed to reduce pollutants, organic matter, and hazardous substances to levels that are safe for discharge into natural water bodies or for reuse in non-potable applications. The treatment process is generally divided into different stages, each targeting specific types of contaminants, thereby ensuring a comprehensive approach to wastewater management. These stages include physical, chemical, and biological processes, each with its own mechanisms and purposes in treating wastewater. Wastewater treatment methods have evolved significantly over the years, driven by a need to address the growing concerns over water pollution and the scarcity of clean water resources. The development of these conventional processes has been guided by regulatory standards that ensure treated water meets the quality requirements for environmental and public health safety. Conventional methods often serve as a baseline for many modern and hybrid treatment approaches, which aim to enhance the efficiency and effectiveness of the treatment process. The integration of different stages allows for a more targeted removal of various pollutants, adapting the treatment strategy to the specific characteristics of the wastewater being treated.

These conventional methods focus primarily on the removal of suspended solids, organic and inorganic pollutants, and pathogenic microorganisms. As wastewater characteristics vary greatly depending on their source, the selection and combination of treatment methods must be carefully tailored to achieve optimal results. Each method, whether physical, chemical, or biological, is selected based on its capacity to handle specific types of contaminants and its compatibility with other stages in the treatment process. This integrated approach helps in achieving higher treatment efficiencies and compliance with environmental discharge standards, while also ensuring the conservation and reuse of water resources. Despite their widespread use, conventional wastewater treatment processes face challenges such as energy consumption, the production of residual sludge, and the need for specialized infrastructure. These challenges have spurred the search for more sustainable and efficient alternatives, including the integration of biological treatment systems like fungal and algal processes. Such innovations aim to complement traditional methods, offering a promising path towards more effective and environmentally friendly wastewater management solutions. Conventional methods are divided into three categories, which include physical, chemical and biological treatment methods.

Physical treatment methods are designed to remove solid materials from wastewater through mechanical means. These processes are often the first step in wastewater treatment and aim to reduce the load on subsequent treatment stages. The main physical treatment processes include screening, sedimentation and filtration as shown in Fig. 10.1. Screening involves the use of screens or mesh to remove large debris, such as sticks, leaves, and plastic, from the incoming wastewater. This process prevents damage to equipment and clogging of pipes. Screens can vary in size and type, including coarse screens for larger objects and fine screens for smaller particles. Sedimentation, or settling, is used to remove suspended solids by allowing

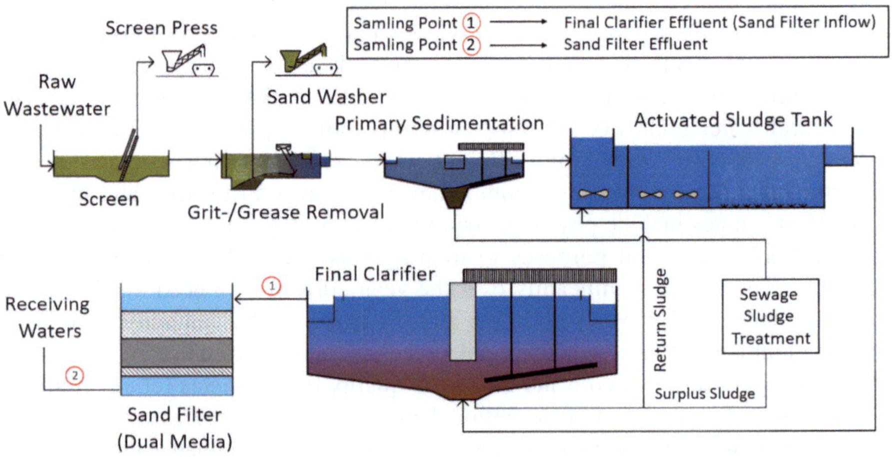

Fig. 10.1 Flow diagram of a municipal wastewater treatment plant with a downstream sand filter, indicating the sampling locations: point 1 before the sand filter and point 2 after the sand filter. Reproduced from Wolff et al. (2020). Licensee MDPI, Basel, Switzerland

them to settle out of the wastewater under the force of gravity (Engineering, n.d.). This is typically achieved in sedimentation tanks or clarifiers, where the flow rate is reduced to allow particles to settle at the bottom as sludge (Engineering, n.d.). Sedimentation is effective for removing larger particulate matter but less effective for fine particles or dissolved substances. Filtration is employed to further remove smaller suspended particles that remain after sedimentation. Various types of filters, including sand filters, membrane filters, and activated carbon filters, are used to trap particles and contaminants as water passes through them (Cescon, n.d.). Filtration can significantly improve the clarity and quality of the treated water.

Chemical treatment processes involve the addition of chemicals to wastewater to remove or neutralize contaminants through chemical reactions. These processes are often used in conjunction with physical treatment methods to address specific pollutants. The key chemical treatments as shown in Fig. 10.2 include coagulation and flocculation, disinfection and chemical oxidation/reduction. Coagulation involves

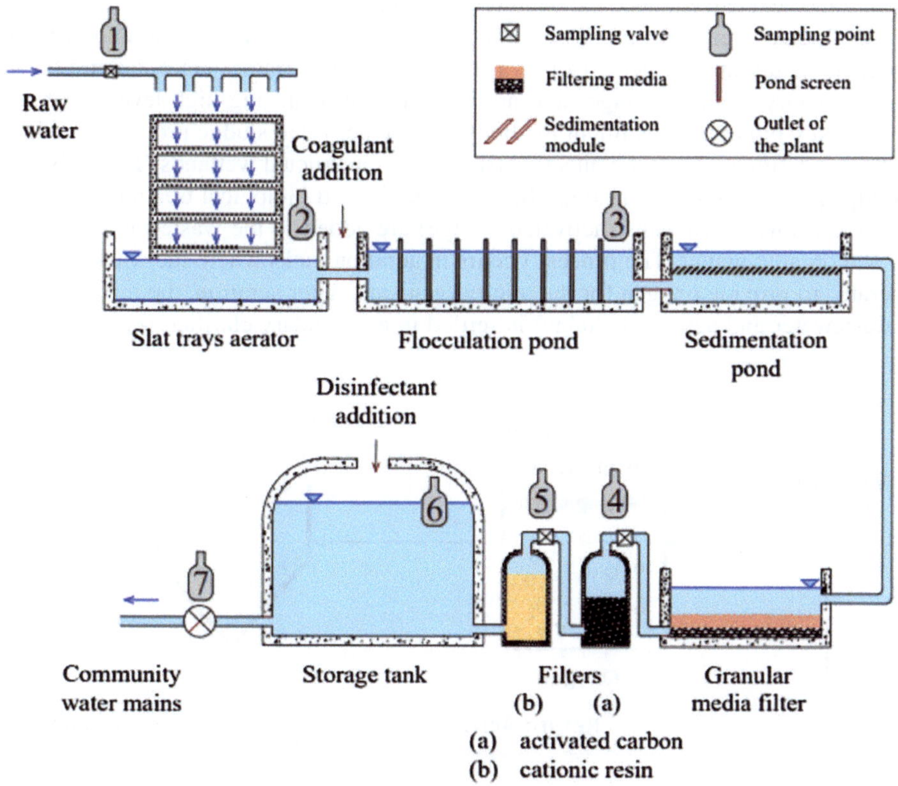

Fig. 10.2 Illustrative images depicting key stages of water treatment: coagulation and flocculation, where particles are clumped together for easier removal, and disinfection, the final process to eliminate harmful pathogens and ensure water safety. Reproduced from Mendez-Ruiz et al. (2023). Licensee MDPI, Basel, Switzerland

the addition of chemicals called coagulants (e.g., alum, ferric chloride) to destabilize colloidal particles and form flocs. Flocculation follows coagulation and involves gentle mixing to aggregate the destabilized particles into larger flocs, which can then be removed through sedimentation or filtration (Tzoupanos, 2008). This process is effective for removing suspended solids, organic matter, and some contaminants. Disinfection aims to eliminate pathogenic microorganisms from wastewater to ensure that it is safe for discharge or reuse. Common disinfection methods include chlorination, which uses chlorine or chlorine compounds to kill bacteria and viruses, and ultraviolet (UV) disinfection, which uses UV light to inactivate microorganisms (Lanrewaju et al., 2022). Disinfection is crucial for protecting public health and preventing the spread of waterborne diseases. Chemical oxidation or reduction processes involve the use of oxidizing or reducing agents to chemically transform or degrade pollutants. For example, ozonation uses ozone to oxidize organic compounds and break down complex molecules, while hydrogen peroxide can be used to degrade contaminants through advanced oxidation processes (AOPs) (Muruganandham et al., 2014). These methods are effective for treating specific pollutants and can enhance the overall treatment efficiency.

Biological treatment processes utilize microorganisms to degrade and remove organic matter and nutrients from wastewater. These processes are often employed in secondary treatment stages and are essential for achieving high levels of treatment. Key biological treatment methods include activated sludge process, trickling filters, rotating biological contactors (rbcs) and constructed wetlands. The activated sludge process as shown in Fig. 10.3 is a widely used biological treatment method in which microorganisms (activated sludge) are added to the wastewater to break down organic matter. The process occurs in aeration tanks where the wastewater is aerated to provide oxygen for the microorganisms. After aeration, the mixed liquor (wastewater and activated sludge) is settled in a secondary clarifier, separating the

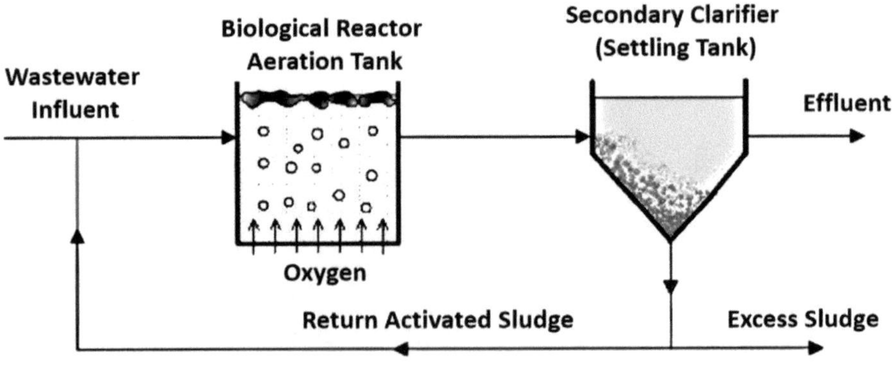

Fig. 10.3 Illustration of the activated sludge process, where wastewater is aerated to promote microbial growth that breaks down organic pollutants, followed by settling to separate treated water from the activated sludge. Reproduced from Conferences (n.d.) published by EDP Sciences, 2016

treated water from the excess sludge. This method is effective for reducing biochemical oxygen demand (BOD) and removing organic contaminants (Rjwave.Org, 2016). Trickling filters consist of a rotating bed of media (e.g., plastic or rock) on which microorganisms grow. As wastewater is distributed over the media, microorganisms attached to the media degrade organic matter through biological processes. The treated water then flows through a clarifier to remove any remaining solids (Deng, 2018). Trickling filters are used for secondary treatment and can handle varying flow rates and contaminant loads. RBCs as shown in Fig. 10.4 are similar to trickling filters but use a series of rotating disks or drums covered with a biofilm of microorganisms. The disks rotate through the wastewater, allowing microorganisms to degrade organic matter. RBCs are effective for treating wastewater with varying organic loads and can be used as a secondary treatment or polishing step (Hochheimer, n.d.).

Constructed wetlands are engineered systems that mimic natural wetlands to treat wastewater through physical, chemical, and biological processes. These systems use a combination of wetland plants, microorganisms, and substrate to remove pollutants from wastewater. Constructed wetlands are particularly effective for treating nutrients, organic matter, and suspended solids and offer the added benefits of habitat creation and aesthetic value (Sustainability, 2019).

Despite the effectiveness of conventional wastewater treatment methods, they often face challenges in managing fungal contaminants, including yeasts, molds, and filamentous fungi, due to their resilience and ability to form complex structures (Novak Babič et al., 2017). The limitations of conventional methods in addressing fungal contaminants include resistance to chemical treatments, as fungi are often resistant to chlorination and oxidation, which may not fully inactivate fungal spores

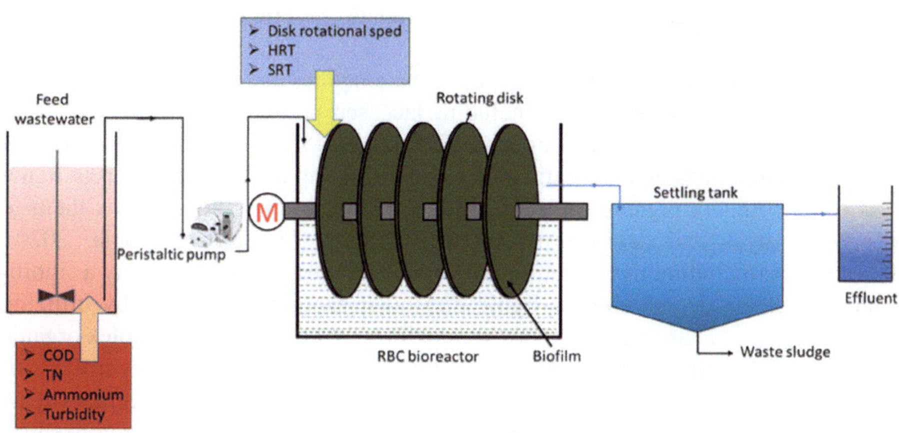

Fig. 10.4 Schematic diagram of a rotating biological contactor (RBC) bioreactor, illustrating the rotating disk system used for biological treatment, where microorganisms grow on the surfaces of the discs to break down organic pollutants in wastewater. Reproduced from Waqas et al. (2023). Licensee MDPI, Basel, Switzerland

or degrade fungal metabolites, resulting in incomplete removal and potential reoccurrence (Novak Babič et al., 2017). Additionally, fungi can form biofilms on surfaces within treatment systems such as pipes, tanks, and filters (Afonso & Simões, n.d.), impeding treatment processes, reducing system efficiency, and increasing maintenance requirements. Conventional physical and chemical methods may not effectively disrupt or remove these biofilms. Furthermore, fungi have specific nutrient requirements that may not be adequately addressed by conventional treatment processes. The presence of complex organic compounds and limited nutrient availability in wastewater can affect the growth and activity of fungal microorganisms, making them difficult to manage. Finally, fungal contaminants pose health risks, including allergic reactions and infections, particularly for sensitive populations, and their presence in treatment systems can lead to operational issues such as clogging, reduced treatment efficiency, and increased maintenance needs.

While conventional wastewater treatment methods play a crucial role in managing various contaminants, they often face limitations in effectively addressing fungal contaminants. This chapter will explore how integrating fungal and algal treatment systems can offer a promising solution to these challenges, enhancing the overall effectiveness and sustainability of wastewater treatment processes.

10.3 Fungal Treatment Systems

Fungal treatment systems leverage the unique metabolic capabilities of fungi to degrade and remove pollutants from wastewater. Fungi are renowned for their ability to break down a wide range of organic compounds, including complex polymers, dyes, and pharmaceuticals, through their enzymatic activities. Understanding the mechanisms by which fungi achieve this degradation is crucial for optimizing their application in wastewater treatment.

Fungi secrete a diverse array of extracellular enzymes that are instrumental in the degradation of organic pollutants, including laccases, peroxidases, cellulases, proteases, and lipases. Laccases are copper-containing enzymes that oxidize a variety of phenolic compounds and aromatic amines, contributing to the breakdown of lignin and other complex organic molecules; their high redox potential allows them to decolorize dyes and degrade persistent pollutants (Singh & Biologia, 2020). Peroxidases, including lignin peroxidase and manganese peroxidase, play a significant role in the degradation of lignin and other aromatic polymers by utilizing hydrogen peroxide as an electron donor to oxidize and break down complex organic structures (Chowdhary et al., 2019). Cellulases are involved in the hydrolysis of cellulose, a major component of plant biomass, breaking it down into simpler sugars that fungi can utilize for growth and metabolism, thereby reducing the concentration of organic matter in wastewater (Zerin Luna et al., 2016). Additionally, proteases degrade proteins, while lipases hydrolyze lipids, aiding in the breakdown of proteins and fats, which are common components of wastewater from food processing and other industries (Technology, 2012).

Fungal mycelium, the vegetative part of a fungus, forms an extensive network of hyphae that can colonize and integrate into wastewater treatment systems (Garg, 2011). Acting as a biofilm, the mycelium provides a surface for the attachment of microorganisms and enhances the degradation of pollutants. This biofilm formation offers several advantages, including increased surface area for microbial attachment and enzyme activity, leading to improved degradation rates of organic contaminants. Additionally, the network of hyphae can access and utilize nutrients more efficiently, supporting the growth and metabolic activity of fungi in wastewater environments. Furthermore, the biofilm structure protects fungi from harsh environmental conditions and toxic compounds, allowing them to persist and function effectively within treatment systems.

Various types of fungi have been identified and utilized in wastewater treatment due to their specific metabolic capabilities and adaptability. Key fungal species and groups include white-rot fungi, brown-rot fungi, yeasts and molds. White-rot fungi shown in Fig. 10.5a, such as *Phanerochaete chrysosporium* and *Trametes versicolor*, are renowned for their ability to degrade lignin and other complex organic compounds (Singh & Chen, 2008). These fungi are particularly effective in the treatment of lignin-rich wastewater from paper and pulp industries. Their production of lignin peroxidase and manganese peroxidase facilitates the breakdown of recalcitrant pollutants and contributes to the decolorization of dyes (Sellami et al., n.d.). Brown-rot fungi as shown in Fig. 10.5b, including species such as *Gloeophyllum trabeum* and *Postia placenta*, are effective in degrading cellulose and hemicellulose. These fungi are useful in treating wastewater with high cellulose content, such as that from agricultural and food processing industries (Kantharaj & Boobalan, 2017).

Fig. 10.5 (**a**) White rot fungi, nature's efficient decomposer, breaks down both lignin and cellulose, playing a vital role in the recycling of organic matter on the forest floor generated by AI using lexica.art. (**b**) Nature's silent recyclers: brown rot fungi decompose wood, leaving behind a unique pattern of cracked cellulose while enriching the forest floor generated by AI using lexica.art

Yeasts, such as *Candida* and *Saccharomyces* species, are versatile microorganisms used in wastewater treatment due to their ability to metabolize a range of organic substrates (Conference, 2014). Yeasts are employed in the treatment of wastewater containing high concentrations of sugars and alcohols, as well as in the removal of pollutants through biosorption and bioaccumulation (Nicula et al., 2023). Molds, including species like *Aspergillus* and *Penicillium*, are utilized for their ability to degrade proteins, lipids, and other organic compounds (Plascencia-Jatomea & Susana, n.d.). These fungi are effective in treating wastewater from food processing and dairy industries, where they contribute to the breakdown of complex organic matter.

Several factors influence the effectiveness of fungal treatment systems, including environmental conditions, nutrient availability, and the presence of other microorganisms. Understanding these factors is essential for optimizing fungal treatment processes. Fungi have specific pH and temperature requirements for optimal growth and enzymatic activity. Most fungi thrive in neutral to slightly acidic conditions (pH 5–7) and at moderate temperatures (20–30 °C) (Gleason & Daynes, n.d.). Deviations from these conditions can affect fungal metabolism and treatment efficiency. Aerobic fungi require adequate oxygen for their metabolic processes. In aerobic treatment systems, ensuring proper aeration is crucial for maintaining fungal activity and degradation rates (Sankaran et al., 2010). Anaerobic fungi, on the other hand, can function in low-oxygen environments and may be used in specific treatment applications.

Fungi require carbon sources for growth and metabolism, and the availability of organic carbon in wastewater can significantly influence fungal activity and the rate of pollutant degradation (Sankaran et al., 2010). In cases where carbon sources are limited, supplemental carbon may be needed to support fungal growth (Ene & Cheng, 2013). Additionally, nitrogen and phosphorus are essential for protein synthesis and cellular function in fungi (Marzluf, 1997). The availability of these nutrients in wastewater can impact fungal performance and treatment efficiency, making it crucial to balance nutrient levels for optimizing fungal growth and pollutant removal (Chu et al., 2001). Fungal treatment systems may involve interactions with other microorganisms present in the wastewater. Competition for nutrients and space can affect fungal activity, while synergistic interactions with bacteria and algae may enhance treatment outcomes. Understanding these interactions can help in designing effective treatment systems.

Several case studies highlight the successful application of fungal treatment systems in various wastewater contexts. A case study of *Trametes versicolor* in the treatment of textile industry wastewater demonstrated the effectiveness of white-rot fungi in degrading azo dyes and reducing chemical oxygen demand (COD). The study highlighted the ability of *T. versicolor* to decolorize dyes and remove color from wastewater, showcasing its potential for treating industrial effluents with high dye concentrations (Zuleta-Correa, n.d.). In another study, *Phanerochaete chrysosporium* was utilized for treating wastewater from the pulp and paper industry. The white-rot fungus was found to effectively degrade lignin and other complex organic compounds, reducing the toxicity and improving the overall quality of the treated effluent (Kumar et al., 2020). The study demonstrated the potential of *P. chrysosporium* in addressing the challenges associated with lignin-rich wastewater (Raychaudhuri, n.d.). A case study involving *Aspergillus niger* for the treatment of agricultural and food processing wastewater highlighted the mold's ability to degrade proteins and lipids (Singh & Singh, 2022). The study reported significant reductions in biochemical oxygen demand (BOD) and total suspended solids (TSS), demonstrating the effectiveness of *A. niger* in managing organic waste from food-related industries (Green, n.d.). The use of *Penicillium* species in treating dairy industry wastewater was explored in a case study that emphasized the mold's capability to degrade fats and proteins (Awasti & Anand, 2020). The study revealed improvements in the removal of organic matter and reduction in pollution levels, showcasing the potential of *Penicillium* for treating wastewater with high fat content (Sankaran et al., 2010).

The fungal treatment systems offer a promising approach for addressing various contaminants in wastewater through their unique enzymatic capabilities and biofilm formation. The selection of appropriate fungal species, understanding influencing factors, and reviewing successful case studies can guide the effective implementation of fungal treatment systems in diverse wastewater contexts.

10.4 Integration of Fungal and Algal Systems

The integration of fungal and algal systems in wastewater treatment represents a significant advancement in addressing the complex challenges associated with managing pollutants, including fungal contaminants. Both fungi and algae offer distinct advantages in wastewater treatment: fungi are adept at breaking down complex organic substances and forming stable biofilms (Pandit et al., 2020), while algae excel in nutrient removal and oxygen production (Teles et al., n.d.). By combining these systems, it is possible to harness their complementary strengths, leading to enhanced treatment efficiency and sustainability.

Fungi possess a remarkable ability to degrade a wide range of organic pollutants through their extracellular enzymes (Vishnoi, 2019). Their capability to handle complex and recalcitrant compounds, such as lignin and synthetic dyes, makes them invaluable in treating industrial wastewater. Additionally, fungal biofilms can improve nutrient removal and provide a stable environment for microbial interactions (Pandit et al., 2020). Algae, including both microalgae and macroalgae, are highly effective at removing nutrients like nitrogen and phosphorus from wastewater (Nguyen et al., 2022). Through photosynthesis, algae produce oxygen, which supports aerobic microbial processes and enhances overall treatment performance. Algae also sequester carbon dioxide and can produce valuable by-products, such as biofuels and bioplastics (Agarwal et al., 2022).

Integrating fungal and algal systems can lead to enhanced pollutant removal, as fungi degrade complex organic compounds while algae address nutrient pollution, contributing to overall water quality improvement. This dual-action approach ensures that multiple types of pollutants are effectively managed (Banerjee & Choudhury, n.d.). Additionally, the combined use of fungi and algae can increase system stability, with fungal biofilms providing a stable environment for algae, and algae's oxygen production supporting fungal metabolism. This synergy leads to more reliable and robust treatment processes (Yu et al., n.d.). Moreover, the integration of fungal and algal systems facilitates resource recovery and promotes sustainability, with algae producing biomass for bioenergy and fungi enhancing the degradation of organic waste, thereby reducing the need for additional treatment steps. This approach aligns with the principles of circular economy and sustainable resource management (Ahmad et al., 2021).

Several strategies and methods can be employed to integrate fungal and algal systems effectively. These strategies vary based on the specific treatment goals, wastewater characteristics, and operational requirements. Co-cultivation involves growing fungi and algae together in a shared environment, and this method can be implemented in several ways. In mixed reactors, fungi and algae are cultured together in a single reactor, with the reactor design supporting the growth of both organisms by providing appropriate conditions for fungal biofilm formation and algal photosynthesis (Le et al., 2019). This approach allows for simultaneous pollutant degradation and nutrient removal. Alternatively, sequential cultivation involves separate stages for fungal and algal treatment, where wastewater may first undergo fungal treatment to degrade complex organic compounds and then be transferred to an algal system for nutrient removal and oxygenation (Sankaran et al., 2010). This method optimizes the performance of each system by targeting specific pollutants at different stages.

Sequential treatment processes utilize distinct stages for fungal and algal treatment, allowing for targeted pollutant removal and system optimization. In this method, fungi are used as a pre-treatment step to degrade complex organic matter and reduce the load on subsequent algal treatment (Collado et al., n.d.). The pre-treated effluent is then introduced to an algal system for nutrient removal and further polishing. This approach enhances the overall treatment efficiency and reduces the risk of clogging and fouling in algal systems. Alternatively, algae can be used as a post-treatment step after fungal degradation. Fungal treatment reduces the concentration of complex pollutants, making the wastewater more amenable to algal nutrient removal and oxygenation (Sankaran et al., 2010). This sequential approach can improve the quality of treated effluent and enhance resource recovery.

Hybrid systems combine elements of both co-cultivation and sequential treatment to optimize performance and address specific treatment challenges. In integrated biofilm systems, fungal biofilms serve as a support matrix for algal growth, providing a stable surface for algae to attach and grow, while the algae contribute to nutrient removal and oxygen production (Sankaran et al., 2010). This integration of biofilm and algal technologies enhances treatment efficiency and system stability. Additionally, modular hybrid systems incorporate separate modules for fungal and algal treatment, allowing for flexible configuration and optimization (Brar et al., 2022). Each module is designed to target specific pollutants or treatment goals, and the modular design enables easy adjustments based on wastewater characteristics and treatment requirements (Brar et al., 2022).

The integration of fungal and algal systems can lead to several synergistic effects that enhance overall treatment performance. The combination of fungal and algal systems can result in more efficient pollutant degradation. Fungi degrade complex organic substances, reducing the concentration of pollutants that may otherwise inhibit algal growth (Silva, Delerue-Matos, & Figueiredo, n.d.). Algae, in turn, remove nutrients and produce oxygen, creating favorable conditions for continued fungal activity. This synergy results in comprehensive treatment of various types of pollutants.

Algae excel in removing nutrients such as nitrogen and phosphorus from wastewater. When combined with fungal systems that degrade organic matter, the overall nutrient removal efficiency is improved (Silva, Magano, et al., n.d.). The reduction in nutrient concentrations enhances water quality and reduces the risk of eutrophication in receiving water bodies (Yang et al., 2008). The integration of fungi and algae can lead to increased system efficiency by leveraging the strengths of both organisms. Fungal treatment reduces the load on algal systems by breaking down complex pollutants (Pandit et al., 2020), while algae enhance overall treatment performance through nutrient removal and oxygenation (Teles et al., n.d.). This combined approach can lead to more efficient and cost-effective wastewater treatment.

While the integration of fungal and algal systems offers numerous benefits, several challenges must be addressed to ensure successful implementation. Designing and managing integrated systems requires careful consideration of the needs and interactions of both fungi and algae. Factors such as pH, temperature, and nutrient levels must be optimized to support the growth and activity of both organisms (Russell, 2016), and advanced control and monitoring systems may be necessary to maintain optimal conditions. Additionally, compatibility issues may arise, as fungal and algal systems can have different growth requirements and sensitivities. Ensuring compatibility and minimizing negative interactions between fungi and algae is crucial for maintaining system performance. Pilot studies and lab-scale experiments can help identify and address potential issues before full-scale implementation (Zhu et al., 2019).

The integration of fungal and algal systems may involve higher initial investment costs compared to conventional treatment methods. Costs associated with system design, construction, and operation must be evaluated in the context of long-term benefits and resource recovery (Duan et al., 2019). Additionally, integrated systems may require additional resources, such as nutrients and energy, to support the growth and activity of fungi and algae. Efficient resource management and optimization are essential to ensure the sustainability and cost-effectiveness of integrated treatment systems (Engineering, n.d.). Scaling up integrated fungal-algal systems from laboratory or pilot-scale to full-scale applications can present challenges. Factors such as system stability, performance consistency, and operational complexity must be addressed to achieve successful implementation at larger scales (Meerbergen, 2018). Additionally, regulatory frameworks and technical standards for integrated systems may be less established compared to conventional treatment methods

(Schellenberg et al., 2020). Collaboration with regulatory agencies and industry experts can help address these barriers and promote the adoption of integrated technologies.

The integration of fungal and algal systems offers a promising approach to enhancing wastewater treatment through the combined strengths of both organisms. By leveraging their complementary capabilities, addressing challenges, and optimizing system performance, it is possible to achieve more effective and sustainable treatment solutions. The exploration of various integration strategies, understanding synergistic effects, and overcoming operational challenges will be key to advancing the application of fungal-algal systems in wastewater management.

10.5 Factors Affecting the Efficiency of Integrated Fungal and Algal Treatment Systems

Integrated fungal and algal systems for wastewater treatment exhibit considerable promise in addressing environmental pollution and resource recovery challenges. However, the efficiency of these systems can be influenced by various factors. This section provides an in-depth analysis of the key factors affecting the performance of integrated fungal and algal treatment systems, including environmental conditions, system design, biological factors, and operational parameters.

Temperature is a critical factor affecting both fungal and algal growth and activity, with each microorganism having an optimal temperature range within which it functions most effectively. For fungal systems, fungi generally thrive in moderate temperature ranges, typically between 20 °C and 30 °C (Maheshwari et al., 2000). Extreme temperatures can inhibit fungal growth or affect enzymatic activity, thereby reducing degradation efficiency. High temperatures can increase metabolic rates, leading to faster degradation of organic matter, but can also increase the risk of system instability and competition from undesired microorganisms (Maheshwari et al., 2000). In algal systems, algae also have specific temperature requirements, with most species thriving in temperatures ranging from 15 °C to 35 °C (Lürling et al., 2018). Temperature influences algal photosynthesis rates, nutrient uptake, and biomass productivity. Extreme temperatures can affect the chlorophyll content and overall health of algal cultures, leading to reduced nutrient removal efficiency (Lürling et al., 2018).

Light is essential for algal photosynthesis and, consequently, for nutrient removal in integrated systems. In algal systems, adequate light intensity is crucial for algal growth and productivity. In photobioreactors, light intensity must be optimized to ensure that all algal cells receive sufficient light for photosynthesis (Sartori & Deprá, n.d.). Too high or too low light levels can affect algal growth rates and nutrient uptake. Techniques such as light-emitting diodes (LEDs) and light diffusion systems are often used to control light conditions in algal systems (Sartori & Deprá, n.d.). Although fungi do not rely on light for growth, the interaction between light and algal systems can impact the overall efficiency of integrated treatment systems. Proper management of light in hybrid systems ensures optimal performance of both fungal and algal components (Sartori & Deprá, n.d.).

pH levels influence the solubility of nutrients and pollutants, as well as microbial activity. In fungal systems, fungi generally prefer neutral to slightly acidic pH levels, typically between 4.5 and 7.5 (Thesis et al., 2018). Deviations from this range can affect fungal growth and enzymatic activity, potentially reducing the efficiency of organic matter degradation. Acidic or alkaline conditions can inhibit fungal growth and affect the breakdown of pollutants (Liu et al., 2016). In algal systems, algae can tolerate a broader pH range but generally perform best in neutral to slightly alkaline conditions (pH 6.5–8.5). pH affects nutrient availability and algal metabolism, and extreme pH levels can lead to reduced algal growth and nutrient uptake efficiency (Thesis et al., 2018).

Nutrient availability impacts the growth and productivity of both fungi and algae. In fungal systems, fungi require a range of nutrients for growth, including carbon, nitrogen, and minerals (Dighton, 2018). In wastewater treatment, the availability of these nutrients can affect fungal activity and efficiency. For optimal performance, nutrient levels should be balanced to support fungal growth and enzymatic activity (Dighton, 2018). In algal systems, algae need nutrients such as nitrogen, phosphorus, and trace elements for photosynthesis and biomass production (Procházková et al., 2014). Nutrient deficiency or excess can affect algal growth and nutrient removal, making efficient nutrient management crucial for maintaining algal health and optimizing performance.

The design of reactors significantly affects the efficiency of integrated fungal-algal systems. For algal systems, photobioreactor design influences light distribution, gas exchange, and nutrient mixing. Reactor types shown in Fig. 10.6 such as tubular, flat-panel, and vertical reactors offer different advantages in terms of light

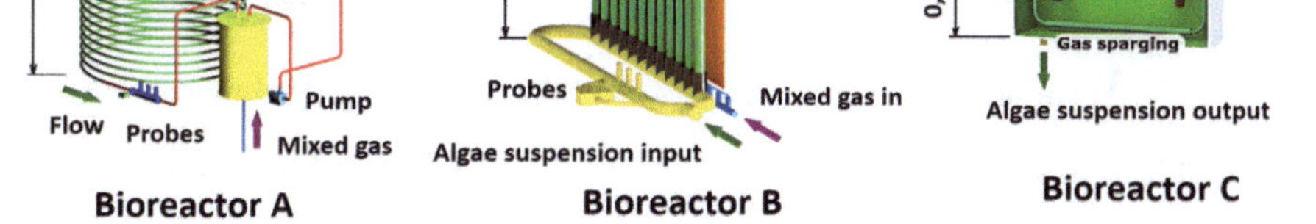

Fig. 10.6 Schematic representation of the design of the three photobioreactors used in this study: Helical-tubular PBR (Bioreactor A), Multi-tubular airlift PBR (Bioreactor B), and Flat panel PBR (Bioreactor C), each offering distinct configurations for optimizing light exposure and gas exchange in algal cultivation. Reproduced from Sukačová et al. (2021). Licensee MDPI, Basel, Switzerland

exposure and space utilization (Carvalho, 2006). Optimizing reactor design to ensure effective light penetration and gas transfer is crucial for maximizing algal productivity.

Fungal bioreactors need to facilitate adequate air circulation, moisture control, and temperature regulation. Designs such as fixed-bed, fluidized-bed, or rotating-drum reactors as shown in Fig. 10.7 offer different benefits depending on the scale and type of wastewater being treated (Manan & Bioeng, 2017). Ensuring proper mixing and aeration can enhance fungal activity and efficiency.

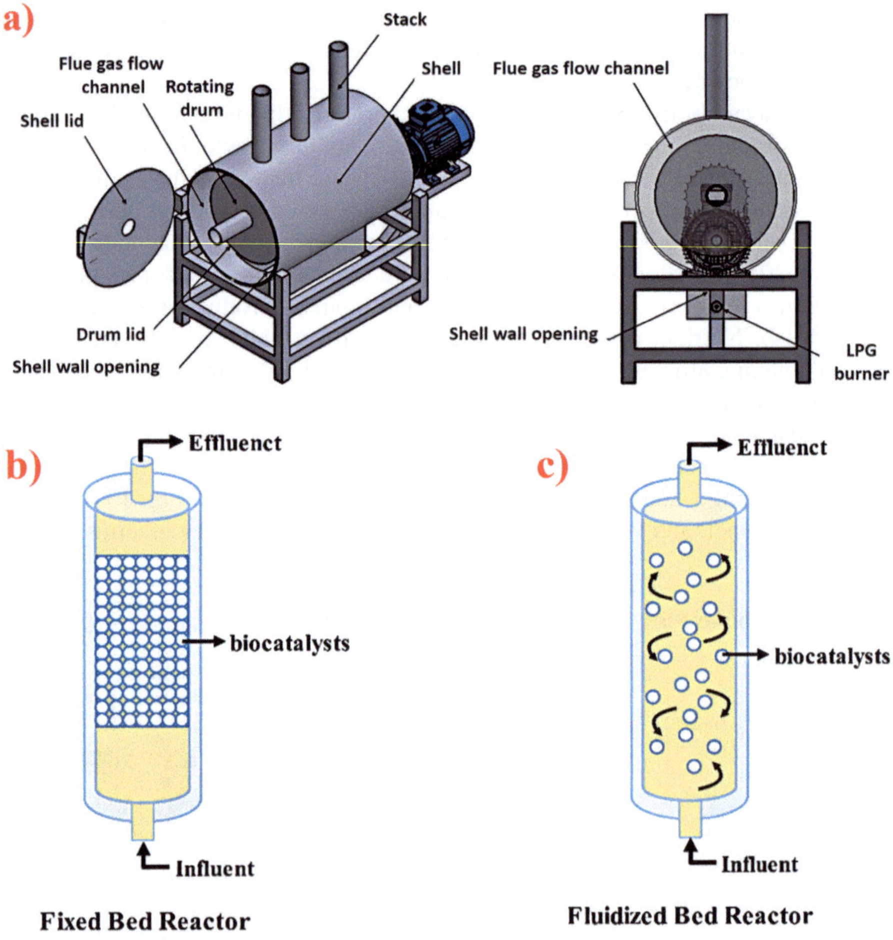

Fig. 10.7 (**a**) Structure of a Small-Scale Rotary Drum Reactor (SS-RDR), optimized for efficient mixing and reaction through the rotation of a cylindrical drum. Reproduced from Soponpongpipat et al. (n.d.) Licensee MDPI, Basel, Switzerland. Schematic diagram comparing two types of reactors: (**b**) Fixed Bed Reactor, where the solid catalyst remains stationary while the fluid passes through, and (**c**) Fluidized bed reactor, where the solid particles are suspended and mixed by the upward flow of fluid, enhancing mass transfer and reaction efficiency. Reproduced from Liu et al. (2022). Licensee MDPI, Basel, Switzerland

The integration of fungal and algal systems requires careful consideration of how the two components interact. In sequential systems, fungal treatment is followed by algal treatment, or vice versa (Thalla & Devika, 2024). The efficiency of nutrient removal depends on the effectiveness of each stage and how well the effluent from one stage supports the performance of the next stage (Fig. 10.8). In combined systems, fungi and algae are co-cultured or operate in parallel (Naseema Rasheed et al., 2023). The design must ensure that both components receive optimal conditions for growth and activity, with balancing the needs of both fungal and algal populations being key to achieving high performance in integrated systems (Naseema Rasheed et al., 2023).

Flow dynamics within reactors affect the distribution of microorganisms, nutrients, and pollutants. Hydraulic retention time (HRT) influences the contact time between wastewater and treatment microorganisms, with longer retention times generally improving treatment efficiency but potentially requiring larger reactor volumes (Martín-Pascual et al., 2016). Optimizing HRT ensures that pollutants are adequately treated and removed. Additionally, effective mixing and mass transfer are essential for ensuring uniform distribution of nutrients and pollutants. Poor

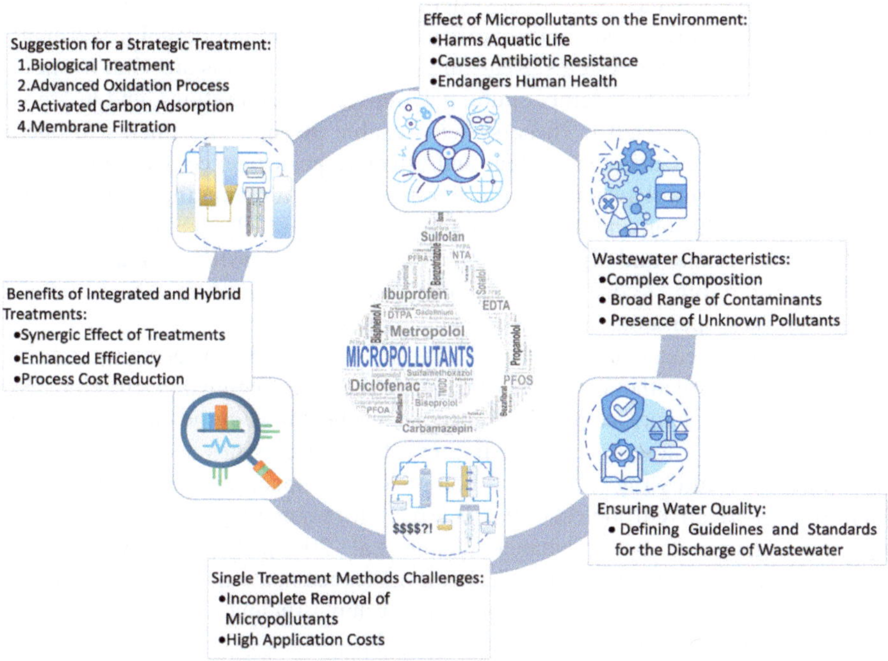

Fig. 10.8 Overview of strategies for addressing micropollutants in wastewater treatment, highlighting the effects on the environment, wastewater characteristics, benefits of integrated treatments, challenges of single treatment methods, and the importance of ensuring water quality through defined guidelines. Reproduced from Asheghmoalla and Processes (2024). Licensee MDPI, Basel, Switzerland

mixing can lead to localized inefficiencies and reduced overall treatment performance.

Interactions between fungal and algal species can influence system performance. In integrated systems, fungi and algae may exhibit synergistic effects, where the presence of one enhances the performance of the other. For example, fungal degradation of organic matter can improve nutrient availability for algae, leading to enhanced nutrient removal (Hu et al., 2019). Conversely, competition for nutrients or space can occur between fungal and algal species. Ensuring balanced microbial populations and optimizing conditions to minimize competition is important for maintaining system efficiency (Hu et al., 2019).

Fungi and algae have the potential to adapt to changing conditions over time, which can influence the performance of wastewater treatment systems (Rani et al., 2020). Evolutionary adaptations among microbial species may occur, allowing them to better tolerate specific environmental conditions or pollutants (Bleuven & Landry, 2016). Understanding these evolutionary changes is crucial for selecting and optimizing strains tailored to particular applications (Dragosits, 2013). To address these dynamics, implementing adaptive management strategies is essential. This approach involves adjusting operational parameters based on the observed performance of microorganisms and fluctuating environmental conditions, thereby ensuring the maintenance of optimal system efficiency (Nsair et al., n.d.).

Effective monitoring and control are crucial for sustaining optimal performance in wastewater treatment systems. Real-time monitoring systems are vital for tracking key parameters such as temperature, pH, nutrient concentrations, and microbial activity (Park & Kim, n.d.). These systems enable prompt adjustments to operational conditions, ensuring consistent and effective performance. Complementing this, automated control systems play a significant role in regulating environmental factors, nutrient dosing, and flow rates (Beltrán et al., n.d.). By maintaining optimal conditions for fungal and algal growth, advanced control systems enhance overall efficiency and minimize the need for manual intervention, leading to more reliable and streamlined operations (Dębowski et al., 2022).

Regular maintenance and troubleshooting are essential for ensuring the reliability of wastewater treatment systems. Preventive maintenance involves routine tasks such as cleaning reactors, calibrating sensors, and checking equipment functionality (Lee et al., 2020). These measures help prevent potential issues and ensure smooth operation by addressing problems before they impact system performance. Additionally, effective troubleshooting is crucial for managing challenges like fouling, clogging, or microbial imbalances (Jeong et al., 2018). By promptly addressing these issues and implementing troubleshooting protocols, along with having contingency plans in place, operators can maintain system efficiency and address operational challenges effectively (Hale, n.d.).

Scaling up from laboratory to full-scale operations presents several challenges that require careful adaptation of operational parameters. Key challenges include addressing issues related to reactor design, flow dynamics, and nutrient management, which can differ significantly between laboratory and larger-scale systems (Joshi & Nandakumar, 2015). Pilot studies and pilot-scale testing play a crucial role in identifying and resolving these scale-up issues, providing valuable insights for successful implementation (McNeil et al., 2017). To achieve consistent performance across different scales and wastewater types, it is essential to adapt operational parameters accordingly. Implementing flexible and adaptive management strategies can help optimize integrated systems and ensure their effective scaling (Williams, 2014).

10.6 Future Directions and Innovations

The integration of fungal and algal systems in wastewater treatment represents a promising frontier in environmental technology. However, the field is still evolving, and several areas offer potential for innovation and advancement. This section explores emerging trends, research directions, and technological innovations that could shape the future of integrated fungal-algal wastewater treatment systems. Synthetic biology and genetic engineering are opening new avenues for enhancing the performance of fungi and algae in wastewater treatment. By modifying the genetic makeup of these organisms, researchers can significantly improve their effectiveness in several ways. For fungi, genetic engineering can enhance enzymatic capabilities and metabolic pathways (Dare Asemoloye et al., 2021), such as introducing genes for novel enzymes or boosting the expression of existing ones to increase pollutant degradation efficiency (Díaz, 2004). Additionally, engineering fungi to withstand harsh environmental conditions or high pollutant concentrations can make them more versatile in various wastewater contexts (Aragaw, 2021). Similarly, algae can be genetically modified to improve nutrient removal, increase biomass production, or generate valuable by-products (Bansal et al., 2018). For example, algae engineered to express genes involved in carbon fixation or nutrient uptake can become more efficient at removing nitrogen and phosphorus from wastewater (Wang & Reports, n.d.). Furthermore, algae can be tailored to produce biofuels, bioplastics, or other high-value products, thereby adding economic value to the wastewater treatment process.

Microbial fuel cells (MFCs) as shown in Fig. 10.9a and bioelectrochemical systems (BESs) as shown in Fig. 10.9b represent innovative approaches to integrating fungal and algal systems with energy generation and pollutant removal (Angelaalincy et al., 2018). These technologies harness the metabolic activities of microorganisms to produce electricity and treat wastewater simultaneously. Fungal-based MFCs utilize fungal mycelium as a bioelectrode or mediator to facilitate electron transfer during microbial oxidation processes. The integration of fungi with MFCs can enhance pollutant degradation and energy production, offering a sustainable solution for wastewater treatment and power generation. Algal-based BESs leverage the photosynthetic capabilities of algae to generate electrical current while removing pollutants. Algae can act as the anode or cathode in bioelectrochemical systems, and their photosynthetic activity contributes to the overall energy production (Gude,

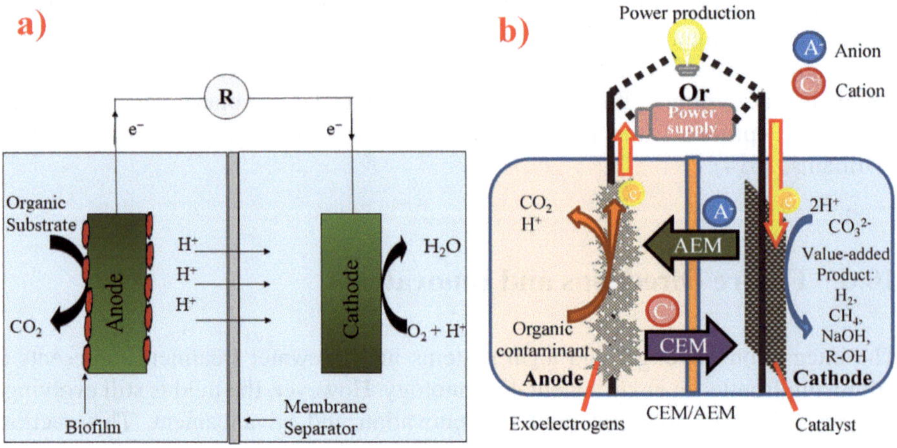

Fig. 10.9 (**a**) A typical microbial fuel cell (MFC), illustrating the system where microorganisms convert organic matter into electricity through biochemical reactions, highlighting the anode and cathode compartments, and the flow of electrons and ions reproduced from Bipin Patwardhan et al. (2021) Copyright © 2021 by the authors. Licensee MDPI, Basel, Switzerland. (**b**) Schematic of a bioelectrochemical system (BES), featuring cation exchange membrane (CEM) and anion exchange membrane (AEM), illustrating the separation of ions and the flow of electrical current during biochemical processes. Reproduced from Lu et al. (2015). licensee MDPI, Basel, Switzerland

n.d.). This approach can enhance nutrient removal and energy recovery in wastewater treatment.

The design and engineering of integrated photobioreactors and bioreactors play a crucial role in optimizing the performance of fungal-algal systems. Advances in reactor design can improve nutrient and pollutant removal, enhance system efficiency, and reduce operational costs. Innovations in photobioreactor design include improved light distribution, enhanced gas exchange, and efficient nutrient delivery (Carvalho, 2006). These advancements support the growth and activity of both fungi and algae, leading to more effective wastewater treatment. In hybrid bioreactor systems, fungal and algal treatment processes are integrated within a single reactor or sequential modules. Innovations in hybrid reactor design include modular configurations, advanced mixing and aeration systems, and optimized flow dynamics (Thalla & Devika, 2024). These systems enhance the interaction between fungi and algae and improve overall treatment performance.

The development of smart monitoring and control systems significantly enhances the efficiency and reliability of integrated fungal-algal systems by leveraging advanced technology (Neethirajan et al., n.d.). These systems employ sophisticated sensors to provide real-time data on critical parameters like pH, temperature, dissolved oxygen, and nutrient concentrations, allowing for immediate adjustments to maintain optimal conditions (Su et al., 2020). Data analytics and machine learning further elevate system performance by analyzing large volumes of data to identify trends, predict outcomes, and fine-tune treatment processes (Aragaw, 2021).

Machine learning algorithms can create predictive models that help in managing and controlling integrated systems more effectively. Additionally, automation plays a crucial role by streamlining operations such as nutrient dosing, aeration, and mixing (Pöysä et al., 2022). Automated control systems improve consistency, minimize manual intervention, and enhance the overall reliability and efficiency of wastewater treatment processes.

The integration of fungal and algal systems not only enhances wastewater treatment but also creates opportunities for resource recovery and the production of high-value by-products, contributing to a circular economy. Algae, for instance, can generate biofuels such as biodiesel or bioethanol from their biomass, while fungi can be used to produce bioplastics and other biodegradable materials (Coppola et al., 2021). This integration enables the simultaneous treatment of wastewater and the generation of valuable by-products. Additionally, algae play a crucial role in sequestering nutrients like nitrogen and phosphorus, which can be recovered for use as fertilizers or feed additives (Zabochnicka et al., 2022). By combining fungal and algal systems, nutrient removal and recovery are optimized, leading to more sustainable wastewater management and enhanced resource utilization.

The integration of fungal and algal systems significantly contributes to waste reduction and management by transforming wastewater contaminants into valuable products and by-products. Fungi play a crucial role in degrading complex organic matter, which reduces both the volume of waste and its environmental impact (Sánchez, 2008). Following fungal treatment, algae further process the residual organic matter, converting it into useful biomass or energy. This waste-to-resource approach minimizes the need for additional waste disposal and aligns with sustainable practices (Makepa et al., 2024). Furthermore, the combined fungal and algal treatment processes reduce the overall environmental impact of wastewater management. By lowering residual sludge production, reducing chemical usage, and minimizing pollution, these integrated systems support more sustainable and eco-friendly wastewater management practices (Poblete & Araujo, n.d.).

The adoption of integrated fungal-algal systems in wastewater treatment necessitates updates to existing regulatory frameworks and standards. Policymakers and regulators must address the unique requirements and considerations associated with these innovative technologies (Bergis et al., n.d.). Developing comprehensive standards and guidelines is essential to ensure the safe and effective operation of fungal-algal treatment systems. These standards should encompass aspects such as system design, monitoring practices, and performance reporting to guarantee environmental protection and operational efficiency (Silva, Magano, et al., n.d.). Additionally, to foster the adoption of these advanced technologies, governments and regulatory bodies can offer various forms of support. Financial incentives, research grants, and technical assistance can play a crucial role in encouraging the development and implementation of integrated fungal-algal systems, facilitating their broader adoption and integration into wastewater management practices (Chu et al., 2001).

Raising public awareness and engaging stakeholders are crucial for advancing the adoption of integrated fungal-algal systems in wastewater treatment. Educational programs and outreach efforts play a significant role in highlighting the benefits and advantages of these technologies. By offering educational programs and workshops, stakeholders can gain a better understanding of fungal-algal systems, learning about their potential through success stories, research findings, and best practices. Additionally, involving various stakeholders—including industry representatives, environmental organizations, and community members—can foster broader acceptance and support. Collaborative efforts can address concerns, build advocacy, and promote the integration of these innovative solutions into wastewater management practices (Kumar & Chaturvedi, n.d.).

The future of integrated fungal-algal wastewater treatment systems holds significant promise for advancing environmental technology and sustainability. By exploring advanced cultivation techniques, enhancing system design, recovering valuable resources, and addressing policy considerations, the field can continue to evolve and deliver more effective and sustainable solutions for wastewater management. The ongoing research, innovation, and collaboration in this area will be crucial for realizing the full potential of integrated fungal-algal systems and achieving long-term environmental and economic benefits.

10.7 Economic and Environmental Implications

The integration of fungal and algal systems in wastewater treatment offers significant economic and environmental benefits. This section explores these implications in detail, assessing the cost-effectiveness, resource efficiency, and environmental impact of these integrated systems. It also considers the broader implications for sustainability and the circular economy. The initial investment for integrated fungal-algal systems involves various costs associated with system design, construction, and installation. Key components, such as fungal bioreactors and algal photobioreactors or wetlands, require significant capital expenditure, with costs varying based on the system's scale, the type of wastewater being treated, and the complexity of the design (Arora et al., n.d.). For fungal bioreactors, expenses include constructing the bioreactor vessels, inoculating fungal strains, and establishing environmental controls such as temperature and pH regulation (Ganeshan & Kim, 2021). Advanced systems that incorporate automated controls and real-time monitoring tend to have higher initial costs. Similarly, algal systems, whether they are photobioreactors or wetlands, involve costs related to reactor construction, lighting (if needed), and nutrient management. Additionally, maintaining optimal growth conditions and harvesting algal biomass contribute to the overall investment (Barry et al., 2015).

Operational costs for integrated fungal-algal systems encompass energy, labor, and materials required for the system's ongoing operation, as well as maintenance expenses for equipment and monitoring systems (Raj et al., n.d.). Energy costs can be significant, especially for systems utilizing artificial lighting or aeration. Despite these costs, they may be partially offset by the energy recovery potential of the system, such as through the production of biofuels (Beal et al., 2012). Labor costs are another important factor, as skilled personnel are needed for system operation, monitoring, and maintenance. Training staff and ensuring adequate staffing levels can impact overall operational expenses. Maintenance costs are crucial for ensuring system performance. These expenses may include cleaning reactors, calibrating sensors, and addressing issues such as fouling or clogging to maintain optimal operation (Gkotsis et al., 2014).

Integrated fungal-algal systems offer several economic benefits. These systems can generate revenue through resource recovery by producing valuable by-products such as biofuels, bioplastics, or recovered nutrients. Algal biomass, for instance, can be utilized in biofuel production, while recovered nutrients can be used as fertilizers or feed additive (Zabochnicka et al., 2022). Additionally, by effectively removing pollutants and nutrients, these systems reduce the need for chemical treatments. This leads to cost savings and decreases expenses related to chemical handling and disposal (Aiyuk et al., 2004). Moreover, integrated systems typically generate less residual sludge compared to conventional treatment methods. This reduction in sludge production translates to lower disposal costs and a diminished environmental impact (Ferrentino et al., 2023). Integrated fungal-algal systems are effective in removing nutrients such as nitrogen and phosphorus from wastewater, which is crucial for preventing eutrophication in receiving water bodies. Eutrophication can lead to harmful algal blooms and degradation of aquatic ecosystems(Sellner et al., 2008). In these systems, fungi play a key role in the pre-treatment phase by degrading complex organic matter and reducing nutrient concentrations in wastewater (Collado et al., n.d.). This initial fungal treatment enhances the effectiveness of the subsequent algal treatment.

Following fungal treatment, algae contribute to nutrient removal through photosynthesis. Algae absorb and assimilate nitrogen and phosphorus, converting these nutrients into biomass. This process significantly reduces the overall nutrient load in the treated effluent (Al-Jabri et al., 2020). Fungal and algal systems are capable of degrading a wide range of pollutants, including organic contaminants, heavy metals, and pharmaceuticals (Ghosh et al., 2023). Fungi effectively break down complex organic compounds through various enzymatic processes, which helps in the degradation of organic pollutants. Algae further enhance this process by adsorbing and assimilating residual organic matter and pollutants (Shahid et al., n.d.). In addition to organic pollutants, certain fungal and algal species have the ability to accumulate and detoxify heavy metals from wastewater. This capability plays a crucial role in reducing metal contamination in the treated effluent (Kumar & Chaturvedi, n.d.).

Integrated fungal-algal systems contribute to environmental sustainability in several key ways. By harnessing natural processes and minimizing reliance on chemical treatments, these systems achieve a reduced carbon footprint compared to conventional wastewater treatment methods. The effective removal of pollutants and nutrients helps protect aquatic ecosystems and supports biodiversity by reducing nutrient loading and pollutant concentrations, thereby promoting healthier water bodies and maintaining ecological balance. Additionally, the recovery of valuable resources, such as biofuels and nutrients, enhances resource efficiency and supports a circular economy. By converting waste into useful products, these systems not only manage waste more effectively but also contribute to overall sustainability.

Integrated fungal-algal systems align with the principles of the circular economy by promoting resource recovery and minimizing waste. These systems facilitate waste-to-resource strategies, where wastewater and by-products are transformed into valuable products such as biofuels, bioplastics, and recovered nutrients. This approach not only reduces the overall environmental impact but also supports a closed-loop system where waste materials are converted into useful resources. By recovering and reusing resources, these systems contribute to more sustainable resource management practices and lessen the reliance on virgin materials, thereby enhancing overall sustainability.

The implementation of integrated fungal-algal systems significantly supports several Sustainable Development Goals (SDGs). These systems contribute to SDG 6: Clean Water and Sanitation by enhancing wastewater treatment and nutrient removal, thereby improving access to clean water and sanitation. They also align with SDG 12: Responsible Consumption and Production through effective resource recovery and waste reduction practices, promoting sustainable consumption and production patterns. Furthermore, the reduced carbon footprint and improved resource efficiency of these systems advance SDG 13: Climate Action, supporting efforts to combat climate change and enhance environmental sustainability (Ileana et al., 2021). Government policies and incentives are crucial for promoting the adoption of integrated fungal-algal systems. Supportive policies can drive research, development, and implementation of these technologies. Research grants provide essential funding for innovation and performance improvement, supporting pilot projects and scale-up efforts. Additionally, regulatory incentives such as tax credits or subsidies can encourage the adoption of sustainable technologies, facilitating the implementation of integrated fungal-algal systems in wastewater treatment.

The adoption of integrated systems may face regulatory challenges, including the need for developing comprehensive standards and guidelines. Establishing clear standards for the operation and performance of fungal-algal systems is crucial to ensure their safety and effectiveness. Regulatory frameworks must address the specific requirements of these innovative technologies. Additionally, ensuring compliance with environmental regulations and monitoring system performance are critical for successful implementation. Collaboration between regulators and operators is essential to address issues related to system operation and effluent quality (Kagan & Gunningham, n.d.). The integrated fungal-algal systems offer significant economic and environmental benefits, contributing to more sustainable and efficient wastewater treatment. By addressing cost considerations, pollution reduction, and broader sustainability implications, these systems provide valuable solutions for managing wastewater and promoting a circular economy. Continued research, policy support, and practical implementation will be essential for realizing the full potential of integrated fungal-algal systems and advancing towards a more sustainable future.

10.8 Conclusion

The integration of fungal and algal treatment systems represents a significant advancement in wastewater management, offering a dual approach to pollutant degradation and nutrient removal. This chapter has explored various factors influencing the efficiency of these systems, including environmental conditions, system design, biological interactions, and operational parameters. Fungal and algal systems, when effectively combined, harness the complementary strengths of each microorganism, leading to enhanced treatment performance and resource recovery. Fungi contribute to the breakdown of complex organic pollutants, while algae excel in nutrient uptake and biomass production through photosynthesis. The success of these integrated systems hinges on careful optimization of conditions such as temperature, light intensity, pH, and nutrient availability, as well as the design and configuration of reactors. Despite the promising potential, several challenges remain, including scaling up from laboratory to full-scale applications, ensuring adaptability to diverse wastewater types, and addressing operational complexities. Future research should focus on advancing strain selection, reactor design, and process integration, while also addressing environmental impact and economic feasibility. The integrated fungal-algal systems offer a sustainable and innovative solution to wastewater treatment challenges. Continued exploration and refinement of these technologies will contribute to more efficient, cost-effective, and environmentally friendly wastewater management practices, paving the way for their broader adoption and implementation.

References

Afonso, T., & Simões, L. (n.d.). *Occurrence of filamentous fungi in drinking water: Their role on fungal-bacterial biofilm formation*. Elsevier. Retrieved August 24, 2024, from https://www.sciencedirect.com/science/article/pii/S0923250820301042

Agarwal, P., Soni, R., Kaur, P., Madan, A., Mishra, R., Pandey, J., Singh, S., & Singh, G. (2022). Cyanobacteria as a promising alternative for sustainable environment: Synthesis of biofuel and biodegradable plastics. *Frontiers in Microbiology, 13*. https://doi.org/10.3389/FMICB.2022.939347/FULL

Ahmad, A., Banat, F., Hasan Khalifa, S. W., Alsafar, H., & Hasan, S. W. (2021). *Algae biotechnology for industrial wastewater treatment, bioenergy production, and high-value bioproducts*. Elsevier. https://doi.org/10.1016/j.scitotenv.2021.150585

Ahmed, S., Mofijur, M., & Nuzhat, S. (n.d.). *Recent developments in physical, biological, chemical, and hybrid treatment techniques for removing emerging contaminants from wastewater*. Elsevier. Retrieved August 22, 2024, from https://www.sciencedirect.com/science/article/pii/S0304389421008761

Aiyuk, S., Amoako, J., Raskin, L., Van Haandel, A., & Verstraete, W. (2004). *Removal of carbon and nutrients from domestic wastewater using a low investment, integrated treatment concept* (Vol. 38, pp. 3031–3042). Elsevier. https://doi.org/10.1016/j.watres.2004.04.040

Al-Jabri, H., Das, P., Khan, S., Thaher, M., & Water, M. A. (2020). Treatment of wastewaters by microalgae and the potential applications of the produced biomass—A review. *Water*. https://doi.org/10.3390/w13010027

Angelaalincy, M. J., Navanietha Krishnaraj, R., Shakambari, G., Ashokkumar, B., Kathiresan, S., & Varalakshmi, P. (2018). Biofilm engineering approaches for improving the performance of microbial fuel cells and bioelectrochemical systems. *Frontiers in Energy Research, 6*. https://doi.org/10.3389/FENRG.2018.00063/FULL

Angelakis, A., Capodaglio, A., & Water, E. D. (2022). Wastewater management: From ancient Greece to modern times and future. *Water*. https://doi.org/10.3390/w15010043

Aragaw, T. A. (2021). Functions of various bacteria for specific pollutants degradation and their application in wastewater treatment: A review. *International Journal of Environmental Science and Technology, 18*(7), 2063–2076. https://doi.org/10.1007/s13762-020-03022-2

Arora, K., Kaur, P., & Kumar, P. (n.d.). *Valorization of wastewater resources into biofuel and value-added products using microalgal system*. Frontiersin.Org. Retrieved September 29, 2024, from https://www.frontiersin.org/articles/10.3389/fenrg.2021.646571/full

Asheghmoalla, M., & Processes, M. M. (2024). Integrated and hybrid processes for the treatment of actual wastewaters containing micropollutants: A review on recent advances. *Processes*. https://doi.org/10.3390/pr12020339

Awasti, N., & Anand, S. (2020). The role of yeast and molds in dairy industry: An update. In *Dairy processing: Advanced research to applications* (pp. 243–262). Springer. https://doi.org/10.1007/978-981-15-2608-4_12

Banerjee, A., & Choudhury, M. (n.d.). Microbiological degradation of organic pollutants from industrial wastewater. In *Nanobiotechnology for green environment*. CRC Press. https://doi.org/10.1201/9780367461362-4

Bansal, A., Shinde, O., & Biodegrad, S. S.-J. B. (2018). Industrial wastewater treatment using phycoremediation technologies and co-production of value-added products. *Journal of Bioremediation & Biodegradation*. https://doi.org/10.4172/2155-6199.1000428

Barry, A. N., Starkenburg, S. R., & Sayre, R. T. (2015). Strategies for optimizing algal biology for enhanced biomass production. *Frontiers in Energy Research, 3*. https://doi.org/10.3389/FENRG.2015.00001/FULL

Beal, C. M., Stillwell, A. S., King, C. W., Cohen, S. M., Berberoglu, H., Bhattarai, R. P., Connelly, R. L., Webber, M. E., & Hebner, R. E. (2012). Energy return on investment for algal biofuel production coupled with wastewater treatment. *Wiley Online Library, 84*(9), 692–710. https://doi.org/10.2175/106143012X13378023685718

Beltrán, S., Irizar, I., & Plants, E. A.-S. T. (n.d.). *Instrumentation, monitoring and real-time control strategies for efficient sewage treatment plant operation*. Library.Oapen.Org. Retrieved October 2, 2024, from https://library.oapen.org/bitstream/handle/20.500.12657/43789/1/external_content.pdf#page=209

Bergis, H., Betts, G., & Binet, R. (n.d.). *P3-160 challenge tests to study inactivation potential and kinetic parameters (ISO 20976-2: 2022)*. Library.Wur.Nl. Retrieved October 1, 2024, from https://library.wur.nl/WebQuery/wurpubs/fulltext/645250

Bipin Patwardhan, S., Savla, N., Pandit, S., Kumar Gupta, P., Singh Mathuriya, A., Lahiri, D., Jadhav, D. A., Kumar Rai, A., Rani Ray, R., Singh, V., Kumar, V., Prasad, R., Treatment Appl Sci, W., & Fernandez-Morales, J. (2021). Microbial fuel cell united with other existing technologies for enhanced power generation and efficient wastewater treatment. *Applied Sciences*. https://doi.org/10.3390/app112210777

Bleuven, C., & Landry, C. R. (2016). Molecular and cellular bases of adaptation to a changing environment in microorganisms. *Proceedings of the Royal Society B: Biological Sciences, 283*(1841). https://doi.org/10.1098/RSPB.2016.1458

Brar, S., Kumar, P., & Cuprys, A. (2022). *Modular treatment approach for drinking water and wastewater*. https://books.google.com/books?hl=en&lr=&id=wxxoEAAAQBAJ&oi=fnd&pg=PP1&dq=%E2%80%A2+Modular+Systems:+Modular+hybrid+systems+incorporate+separate+modules+for+fungal+and+algal+treatment,+allowing+for+flexible+configuration+and+optimization&ots=De-SaGjgy0&sig=MpSVJ1fUm9O7YCGF1jS7f04Tfps

Carvalho, A. (2006). Microalgal reactors: A review of enclosed system designs and performances. *Biotechnology Progress, 22*(6), 1490–1506. https://doi.org/10.1021/bp060065r

Cescon, A. (n.d.). Filtration process and alternative filter media material in water treatment. *Water*. https://doi.org/10.3390/w12123377

Chowdhary, P., Shukla, G., Raj, G., Ferreira, F. R., & Bharagava, R. N. (2019). Microbial manganese peroxidase: A ligninolytic enzyme and its ample opportunities in research. *SN Applied Sciences, 1*(1), 45. https://doi.org/10.1007/s42452-018-0046-3

Chu, R., Li, S., Zhu, L., Yin, Z., Hu, D., & Liu, C. (2001). A review on co-cultivation of microalgae with filamentous fungi: Efficient harvesting, wastewater treatment and biofuel production. *Renewable and Sustainable Energy Reviews, 139*, 110689.

Collado, S., Oulego, P., Suárez-Iglesias, O., Díaz, M., Clavería, J., & Oviedo, S. (n.d.). Leachates and natural organic matter. A review of their biotreatment using fungi. *Waste Management, 96*, 108–120.

Conference, M. T.-P. (2014). *Yeast biotechnology: Diversity and applications*. Researchgate.Net. https://www.researchgate.net/profile/Mustafa-Tuerker-3/publication/285598626_Yeast_Biotechnology_Diversity_and_Applications/links/5661d36408ae4931cd5c05ae/Yeast-Biotechnology-Diversity-and-Applications.pdf.

Conferences, P. F.-E. W. (n.d.). Nonlinear modeling of activated sludge process using the Hammerstein-Wiener structure. https://doi.org/10.1051/2016

Coppola, G., Gaudio, M. T., Lopresto, G., Calabro, V., Curcio, S., & Chakraborty, S. (2021). Bioplastic from renewable biomass: A facile solution for a greener environment. *Earth Systems and Environment, 5*(2), 231–251. https://doi.org/10.1007/s41748-021-00208-7

Dare Asemoloye, M., Marchisio, M. A., Gupta, V. K., & Pecoraro, L. (2021). Genome-based engineering of ligninolytic enzymes in fungi. *Microbial Cell Factories, 20*(1). https://doi.org/10.1186/s12934-021-01510-9

Dębowski, M., Świca, I., Kazimierowicz, J., Energies, M. Z.-, & 2022. Large scale microalgae biofuel technology—Development perspectives in light of the barriers and limitations. Energies. https://doi.org/10.3390/en16010081.

Deng, Y. (2018). *Improvements to the performance of trickling filters by inclusion of alternative surface-active media*. https://core.ac.uk/download/pdf/288358191.pdf.

Díaz, E. (2004). *Bacterial degradation of aromatic pollutants: A paradigm of metabolic versatility*. https://digital.csic.es/handle/10261/2134.

Dighton,J.(2018).*Fungiinecosystemprocesses*.CRCPress.https://doi.org/10.1201/9781315371528

Dragosits, M. (2013). Adaptive laboratory evolution–principles and applications for biotechnology. *Microbial Cell Factories, 12*(1), 64. https://doi.org/10.1186/1475-2859-12-64

Duan, M., O'Dwyer, E., & Stuckey, D. (2019). Wastewater to resource: Design of a sustainable phosphorus recovery system. *ChemistryOpen, 8*(8), 1109–1120. https://doi.org/10.1002/open.201900189

Egbuta, M., & Mwanza, M. (1995). Health risks associated with exposure to filamentous fungi. *International Journal of Environmental Research and Public Health, 8*(2), 161–179.

Ene, I., & Cheng, S. (2013). Growth of Candida albicans cells on the physiologically relevant carbon source lactate affects their recognition and phagocytosis by immune cells. *American Society for Microbiology, 81*(1), 238–248. https://doi.org/10.1128/IAI.01092-12

Engineering. (n.d.). *Integrating the three E's in wastewater treatment: Efficient design, economic viability, and environmental sustainability*. Elsevier. Retrieved August 31, 2024, from https://www.sciencedirect.com/science/article/pii/S2211339819300371

Ferrentino, R., Langone, M., & Fiori, L. (2023). Full-scale sewage sludge reduction technologies: A review with a focus on energy consumption. *Water*. https://doi.org/10.3390/w15040615

Ganeshan, S., & Kim, S. (2021). Scaling-up production of plant endophytes in bioreactors: Concepts, challenges and perspectives. *Bioprocessing, 8*(1), 63. https://doi.org/10.1186/s40643-021-00417-y

Garg, N. (2011). Arbuscular mycorrhizal networks: Process and functions. *Agriculture, 30*, 581–599.

Ghosh, S., Rusyn, I., Dmytruk, O. V., Dmytruk, K. V., Onyeaka, H., Gryzenhout, M., & Gafforov, Y. (2023). Filamentous fungi for sustainable remediation of pharmaceutical compounds, heavy metal and oil hydrocarbons. *Frontiers in Bioengineering and Biotechnology, 11*. https://doi.org/10.3389/FBIOE.2023.1106973/FULL

Gkotsis, P. K., Ch Banti, D., Peleka, E. N., Zouboulis, A. I., Samaras, P. E., Matis, K. A., & Henson, M. (2014). Fouling issues in membrane bioreactors (MBRs) for wastewater treatment: Major mechanisms, prevention and control strategies. *Processes, 2*, 795–866. https://doi.org/10.3390/pr2040795

Gleason, F., & Daynes, C. (n.d.). *Some zoosporic fungi can grow and survive within a wide pH range*. Elsevier. Retrieved August 25, 2024, from https://www.sciencedirect.com/science/article/pii/S1754504809000683

Green, R. S.-M. (n.d.). *Microbial-assisted remediation of food processing industry waste*. Elsevier. Retrieved August 26, 2024, from https://www.sciencedirect.com/science/article/pii/B978032390452000027X

Gude, V. (n.d.) Beneficial bioelectrochemical systems for energy, water, and biomass production. *Academia.Edu*. Retrieved September 23, 2024, from https://www.academia.edu/download/42561141/Beneficial_Bioelectrochemical_Systems_fo20160210-3976-y8og2h.pdf

Hale, T. (n.d.). Improving supply chain disaster preparedness: A decision process for secure site location. *International Journal of Physical Distribution & Logistics Management*. https://doi.org/10.1108/09600030510594576

Hochheimer, J. (n.d.). *Biological filters: Trickling and RBC design*. Citeseer. Retrieved August 24, 2024, from https://citeseerx.ist.psu.edu/document?repid=rep1&type=pdf&doi=dc151f42f0970635fff4f19e2779ba02fd4e17e7

Hu, Z., Qi, Y., Zhao, L., & Chen, G. (2019). Interactions between microalgae and microorganisms for wastewater remediation and biofuel production. *Waste and Biomass Valorization, 10*(12), 3907–3919. https://doi.org/10.1007/S12649-018-0325-7

Ileana, G., Iacobută, I., Höhne, N., Laura Van Soest, H., & Leemans, R. (2021). Transitioning to low-carbon economies under the 2030 agenda: Minimizing trade-offs and enhancing co-benefits of climate-change action for the SDGs. *Sustainability*. https://doi.org/10.3390/su131910774

Jeong, H. Y., Park, M., Jeong, H. Y., Jun, S.-C., & Cheon, J.-Y. (2018). A review on clogging mechanisms and managements in aquifer storage and recovery (ASR) applications. *Geosciences Journal, 22*(4), 667–679. https://doi.org/10.1007/s12303-017-0073-x

Joshi, J. B., & Nandakumar, K. (2015). Computational modeling of multiphase reactors. *Annual Review of Chemical and Biomolecular Engineering, 6*, 347–378. https://doi.org/10.1146/ANNUREV-CHEMBIOENG-061114-123229

Kagan, R., & Gunningham, N. (n.d.). *Explaining corporate environmental performance: How does regulation matter?* Cambridge.Org. Retrieved September 30, 2024, from https://www.cambridge.org/core/journals/law-and-society-review/article/explaining-corporate-environmental-performance-how-does-regulation-matter/F224359D86E66A6F1CA696F0454E3F6A

Kantharaj, P., & Boobalan, B. (2017). Lignocellulose degrading enzymes from fungi and their industrial applications. *International Journal of Current Research and Review*. https://doi.org/10.7324/IJCRR.2017.9211

Kumar, A., & Chaturvedi, A. (n.d.). *Fungal phytoremediation of heavy metal-contaminated resources: Current scenario and future prospects*. Springer. Retrieved September 30, 2024, from https://link.springer.com/chapter/10.1007/978-3-030-25506-0_18

Kumar, V., Thakur, I. S., & Shah, M. P. (2020). Bioremediation approaches for treatment of pulp and paper industry wastewater: Recent advances and challenges. *Microbial Bioremediation & Biodegradation*. https://doi.org/10.1007/978-981-15-1812-6_1

Lanrewaju, A. A., Enitan-Folami, A. M., Sabiu, S., & Swalaha, F. M. (2022). A review on disinfection methods for inactivation of waterborne viruses. *Frontiers in Microbiology, 13*. https://doi.org/10.3389/FMICB.2022.991856/FULL

Le, T., Tran, D., & Van Do, T. (2019). Design considerations of microalgal culture ponds and photobioreactors for wastewater treatment and biomass cogeneration. In *Microalgae biotechnology for development of biofuel and wastewater treatment* (pp. 535–567). Springer. https://doi.org/10.1007/978-981-13-2264-8_21

Lee, J., Ni, J., Singh, J., Jiang, B., Azamfar, M., & Feng, J. (2020). Intelligent maintenance systems and predictive manufacturing. *Journal of Manufacturing Science and Engineering, 10*(1115/1), 4047856.

Liu, D., Yang, X., Zhang, L., Tang, Y., He, H., Liang, M., Tu, Z., & Zhu, H. (2022). Immobilization of biomass materials for removal of refractory organic pollutants from wastewater. *International Journal of Environmental Research and Public Health*. https://doi.org/10.3390/ijerph192113830

Liu, S.-H., Zeng, G.-M., Niu, Q.-Y., Liu, Y., Zhou, L., Jiang, L.-H., Tan, X.-F., Xu, P., Zhang, C., & Cheng, M. (2016). Bioremediation mechanisms of combined pollution of PAHs and heavy metals by bacteria and fungi: A mini review. *Elsevier*. https://doi.org/10.1016/j.biortech.2016.11.095

Lu, Y., Qin, M., Yuan, H., Abu-Reesh, I. M., He, Z., Angelakis, A. N., Snyder, S. A., & Xagoraraki, I. G. (2015). When bioelectrochemical systems meet forward osmosis: Accomplishing wastewater treatment and reuse through synergy. *Mdpi.Com, 7*, 38–50. https://doi.org/10.3390/w7010038

Lürling, M., Mello, M. M., van Oosterhout, F., Domis, L. de S., & Marinho, M. M. (2018). Response of natural cyanobacteria and algae assemblages to a nutrient pulse and elevated temperature. Frontiers in Microbiology, 9. https://doi.org/10.3389/FMICB.2018.01851/FULL.

Maheshwari, R., Bharadwaj, G., & Bhat, M. K. (2000). Thermophilic fungi: Their physiology and enzymes. *Microbiology and Molecular Biology Reviews, 64*(3), 461–488. https://doi.org/10.1128/MMBR.64.3.461-488.2000

Makepa, D. C., Nocera, F., Kumba, H., Makepa, D. C., Charamba, A. N., & Olanrewaju, O. A. (2024). Towards circular economy: Integrating waste Management for Renewable Energy Optimization in Zimbabwe. *Sustainability*. https://doi.org/10.3390/su16125014

Manan, M., & Bioeng, C. W.-J. A. B. (2017). Design aspects of solid state fermentation as applied to microbial bioprocessing. *Journal of Applied Biotechnology & Bioengineering*. https://doi.org/10.15406/jabb.2017.04.00094

Martín-Pascual, J., Leyva-Díaz, J. C., & Poyatos, J. M. (2016). Treatment of urban wastewater with pure moving bed membrane bioreactor technology at different filling ratios, hydraulic retention times and temperatures. *Annals of Microbiology, 66*(2), 607–613. https://doi.org/10.1007/S13213-015-1145-7

Marzluf, G. A. (1997). Genetic regulation of nitrogen metabolism in the fungi. *Microbiology and Molecular Biology Reviews, 61*(1), 17–32. https://doi.org/10.1128/MMBR.61.1.17-32.1997

McNeil, J., Borg, M., Kennedy, E., Cui, V., Puntha, H., & Rashid, Z. (2017). *Scale-up handbook*. https://aclproject.org.uk/wp-content/uploads/2017/12/SCALE-UP-Handbook-2017-18.pdf

Meerbergen, K. (2018). Decolorization of textile wastewater, with an emphasis on microbial treatment processes. https://lirias.kuleuven.be/retrieve/498295

Mendez-Ruiz, J. I., Barcia-Carreño, M. B., Mejía-Bustamante, L. J., Cornejo-Pozo, Á. K., Salas-Vázquez, C. A., & Valverde-Armas, P. E. (2023). Assessment of the performance of a water treatment plant in Ecuador: hydraulic resizing of the treatment units. *Sustainability*. https://doi.org/10.3390/su15021235

Muruganandham, M., Suri, R. P. S., Jafari, S., Sillanpää, M., Lee, G.-J., Wu, J. J., & Swaminathan, M. (2014). Recent developments in homogeneous advanced oxidation processes for water and wastewater treatment. *Wiley Online Library*. https://doi.org/10.1155/2014/821674

Naseema Rasheed, R., Pourbakhtiar, A., Mehdizadeh Allaf, M., Baharlooeian, M., Rafiei, N., Alishah Aratboni, H., Morones-Ramirez, J. R., & Winck, F. V. (2023). Microalgal co-cultivation-recent methods, trends in omic-studies, applications, and future challenges. *Frontiers in Bioengineering and Biotechnology, 11*. https://doi.org/10.3389/FBIOE.2023.1193424/FULL

Neethirajan, S., Ragavan, V., & Weng, X. (n.d.). Biosensors for sustainable food engineering: Challenges and perspectives. *Biosensors*. Retrieved October 1, 2024, from https://www.mdpi.com/2079-6374/8/1/23

Nguyen, L. N., Aditya, L., Vu, H. P., Hasan Johir, A., Bennar, L., Hoang, N. B., Zdarta, J., Long, A., & Nghiem, D. (2022). Nutrient removal by algae-based wastewater treatment. *Current Pollution Reports, 1*(4), 3. https://doi.org/10.1007/s40726-022-00230-x

Nicula, N.-O., Lungulescu, E.-M., Rîmbu, G. A., Marinescu, V., Corbu, V. M., & Csutak, O. (2023). Bioremediation of wastewater using yeast strains: An assessment of contaminant removal efficiency. International Journal of Environmental Research and Public Health. https://doi.org/10.3390/ijerph20064795.

Novak Babič, M., Gunde-Cimerman, N., Vargha, M., Tischner, Z., Magyar, D., Veríssimo, C., Sabino, R., Viegas, C., Meyer, W., Brandão, J., Ahmed, W., & Beale, D. J. (2017). Fungal contaminants in drinking water regulation? A tale of ecology, exposure, purification and clinical relevance. *International Journal of Environmental Research and Public Health, 14*, 636. https://doi.org/10.3390/ijerph14060636

Nsair, A., Onen Cinar, S., Alassali, A., Qdais, H. A., & Kuchta, K. (n.d.). Operational parameters of biogas plants: A review and evaluation study. *Energies*. https://doi.org/10.3390/en13153761

Pandit, A., Adholeya, A., Cahill, D., Brau, L., & Kochar, M. (2020). Microbial biofilms in nature: Unlocking their potential for agricultural applications. *Journal of Applied Microbiology, 129*(2), 199–211. https://doi.org/10.1111/JAM.14609

Park, J., & Kim, K. (n.d.). Recent advances in information and communications technology (ICT) and sensor technology for monitoring water quality. *Water*. https://doi.org/10.3390/w12020510

Plascencia-Jatomea, M., & Susana, M. (n.d.). *Aspergillus spp. (black mold)*. Elsevier. https://doi.org/10.1016/B978-0-12-411552-1.00003-X

Poblete, I., & Araujo, O. (n.d.). Sewage-Water treatment and sewage-sludge management with power production as bioenergy with carbon capture system: A review. *Processes*. Retrieved October 1, 2024, from https://www.mdpi.com/2227-9717/10/4/788

Pöysä, T., Simo, S. P., Advisors, S., Sc, L., Kuokkanen, A., Sc, M., & Lukkarinen, T. (2022). *Automated quality control and correction of online measurement data in a distributed control system*. https://aaltodoc.aalto.fi/items/19e4958e-2de5-4650-926f-fc96fd0cc9bf.

Procházková, G., Brányiková, I., Zachleder, V., & Brányik, T. (2014). Effect of nutrient supply status on biomass composition of eukaryotic green microalgae. *Journal of Applied Phycology, 26*(3), 1359–1377. https://doi.org/10.1007/S10811-013-0154-9

Raj, T., Sompura, S., & Chandrasekhar, K. (n.d.). *Technology development and challenges for the transformation of municipal solid waste into sustainable energy production*. Elsevier. Retrieved September 29, 2024, from https://www.sciencedirect.com/science/article/pii/S0961953423002647

Rani, S., Singh, R. P., Gunjyal, N., Ojha, C. S. P., & Asce, F. (2020). Review of challenges for algae-based wastewater treatment: Strain selection, wastewater characteristics, abiotic, and biotic factors. *Ascelibrary, 25*(2). https://doi.org/10.1061/(ASCE)HZ.2153-5515.0000578

Raychaudhuri, A. (n.d.). *Microbial degradation of lignin: Conversion, application, and challenges*. Elsevier. Retrieved August 25, 2024, from https://www.sciencedirect.com/science/article/pii/B9780323856577000080

Rjwave.Org. (2016). *Activated sludge treatment process–concept and system design*. https://rjwave.org/ijedr/viewpaperforall.php?paper=IJEDR1602156.

Russell, N. J. (2016). Cold adaptation of microorganisms. *Philosophical Transactions of the Royal Society of London. B, Biological Sciences, 326*(1237), 595–611. https://doi.org/10.1098/rstb.1990.0034

Sánchez, C. (2008). *Lignocellulosic residues: Biodegradation and bioconversion by fungi*. Elsevier. https://doi.org/10.1016/j.biotechadv.2008.11.001

Sankaran, S., Khanal, S., & Jasti, N. (2010). Use of filamentous fungi for wastewater treatment and production of high value fungal byproducts: A review. *Critical Reviews in Environmental Science and Technology, 40*(5), 400–449. https://doi.org/10.1080/10643380802278943

Sartori, R., Deprá, M. (n.d.). *The role of light on the microalgae biotechnology: Fundamentals, technological approaches, and sustainability issues*. Ingentaconnect.com. Retrieved August 31, 2024, from https://www.ingentaconnect.com/content/ben/biot/2024/00000018/00000001/art00005

Schellenberg, T., Subramanian, V., Ganeshan, G., Tompkins, D., & Pradeep, R. (2020). Wastewater discharge standards in the evolving context of urban sustainability–The Case of India. *Frontiers in Environmental Science, 8*. https://doi.org/10.3389/FENVS.2020.00030/FULL

Sellami, K., Couvert, A., & Nasrallah, N. (n.d.). *Peroxidase enzymes as green catalysts for bioremediation and biotechnological applications: A review*. Elsevier. Retrieved August 24, 2024, from https://www.sciencedirect.com/science/article/pii/S0048969721055777

Sellner, K., Doucette, G., & Microbiology, G. K.-J. (2008). Harmful algal blooms: Causes, impacts and detection. *Journal of Industrial Microbiology and Biotechnology*. https://doi.org/10.1504/IJEP.2008.020569

Shahid, A., Malik, S., Zhu, H., Xu, J., Zohaib Nawaz, M., Nawaz, S., Alam, A., & Mehmood, M. A. (n.d.). Multifaceted roles of microalgae in the application of wastewater biotreatment: A review. *Environmental Pollution*. https://doi.org/10.1016/j.scitotenv.2019.135303

Sheriff, B., & Kachalla, B. (n.d.). Sustainable implementation of water and wastewater infrastructures in developing countries: a review. *Journal of Emerging Trends in Engineering and Applied Sciences, 5*, 10.

Silva, A., Delerue-Matos, C., & Figueiredo, S. (n.d.). The use of algae and fungi for removal of pharmaceuticals by bioremediation and biosorption processes: A review. *Water*. https://doi.org/10.3390/w11081555

Silva, C., Magano, J., Moskalenko, A., Nogueira, T., Alzira Pimenta Dinis, M., & Fernando Pedrosa Sousa, H. (n.d.). Sustainable management systems standards (SMSS): Structures, roles, and practices in corporate sustainability. *Sustainability*. https://doi.org/10.3390/su12155892

Singh, A., & Singh, A. (2022). Microbial degradation and value addition to food and agriculture waste. *Current Microbiology, 79*(4). https://doi.org/10.1007/S00284-022-02809-5

Singh, D., & Biologia, N. G. (2020). Microbial laccase: A robust enzyme and its industrial applications. *Biologia, 75*(8), 1183–1193. https://doi.org/10.2478/s11756-019-00414-9

Singh, D., & Chen, S. (2008). The white-rot fungus Phanerochaete chrysosporium: Conditions for the production of lignin-degrading enzymes. *Applied Microbiology and Biotechnology, 81*(3), 399–417. https://doi.org/10.1007/S00253-008-1706-9

Soponpongpipat, N., Nanetoe, S., & Processes, P. C. (n.d.). Thermal and torrefaction characteristics of a small-scale rotating drum reactor. *Processes, 8*, 489.

Su, X., Sutarlie, L., & Loh, X. J. (2020). Sensors, biosensors, and analytical technologies for aquaculture water quality. *Research*. https://doi.org/10.34133/2020/8272705

Sukačová, K., Lošák, P., Brummer, V., Máša, V., Vícha, D., & Zavřel, T. (2021). Perspective design of algae photobioreactor for greenhouses—A comparative study. *Energies*. https://doi.org/10.3390/en14051338

Sustainability, A. S. (2019). The role of constructed wetlands as green infrastructure for sustainable urban water management. *Sustainability*. https://doi.org/10.3390/su11246981

Technology. (2012). Application of lipase in industry. *Asian Journal of Pharmaceutical Technology and Innovation, 2*, 33–37. https://www.indianjournals.com/ijor.aspx?target=ijor:ajpt&volume=2&issue=2&article=001

Teles, I., Cabanelas, D., Ruiz, J., Arbib, Z., Chinalia, F. A., Garrido-Pérez, C., Rogalla, F., Andrade Nascimento, I., & Perales, J. A. (n.d.). Comparing the use of different domestic wastewaters for coupling microalgal production and nutrient removal. *Bioresource Technology*. https://doi.org/10.1016/j.biortech.2012.12.152

Thalla, A. K., & Devika, R. S. (2024). Integrated hybrid biological Systems in wastewater treatment and recycling. In *Biological and hybrid wastewater treatment technology*. Springer. https://doi.org/10.1007/978-3-031-63046-0_1

Thesis, M. P., Abdul, M., & Sarker, Q. (2018). *Water neutral developments: How to successfully integrate micro-algae systems into wastewater management*. http://www.ros.hw.ac.uk/handle/10399/3459

Tzoupanos, N. (2008). *Coagulation-flocculation processes in water/wastewater treatment: The application of new generation of chemical reagents*. Researchgate.Net. https://www.researchgate.net/profile/Anastasios-Zouboulis/publication/229039796_Coagulation-flocculation_processes_in_waterwastewater_treatment_the_application_of_new_generation_of_chemical_reagents/links/0046352a3555433fc5000000/Coagulation-flocculation-processes-in-water-wastewater-treatment-the-application-of-new-generation-of-chemical-reagents.pdf

Vishnoi, N. (2019). *Bioremediation: New prospects for environmental cleaning by fungal enzymes*. Springer. https://doi.org/10.1007/978-3-030-25506-0_2

Wang, X., & Reports, Y. H.-C. P. (n.d.). *Microalgae biofilm and bacteria symbiosis in nutrient removal and carbon fixation from wastewater: A review*. Springer. https://doi.org/10.1007/s40726-022-00214-x

Waqas, S., Yub Harun, N., Sambudi, N. S., Bilad, M. R., Abioye, K. J., Ali, A., & Abdulrahman, A. (2023). A review of rotating biological contactors for wastewater treatment. *Water*. https://doi.org/10.3390/w15101913

Williams, B. (2014). Adaptive management: From more talk to real action. *Environmental Management, 53*(2), 465–479. https://doi.org/10.1007/s00267-013-0205-7

Wolff, S., Weber, F., Kerpen, J., Winklhofer, M., Engelhart, M., & Barkmann, L. (2020). Elimination of microplastics by downstream sand filters in wastewater treatment. *Water*. https://doi.org/10.3390/w13010033

Yang, X., Wu, X., & Hao, H. (2008). Mechanisms and assessment of water eutrophication. *Journal of Zhejiang University Science B, 9*(3), 197–209. https://doi.org/10.1631/jzus.B0710626

Yu, H., Lei, P., Ma, J., Jin, J., Ma, Y., & Fang, Y. (n.d.). *The potential of white-rot fungi for algal control: Mechanisms, strategies, and challenges*. Elsevier. Retrieved August 26, 2024, from https://www.sciencedirect.com/science/article/pii/S0013935123015426

Zabochnicka, M., Krzywonos, M., Romanowska-Duda, Z., Szufa, S., Darkalt, A., & Mubashar, M. (2022). Algae-based biorefinery as a sustainable renewable resource. *Circular Economy and Sustainability*. https://doi.org/10.3390/life12101480

Zerin Luna, I., Nuruzzaman Khan, M., Minhajul Islam, M., Sharmeen, S., Salem, K. S., Rashid, T. U., Zaman, A., Haque, P., & Mizanur Rahman, M. (2016). Cellulases: From lignocellulosic biomass to improved production. *Energies*. https://doi.org/10.1016/B978-0-444-63507-5.00021-6

Zhu, S., Huo, S., & Feng, P. (2019). Developing designer microalgal consortia: A suitable approach to sustainable wastewater treatment. In *Microalgae biotechnology for development of biofuel and wastewater treatment* (pp. 569–598). Springer. https://doi.org/10.1007/978-981-13-2264-8_22

Zuleta-Correa, A. (n.d.) *Use of white rot fungi in the degradation of an azo dye from the textile industry*. Scielo.Org.Co. Retrieved August 25, 2024, from http://www.scielo.org.co/scielo.php?pid=S0012-73532016000400015&script=sci_arttext&tlng=en

Chapter 11
Smart Materials and Nanotechnology for Fungal-Based Treatment Systems for Fungi in Waste Water Treatment

Sharmi Ganguly and Joydip Sengupta

Abstract The environment is at risk from environmental contamination caused by both organic and inorganic contaminants. In recent times, there has been a notable surge in the frequency and endurance of pollutants. There have been several methods used to improve naturally occurring water quality to a level fit for human consumption. Within this framework, the use of state-of-the-art nanotechnology to traditional process engineering opens the door for technological advances in advanced water and wastewater technology. The wastewater originating from various key sectors, including agriculture, industries, and households, is diverse and requires appropriate treatment prior to release into the environment. Untreated wastewater has a high concentration of dissolved solids, both harmful and innocuous, which can harm aquatic life in freshwater if they are not properly handled. Wastewater needs to be properly handled because of all the negative effects it has on the environment. The distinct physical and chemical characteristics of nanomaterials increase their appeal for use in various applications. On the other hand, biotechnology is a vital component of most established companies, providing an environmentally sustainable pathway for the production of active ingredients. The growth in heavy metal ion pollution in effluents, which is brought on by growing urbanisation and human activity, is one of the most dangerous environmental contaminants. The sorption method's low cost, design and operation flexibility, reproducibility, and notable performance make it a popular choice. To get rid of metal ions, more eco-friendly sorbents are therefore needed now more than ever. The use of fungus as adsorbent has drawn a lot of interest because of its special qualities, which include its high absorption capacity, availability, affordability, environmental friendliness, and the presence of chitin and chitosan in its cell wall. The application of certain process-induced microorganisms or the use of microorganism metabolites for synthesis and production processes in order to harness "bio-in-nanotechnology"

S. Ganguly
Electronics and Communication Engineering Department, KPRIET, Coimbatore, India
e-mail: sharmiganguly@kpriet.ac.in

J. Sengupta (✉)
Department of Electronic Science, Jogesh Chandra Chaudhuri College, Kolkata, India

C. M. Hussain et al. (eds.), *Fungi in Waste Water Treatment*, Springer Water,
https://doi.org/10.1007/978-3-032-12693-1_11

or "nano-in-biotechnology." Due to their ability to produce vital enzymes, acids, and biosurfactants as well as to control some pest illnesses and crops on a wide scale, fungi have garnered a great deal of scientific attention. Fungi have been effectively used in nanotechnology to produce metal and metal oxide nanoparticles, including Ag, Au, Se, Fe, Cu, ZnO, and CuO. The goal of this chapter is to illustrate the most recent developments in fungal nanobiotechnology by outlining the primary group of fungi that can be used to produce nanomaterials sustainably and their potential uses in biotechnological fields. In order to overcome this disparity, this review emphasises the potential for fungal biorefinery in specific industrial wastewaters, especially those with low pH levels. The financial effects of chemical separation and biomass recovery are also highlighted. Using a thorough examination of several industrial sectors, including paper pulp, textiles, food processing, and pharmaceuticals, this study investigates how filamentous fungus can efficiently capture wastewater's nutrient-rich content to create useful resources.

Keywords Biosurfactants · Sorption · Microorganisms · Biotechnology · Fungi

11.1 Introduction

On Earth, water makes up over two-thirds of the surface. For many years, the world has struggled with the scarcity of clean water. Nature has a system in place to recycle water and supply us with a sufficient amount of pure water. Unplanned industrialization and unchecked population increase have, however, interfered with the natural purification processes, resulting in a scarcity of drinkable water. Drinking contaminated water is the source of around 90% of all illnesses in the majority of underdeveloped nations. The following factors contribute to the ongoing rise in water demand: (1) population growth and migration to drought-prone areas; (2) rapid industrial development and rising per capita water use; and (3) climate change, which is causing altered weather patterns in populated areas (Pearce, 2008). Because untreated wastewater or man-made wastes are released into the environment, a significant percentage of drinking water sources around the world are found to be contaminated with different toxins and harmful bacteria (Fishman, 2011). However, a lot of viruses (Rizzo et al., 2013) and man-made chemicals that are getting into both urban and rural water bodies (The Challenge of Micropollutants in Aquatic Systems, n.d.) are threatening the quality of the water. It has been acknowledged that one of the main causes of aquatic contamination worldwide is wastewater discharge from industrial and municipal treatment facilities (Hollender et al., 2008). The majority of wastewater from homes and businesses in many developing nations is dumped straight into streams of water either without any treatment steps or merely after first treatment (Kaudare et al., 2019). Therefore, safeguarding our ecology depends heavily on waste water being properly treated before being released. Businesses all over the world have implemented a number of plans to treat wastewater before releasing it into the environment. Large plant areas, high energy requirements, high capital investment, and operation and maintenance costs are all associated with current wastewater treatment systems (Helmer & Hespanhol, 1997). Because developing-nation industries have low profit margins, they are unable to buy such costly technologies, and as a result, their adherence to environmental laws and norms is generally minimal. According to this, the majority of wastewater finds its way into untreated natural water bodies. The average composition of untreated waste water, together with some of its main pollutants, is depicted in Fig. 11.1.

Several environmental issues will arise from the discharge of untreated wastewater into water bodies without any treatment procedures, including: Untreated wastewater with a high organic matter content will use up dissolved oxygen to meet wastewater's biochemical oxygen demand (BOD), which will reduce the amount of dissolved oxygen in the stream needed by aquatic life; Untreated wastewater typically contains a high concentration of pathogenic, or disease-causing, microorganisms and toxic compounds that can enter the human digestive tract and endanger human health; Lakes and streams may become eutrophic due to the presence of nutrients in wastewater that can promote the growth of aquatic plants and algal blooms; The decomposition of the organic compounds present in wastewater can lead to the production of large quantities of malodorous gases (Topare et al., 2011). Therefore, before wastewater is allowed to enter natural water bodies, it must be treated. Industrial wastewater must go through treatment procedures before being recycled or released into the environment in order to guarantee that the effluent water is as pure as possible and to remove any solids that could endanger human health and the environment (More et al., 2010). For the water to fulfil the required requirements for cleanliness, this procedure is crucial. Traditional techniques (physical and chemical) for treating wastewater approaches), with drawbacks like high

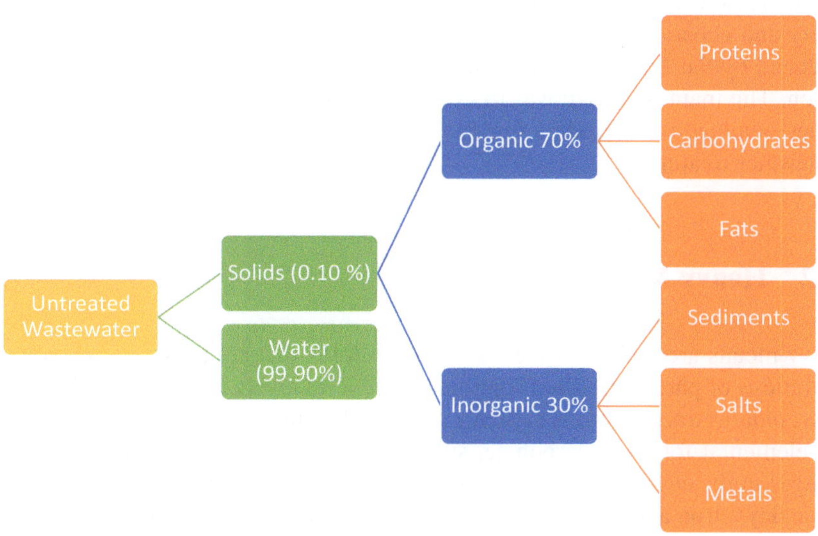

Fig. 11.1 Typical composition of untreated wastewater

expenses and energy usage, along with the creation of dangerous byproducts as a result of chemical use that impact the elimination of particular pollutants from industrial sewage (Dhanavade & Patil, 2023).

Using inexpensive, environmentally friendly, and safe methods like biological degradation, remediation, or decolorisation with different kinds of microorganisms is one way to solve the problem. The industrial sector commonly uses bacteria, fungi, and algae for biological wastewater treatment because of their low cost, simple operating systems, high biomass volume, energy efficiency, and capacity to generate value-added products that can be used for energy production and other purposes (Negi & Das, 2023). The most common microorganisms used in biological wastewater treatment systems to remove organic contaminants are bacterial strains. Contrarily, filamentous fungi have gotten far less attention than bacteria when it comes to treating wastewater, even though certain fungal strains can produce a number of degradative enzymes and surface proteins that are crucial for the biodegradation or biosorption of wastewater pollutants (Negi & Das, 2023). The natural decomposition of organic materials by fungi and their capacity to transform organic waste into beneficial molecules are being utilised to create lucrative goods (Singh et al., 2021). Filamentous fungus have evolved a wide variety of enzyme systems to support themselves through parasitic or saprotrophic ways since they are heterotrophic. They are present on a wide variety of substrates and under a wide range of environmental circumstances. Also, compared to other kinds of microorganisms, filamentous fungus have more benefits. Filamentous fungi also have a lot of potential for treating industrial wastewaters, according to new studies. For example, filamentous fungus are more resilient to hostile environments than bacteria, generate more biomass, and are easier to isolate than bacteria (Dhanavade & Patil, 2023).

This review aimed to emphasise the advantages of both environmental sustainability and economic viability by examining the potential of ligninolytic enzymes produced by the fungi in wastewater purification and metal contamination sequestration. The main objective was to give a thorough grasp of the potential and difficulties involved in expanding fungal biorefinery operations for efficient industrial wastewater treatment.

11.2 Heavy Metals

Heavy metals are a class of metals and metalloids that are poisonous even at concentrations of parts per billion (ppb) and have a relatively high density. The Earth's crust contains trace amounts of heavy metals. They may chemically exist in metallic and elemental forms in carbonate, sulphate, oxide, or silicate rocks. Erosion and changes in the environment cause soil, rivers, and groundwater to be washed away. These days, iron and nickel heavy metals make up the majority of the Earth's core. Actually, the Earth's core was formed around 5 billion years ago when heavy metals pierced the liquid mantle. Mines, automotive exhaust, and industrial wastes are some of the human sources of these metals' discharge into the environment.

Unlike other forms of organic pollution, heavy metals do not break down in the body of an organism and instead prefer to build up in the cell tissues of living things. Because of their extreme toxicity, endurance in the ecosystem, and capacity to bioaccumulate in the food chain, heavy metals are one of the biggest public health issues. In addition to essential elements like copper, iron, nickel, and zinc, these metals are also naturally occurring and include lead, cadmium, and mercury. According to Koller et al. (2018) and Yadav et al. (2019), a prolonged buildup of heavy metal ions in the soft tissues of living things would directly impact the nervous, safety, and cardiovascular systems' normal functions. For aquatic animals, particularly fish, heavy metals are substantial contaminants. They can affect marine species at different levels of the food chain and reach marine environments from a range of sources, such as industrial effluents, urban wastewater, agriculture, and air pollution. Three elements are regarded as dangerous metals with no biological value in the European Union: lead (Pb), cadmium (Cd), and mercury (Hg). This means that none of the known biological species are involved in essential physiological processes and are therefore poisonous.

Heavy metal poisoning of soil has grown to be a major environmental issue in many regions of the world. A major risk to human health, toxic heavy metals are both biodegradable and resistant to metabolic breakdown. The three routes of respiration, ingestion, and dermal absorption are how heavy metals enter urban soils, where they build up in adipose tissue before moving on to neurological and endocrine systems, where they disrupt cellular metabolism and the immune system.

Heavy metals are often divided into four types according to their toxicity and danger as:

1. Toxic heavy metals: Cd, As, Cr etc.
2. Radioactive metals: Am, V, Te, Th, Ra, R etc.
3. Metabolism fundamentals and essential metals: Fe, Na, Zn, Ca etc. and
4. Pseudo metals for the detection of biological effective: Se, Si, B, Ge etc.

The effluents must be purified using innovative and effective techniques before being released into the environment (Koller et al., 2018). Heavy metal toxicity may result in either acute or long-term issues. Chronic exposure to heavy metals causes the body to gradually develop conditions like Parkinson's, neurological, muscular, and Alzheimer's.

11.3 Fungi

Fungi are an essential component of the intricate network in the different ecosystems because they are essential in the breakdown of waste materials (Sharma et al., 2020). Due to their capacity for bioaccumulation, biodegradation, and biosorption, numerous genera of filamentous fungi, including Phanerochaete, Pleurotus,

Penicillium, Fusarium, Ganoderma, Lepiota, Trametes, Daldenia, and Aspergillus, are viable options for treating industrial wastewater containing dyes (Abd El-Zaher, 2010; Kunjadia et al., 2016). By producing a thermostable blue laccase, Aspergillusochraceus NCIM-1146, for instance, has been investigated for its enzymatic potential in dye removal (Telke et al., 2010). Reactive Golden Yellow HER, Reactive Navy Blue HER, and Methyl Orange could all be successfully decolored by 56–90% using this strain (Kurade et al., 2023). Cell-free filamentous fungal extract, actively growing fungal cultures, and fungal consortia with other microorganisms can all be utilised to decolorise dyes (Michalska et al., 2021; Ryu et al., 2017).

By producing a thermostable blue laccase, Aspergillusochraceus NCIM-1146, for instance, has been investigated for its enzymatic potential in dye removal. Reactive Golden Yellow HER, Reactive Navy Blue HER, and Methyl Orange could all be successfully decolored by 56–90% using this strain (Telke et al., 2010). Cell-free filamentous fungal extract, actively growing fungal cultures, and fungal consortia with other microorganisms can all be utilised to decolorise dyes (Michalska et al., 2021; Ryu et al., 2017). Though azoreductases and various peroxidases can catalyse the decontamination reactions in the presence of their appropriate cofactors, laccases are suitable for mycoremediation of olive mill waste, coffee pulp waste, and other industrial and agricultural wastes because of their unique advantages, which include cofactor-free activity and non-specific action for oxidation of xenobiotic materials (Dhanavade & Patil, 2023). Mycelial pellets have unique properties that make them useful for sludge dewatering, biosorption, and biomass aggregation. These characteristics include porosity, rapid growth, net-like texture, quick adaptation to various conditions, and efficient dewatering capabilities (Li et al., 2020; Veiter et al., 2018). Using natural filamentous fungus, Penicillium crustosum PWWS-6, immobilised on agar plugs, a low-cost and environmentally acceptable mycoremediation method has been developed recently for the quick removal of a highly toxic industrial Congo red in addition to mycelial pellets. The expression of different extracellular enzymes, including azoreductase, veratryl alcohol oxidase, NADH-DCIP reductase, and lignin peroxidase, at the molecular level during the decolorisation experiment, indicates that the process entails the biosorption of Congo red followed by mineralisation. Statistical methods were used to optimise the procedure, setting the inoculum volume at six fungal discs, the temperature at 27 °C, and the incubation time at 16 hours. The rate of dye removal was significantly increased, and the decolorisation time was shortened, as a consequence of this optimisation. Filamentous fungi can produce value-added products and facilitate clean water restoration by using wastewater and side streams from municipalities, industries, and wastewater treatment plants that are rich in organic and inorganic nutrients is shown in Fig. 11.2. Clean water that can be used again for human activities and value-added goods can be acquired by converting industrial side streams and eliminating micropollutants. Thus, a paradigm shift in how pollution mitigation is perceived can coexist with the realization of a sustainable circular society is shown in Fig. 11.2.

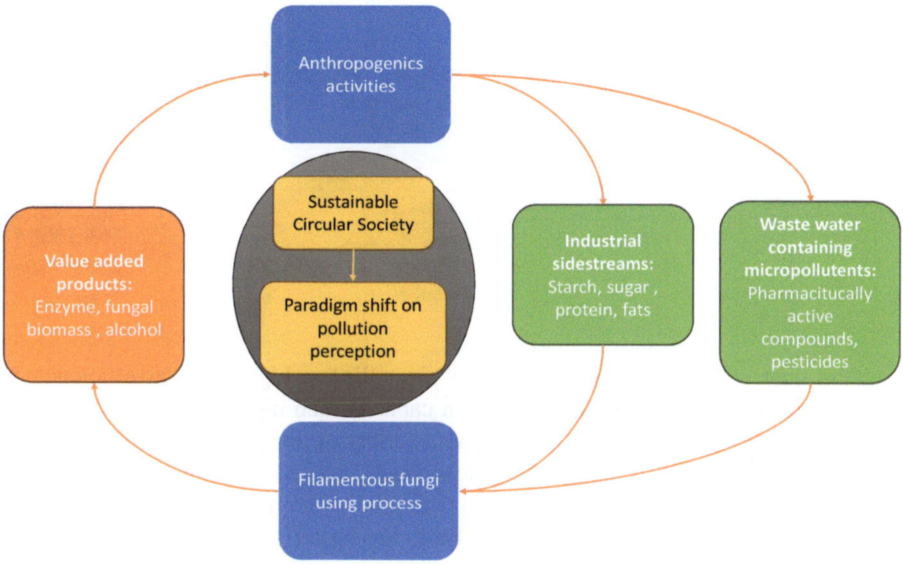

Fig. 11.2 An overview of the part filamentous fungi can play in reducing pollution

Numerous operational factors, including pH level, temperature, pollutant concentration, type and amount of carbon and nitrogen, oxygen, mass transfer, and agitation rate, might influence mycoremediation of industrial effluent wastes. Thus, it is essential to properly modify these parameters in order to guarantee the collection of trustworthy data and outcomes. Among these factors, pH is crucial for mycoremediation and substantially influences the application of filamentous fungi in industry because fungal-mediated systems typically fail quickly in extremely acidic (pH 2.0–4.0) or alkaline (pH 10) environments (Dhanavade & Patil, 2023).

11.4 Bio-Adsorption

An effective method for getting rid of harmful metal ions is bioadsorption, which uses cheap biosorbents. The physical-chemical process of biosorption is used to separate organic or inorganic materials contaminants and substances from solutions utilising biosorbents. This process happens naturally and is not influenced by the metabolism of microorganisms. One of the key steps in biotechnology for environmental protection is biosorption. The inactive adsorption of harmful substances by biologically generated materials, whether living or dead, is the definition of this phenomenon. A different technique for removing contaminants from industrial effluent is the biosorption of heavy metals by microorganisms (Derco & Vrana, 2018; Hussein et al., 2004). One interesting substitute technique for the elimination of heavy metal ions is bioadsorption. When compared to conventional treatments, it offers numerous benefits,

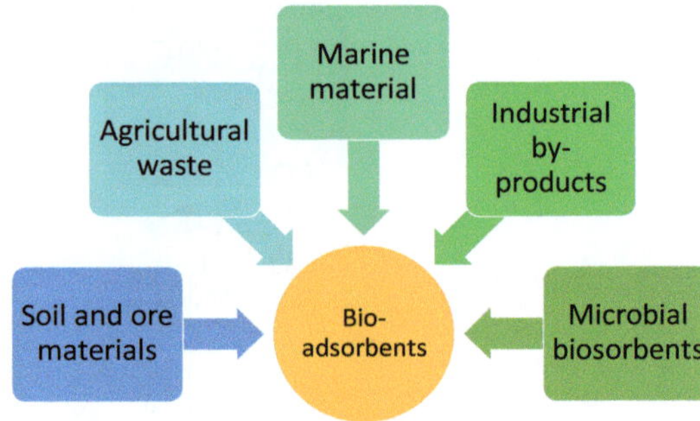

Fig. 11.3 Classification of Bio-adsorbents

such as cost effectiveness, environmental friendliness, high removal efficiency in diluted sewages, high metal ion separation capacity, and high biosorbent reusability (El-Naggar et al., 2018). Figure 11.3 shows a variety of inexpensive biosorbent kinds.

Among the many and inexpensive biosorbents, microorganisms—which include bacteria, fungus, yeasts, and algae—are a crucial group (White et al., 1995). The most prevalent and varied organisms on Earth are bacteria. They are essentially categorised according to their shape as rods, cocci, or spirals. In order to remove harmful metals, bacterial cells have a significant absorption ability and a high specific surface area. Cell walls serve as the conduit for the bacteria's surface connection point and junction strength for metal adsorption. The insertion of metals in the cells is the result of a connection between metallic ions and functional groups on the cell walls during the bioadsorption process (Aryal, 2021; Shamim, 2018). Algae are classified as low plants since they lack actual stem, leaf, and root components. Eukaryotes (brown, red, green, and diatomic algae) and Prokaryotes (green and blue algae) are the two structural classes into which algae are separated. As biosorbents, they have a number of active groups (such as amino, hydroxyl, carboxyl, and sulphate) that can help form chemical interactions with harmful ions (Hallmann, n.d.; Romera et al., 2006; Tamjidi & Ameri, 2020). Among the microbes employed to eliminate heavy metals are yeasts. A lot of research has been done on yeast cells' potential to increase adsorbent capacity and develop novel uses for biomass production. Since yeast cells may be acquired as a by-product of the fermentation process, they are a suitable biomass type for use as a biosorbent (Ojima et al., 2019). Fungi, one of the most popular types of microorganisms, have been thoroughly researched on their own for their ability to remove heavy metals. As illustrated in Fig. 11.4, the most important variables in the uptake process of hazardous metals by fungi are pH, contact time and reaction temperature, adsorbent dosage, and initial concentration of metal ions.

11.4.1 Fungi as Bio-Adsorbent

Cell walls of fungi, which are osmo-heterotrophic eukaryotes, are made of acidic polysaccharides with hydroxyl, phosphate, and amine groups, such as chitin and chitosan. They are crucial to the biosorption of phenolic dyes and heavy metals.

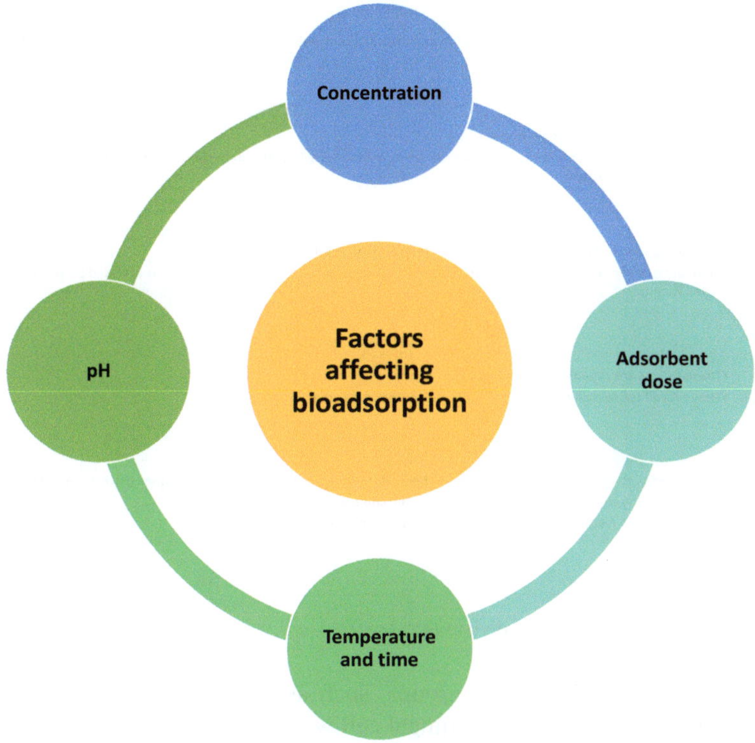

Fig. 11.4 Factors affecting Bio-adsorption

They lack chlorophyll and can have a flamentous or unicellular vegetative structure. They reproduce by sporulation. The traits of the primary fungal categories are depicted in Fig. 11.5. To extract heavy metals from wastewater, numerous researchers use various kinds of fungal adsorbents, such as mycelium, dead fungus, or fungal bullets. The purification of aqueous medium depends heavily on fungal biomass. The effects of combining fungal masses with other microbial masses are not well understood. They can therefore be used in situations that potentially improve elimination efficiency. The compatibility of surfaces with the environmental matrix (air, soil, and water) in which they were generated determines whether stringy fungal species will biosorb their material favourably. According to Dusengemungu et al. (2020), the presence of different functional groups such as carboxyl, amines, and hydroxides influences the adsorption of pollutants on the fungal biomass. By employing biosorption and bio-reduction processes, filamentous fungi may remove heavy metals and other organic and inorganic contaminants. Biosorption is the process by which contaminants are absorbed by living things through adsorption. Through the passive or active trapping of organic and inorganic materials in water by microbial biomass, toxins can be safely and responsibly removed, fostering a clean environment. There are two possible outcomes for the biosorption process: either metabolism-dependent or metabolism-independent (Legorreta-Castañeda

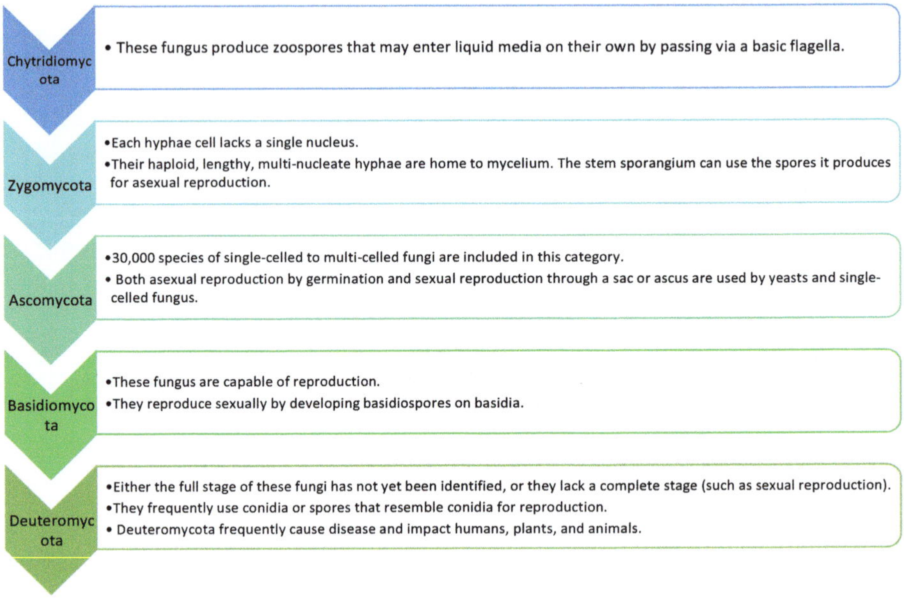

Fig. 11.5 Key attributes of fungal divisions

et al., 2020). Various functional groups, such as hydroxyl, amine, and carboxyl groups, are present in filamentous fungal cell walls and are helpful for biosorption.

Fungal biomass is a suitable material for biosorption because of its high cell wall composition. The capacity of fungi to absorb metals increases with the number of metal-binding functional groups present on the cell wall surface. Due to their high metal tolerance and ability to flourish in low pH environments, filamentous fungi make excellent biosorbents. These environments are frequently found in wastewater containing zinc, such as acid mine drainage and industrial wastewater from battery manufacturing (Legorreta-Castañeda et al., 2020; Negi & Das, 2023). Since filamentous fungus don't require growth, nutrients, or other maintenance, dead cells of these microorganisms are also regarded as common biosorbents for the removal of heavy metals and dyes from wastewater. Furthermore, dead biosorbents are undoubtedly easier to store for an extended period of time than viable biosorbents (Negi & Das, 2023).

11.4.2 The Mechanism of Fungal Biosorption

Adsorption is a complex phenomenon with a complex mechanism. When it comes to fungal species' biosorption of heavy metals, the cell wall structure is effective. The attachment of metallic ions to the cell surface is one of its mechanisms in fungus. Because of their capacity to adsorb into fungal cells via a single mechanism or

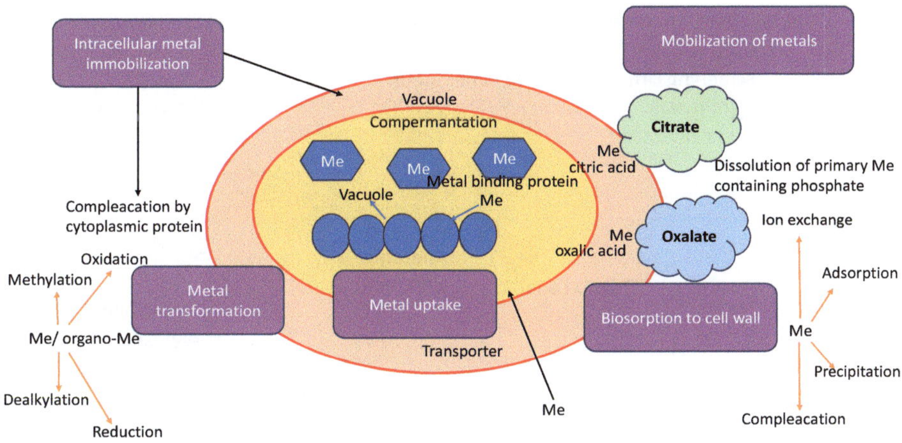

Fig. 11.6 Schematic mechanism for fungi-metal interaction

a combination of processes, these microorganisms are referred to as effective biosorbents (Dusengemungu et al., 2020). Amines, amides, imidazoles, thioether, sulfonates, carbonyls, sulfhydryls, carboxyl, phosphodiesters, phenolics, imines, and phosphate groups are examples of chemical/functional groups that can adsorb and disperse metal ions. Fungi's resistance to heavy metals is characterised by their capacity to withstand metal toxicity by means of a mechanism that is developed in direct reaction to the target metal species. In contrast to other biosorption factors, a significant portion of fungal biomasses are made up of cell wall materials that contribute to the metals' exceptional bonding properties, allowing them to acquire notable concentrations of heavy metals even in the absence of physiological activity (Harms et al., 2011; Kanamarlapudi et al., 2018). There are a number of ways that fungi can interact with metals to detoxify heavy metal ions from contaminated sites, including metal mobilisation, metal biosorption to the cell wall, metal uptake, intracellular metal immobilisation, and metal transformations, all of which are depicted in Fig. 11.6 (Siddiquee et al., 2015).

11.5 Wastewater Treatment

Wastewater treatment is the process of removing contaminants or pollutants from the aqueous phase using a variety of physical and chemical techniques before the water is released into the environment. Although there are numerous methods for treating both residential and nonresidential wastewater, the effluent is often first sent to a municipal sewage treatment facility. In these treatment facilities, a significant amount of the contaminant can be eliminated if it is planned and run correctly. There are three key stages in a normal therapy process: preliminary, primary, and secondary shown in Fig. 11.7. The sewage must have large and/or heavy particles

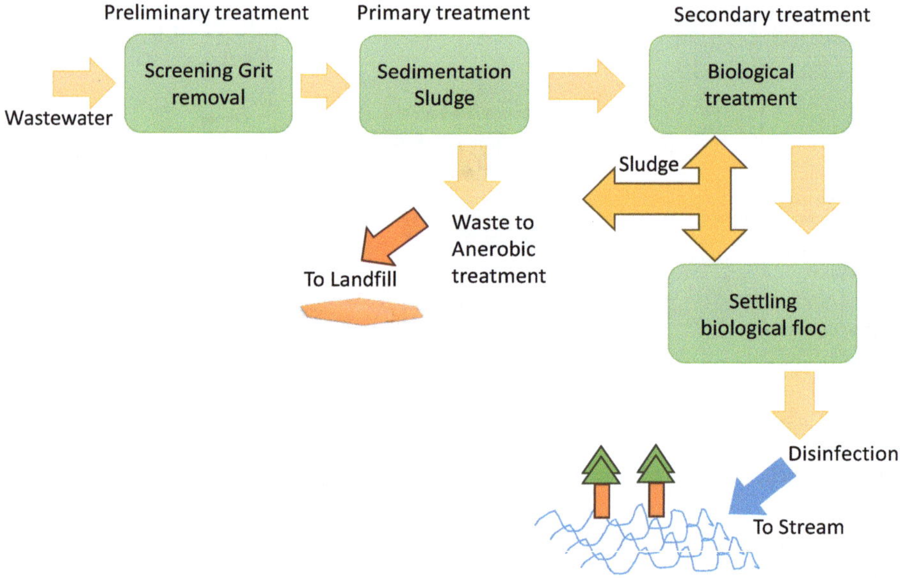

Fig. 11.7 An illustration of the three fundamental steps of the wastewater treatment process at a typical wastewater treatment plant

removed as part of the preliminary treatment process. Typically, this phase includes grit removal and screening. Using screens, big floating debris like rags (about 60%), paper (approximately 25%), and plastics (approximately 5%) are eliminated during the screening process. During the screening process, the leftover debris, known as screens, is typically contaminated with raw faeces, which is why it smells. Screenings are typically gathered in skips and either burned or utilised as filler for land. The next step after screening is to remove the grit. Essentially, grits are heavy inorganic particles that settle in grit channels and are often removed as sand, gravel, and other heavy particulate matter.

Following preliminary treatment, the effluent is sent to the major treatment procedure. Using a sedimentation process, the primary treatment removes a significant amount of the wastewater's suspended solids. The majority of the solid material settles to the bottom of the sedimentation tanks, known as sludge, after the effluent is permitted to pass through them. The effluent remains inside the tanks for a number of hours, allowing the sludge to settle and scum to accumulate on top. Following the removal of the sludge from the bottom and the skimming of the scum from the top, the partially treated wastewater advances to the secondary treatment phase.

Although basic treatment eliminates a sizable portion of hazardous materials from wastewater, it is insufficient to guarantee that all hazardous pollutants have been eliminated. Primary treatment effluent typically has a comparatively high BOD and significant organic compounds. The waste undergoes a secondary treatment, which includes biological processes that can lower the BOD level and most organic matters, as well as a low quantity of suspended solid material, in order to remove the organic matter and the remaining suspended solids. To promote the growth of aerobic bacteria and other microorganisms that oxidize the dissolved organic materials to carbon dioxide and water, the sewage is subjected to intense aeration during this process.

It is crucial to confirm whether the swedge is environmentally safe after treatment. Because the sludge generated during the treatment operations contains a number of dangerous contaminants, handling the sludge following treatment is particularly crucial. Before being released into the environment, treated sewage is typically cleaned using UV (ultra-violet) or chlorination techniques. Anaerobic sludge digestion is used to treat the sludge, which involves promoting the growth of anaerobic bacteria within a digester that holds the sludge. By breaking them down into soluble byproducts and gases, primarily carbon dioxide and methane, these anaerobic bacteria reduce the amount of organic particles in sludge. The cleaned sludge is subsequently delivered to solid waste processing facilities or utilised for land filling. A tertiary treatment stage is added by certain treatment plants in addition to these standard stages in order to use physical and chemical methods to eliminate any leftover organic and inorganic materials as well as microorganisms from the secondary stage effluent (Einschlag, 2011). The water can satisfy drinking water requirements following tertiary treatment. Nevertheless, industries rarely use this level because it is very expensive.

11.5.1 Different Waste Water Sources

The number of wastewater sources has expanded in tandem with the growth of various sectors. Waste waters from various businesses have varying compositions. As a result, different sectors have different purification procedures. Below is a discussion of a few significant waste water generating sectors and the water treatment methods they use.

11.5.1.1 Pesticides

Pesticides are essential to global agricultural production because they guarantee greater crop yields from smaller areas, which is important for sustainable food production and management, especially in light of the growing world population (Münze et al., 2017). Unwantedly, pesticide residues frequently spread well beyond the areas where they are meant to be applied, entering different ecosystems through groundwater, subterranean, overland, and atmospheric channels. The main reason why pesticides enter sewage systems is because of agricultural practices. However, the burden of pesticides in sewage systems is also significantly increased by non-agricultural usage. These sources include pest control methods used in residential houses and gardens, industrial vegetation control along highways and railroads, and grass management techniques used in parks and golf courses (Münze et al., 2017;

Saleh et al., 2020). The problem of pesticide contamination is significant, particularly when using fungus species to remediate wastewater.

According to Münze and his group in the year 2017, effluents released from German Waste Water Treatment Plants included 45 distinct pesticides, which were divided into 5 other metabolites, 14 fungicides, 7 insecticides, and 19 herbicides. Notwithstanding these low quantities, it is impossible to undervalue the possible harm that mixtures of these substances could cause to living things over time (Trellu et al., 2021). For example, long-term exposure to the selective photosynthesis inhibitor atrazine, a pesticide, has been connected to cancer, muscle degradation, retinal degeneration, and cardiovascular problems in humans. According to Saleh et al. (2020), exposure to oxyfluorfen, another pesticide that damages cell membranes irreversibly, might cause liver problems and anaemia in people. The wide pH range of pesticide-polluted water, which can range from extremely acidic (0.5) to highly alkaline (14), as well as the varied influent composition and physical structure of the pesticides, pose a number of difficulties for the treatment of pesticide-contaminated water. Additionally, the BOD in pesticide-laden wastewater can range from 30 to 11,590 mg/L, whereas the COD can range from 150 to 33,750 mg/L. Pesticide concentrations in some water sources may even exceed 107 mg/L (Goodwin et al., 2017; Rodriguez-Narvaez et al., 2017).

Notably, the efficacy of existing Waste Water Treatment Plants in removing pesticides from wastewater is insufficient. As a result, environmental ecosystems commonly contain these dangerous substances, which pose a serious risk to human health as well as the environment (Münze et al., 2017; Rousis et al., 2017; Westlund & Yargeau, 2017). This circumstance emphasises the pressing need for better methods and approaches for treating wastewater, especially ones that can successfully eliminate pesticide residues.

11.5.1.2 Canned Fruits/Vegitables Industries

The increase in population and advancements in production methods have led to a rapid expansion of the worldwide fruit and vegetable processing business, which includes canned and frozen fruits and vegetables as well as processed fruit and vegetable products including wine, sauce, and fruit juices. Environmental contaminants may also come from process effluent from facilities that produce food, beverages, and raw materials (Hubbard et al., 2021). Treatment is crucial for fruit wastewaters because of their high COD levels, phenolic compounds, and organic acid concentration (Chen et al., 2019; Strong & Burgess, 2008; Viuda-Martos et al., 2011). Winery wastewater and agro-industrial wastewater from the processing of fruits with high phenolic content, such grapes, citrus, and pomegranates, can support filamentous fungus (Braho et al., 2024; Satari et al., 2016; Zhang et al., 2008). Another possible substrate for the synthesis of the pectinase enzyme is fruit waste (Kc et al., 2020). Wastewater with a high COD level from sugarcane distilleries is another significant industrial pollutant. The best results in the purification of DSW were obtained by species from the Aspergillus and Trametes genera, according to Chuppa-Tostain and his group (Chuppa-Tostain et al., 2020). Vinasse, a by-product of the molasses-to-ethanol process, is a significant wastewater with high levels of solids and COD, proposed by Nair and Taherzadeh in the year 2016 (Nair & Taherzadeh, 2016). Nair and Taherzadeh (2016), Nitayavardhana et al. (2013), and Karimi et al. (2019) have demonstrated that the edible filamentous fungus *Aspergillus oryzae*, *Rhizopus oligosporus*, and *Neurospora intermedia* can grow on diluted vinasse and significantly aid in the elimination of COD.

11.5.1.3 Pharmaceutical Company

Human activity is connected to the presence of pharmaceuticals in wastewater and, by extension, in the environment. Predicting the possible effects of pharmaceutical compounds on target and non-target species is difficult due to their wide range of variations in origin, modes of action, and resistance to (bio)degradation. Endocrine disruptors, hormones, non-steroidal anti-inflammatory drugs, antimicrobial agents, pharmaceuticals and personal care products, beta-blockers, and lipid-regulating agents are among the many pharmaceutical xenobiotics that are commonly found in wastewaters and the environment (Dong et al., 2023; Khan et al., 2020; Marchlewicz et al., 2023; Nasirabadi et al., 2016; Rezaei et al., 2022; Winker et al., 2008). Instead of being broken down or decomposed by organisms, pharmaceutical chemicals usually go through two different detoxification stages. These mechanisms change them into more polar and water-soluble forms, hence simplifying their removal via urine. During the initial stage, the enzyme system NADPH cytochrome P450 reductase (P450R)/cytochrome P450 mainly catalyses the oxidation, reduction, or hydrolysis of medicines.

These medications and their metabolites are excreted and then sent to Waste Water Treatment Plants, where some of them are partially degraded or undergo further metabolism. Pharmaceutical residues may be found in raw sewage, treated wastewater, and the surrounding environment as a result of this ineffective clearance. Furthermore, because of their greater solubility and bioavailability, these medications' metabolites frequently show higher levels of toxicity than their parent substances. The main goal of pharmaceuticals, which are intentionally created as physiologically active substances, is to provide therapeutic benefits at the lowest feasible dosages (Citkowska et al., 2019; Lawson et al., 2021). As a result, the presence of some medications in environmental matrices is concerning because they may have an effect on different organisms. Long-term, low-dose exposure to these medications may have chronic effects, which adds another level of intricacy. These medications' acute effects on their target creatures are frequently thoroughly researched and understood, but the same cannot be true for non-target organisms. There is still much to learn about the effects of pharmacological exposure on these creatures, especially on a multigenerational timeline. One of the most prevalent metabolites in wastewater and the environment are oxidised metabolites and their conjugates. Ascomycota (Fusarium and Trichoderma) and Mucoromycota (Umbelopsis ramanniana and Mucor rammanianus) contain filamentous fungal species that have the ability to considerably break down pharmaceutical substances like carbamazepine, fluoroquinolones, and acetaminophen (Esterhuizen et al., 2021; Madadi & Bester, 2021; Olicón-Hernández et al., 2017).

11.5.1.4 Paper Industry

The generation of wastewater is a major concern in the pulp and paper business because of the large amount of water used in the manufacturing process. A 2022 study from the International Energy Agency (Tracking Clean Energy Progress, 2023) states that 415 million tons of paper and pulp were produced worldwide annually, the most since 2010. Water consumption throughout the production process can vary greatly, ranging from 5 to 100 m^3 per ton of pulp and paper. Numerous variables, such as the kind of substrate utilised, the method employed, and the characteristics of the finished paper product, affect this variance. Paper mills generally use 10–50 m^3/ton of water on average (Sharma et al., 2022; Toczyłowska-Mamińska, 2017). Several steps are involved in the processing of timber in paper mills, including debarking, chopping, pulping, bleaching, washing, filtering, screening, and either pulp drying or paper production (Kamali et al., 2016). Because of the breakdown of lignin during the bleaching stage, these processes result in the creation of wastewater that contains over 500 distinct chemical compounds. The fact that chlorinated compounds use chlorine dioxide in the bleaching process makes them especially dangerous (Cabrera, 2017). Together with a variety of other chemicals, high-pressure and high-temperature conditions are used in the chemical pulping process. For example, in the Kraft process, woodchips are cooked in a pulping fluid using sodium hydroxide and sodium sulphide to Generate white spirits. As an alternative, some methods use acidic treatments with bisulfide ions and sulphurous acid. Different inorganic substances, including sodium bicarbonate, sodium carbonate, and sodium sulphate, sodium thiosulphate, or sodium sulfite, may also be added. Making green spirits as a result. The 90–95% yield from mechanical processing techniques is higher than the 40–50% yield from chemical processes. However, pulp produced by mechanical techniques typically has shorter fibres and more intense colouration, making it of lesser quality. The component that gives intermediate products their dark colour, lignin, is not completely eliminated from the treated material. This, combined with the cellulose content and byproducts of lignin degradation, contributes to the dark hue of mill effluent, which is referred to as "black liquor." According to Morya et al. (2022), there can be around seven tons of this byproduct produced for every ton of paper pulp. According to several studies (Gao et al., 2018; Morya et al., 2022; Srivastava et al., 2023), black liquor's high organic matter content can make it a reasonable substrate for alternative uses like the production of hydrogen, biogas, bioplastics, biodiesel, and other valuable products.

11.5.1.5 Textile and Dye Industry

An essential part of our daily clothing needs is the worldwide textile industry, which is renowned for its vast array of colours and fabric styles. Although this business contributes significantly to the global economy, it also generates enormous amounts of wastewater, raising serious concerns about the environment and public health. Wastewater from textile and dyeing operations is characterised by high COD and BOD levels, with a 4:1 ratio. The complications of handling such wastewater are highlighted by this disproportionate ratio, which indicates the presence of chemicals resistant to biodegradation. Furthermore, the wastewater has a variety of hazardous dyes and significant levels of suspended particulates, which exacerbates the ecological problem. More than 100,000 different kinds of dyes are currently marketed commercially, and they constitute an essential component of the textile production process. Their astounding 1 million tons of output worldwide has increased, adding to the environmental burden (Gupta & Suhas., 2009; Yaseen & Scholz, 2019). The astounding amount of dye produced and used not only shows how big the problem is, but it also emphasises how urgently we need management and treatment solutions that work to protect the environment and public health.

Although the textile production process is intricate and involves many steps and different treatment techniques, it may be roughly divided into two groups: wet processes and dry processes. In large part, the type of treatment used determines the kind of waste and chemicals that are disposed of.

Because textile manufacturing involves such a wide range of procedures, fibre types, and technology, the resulting effluents contain a diverse variety of pollutants. These may include metal ions, oils, dissolved solids, dissolved solids, dyes, bases, and surfactants (Premaratne et al., 2021). The existence of metal ions among the pollutants is especially worrisome. According to many researchers like Manekar and his group (Manekar et al., 2014), the main elements commonly found in textile wastewater are chromium, arsenic, copper, zinc, lead, iron, mercury, and cobalt. The amount of these metals in textile effluent can vary significantly depending on a number of factors, including geography and the particular production procedures employed. To create efficient and customised wastewater treatment plans, a thorough understanding of the wastewater's composition is therefore necessary.

One of the main environmental issues related to the production of textiles is the use of fabric dyes, especially synthetic ones. These dyes pose serious environmental concerns because of their synthetic nature and resistance to biodegradation. These dyes can be divided into a number of families based on their chemical makeup, including azo, anthraquinone, sulphur, phthalocyanine, and triarylmethane. During the dyeing process, some of these dyes do not stick to the materials and are thus washed out in the washing steps. An estimated 20% of the dyes used may be emitted during the dyeing process or immediately released into the environment. Textile wastewater colour concentrations can vary widely, often ranging from tens to several hundred milligrammes per litre, according to studies (Yaseen & Scholz, 2019). In addition to being hazardous, dyes' ability to adsorb and reflect sunlight can disrupt the natural photosynthesis in damaged habitats, leading to unfavourable environmental effects (Premaratne et al., 2021). White rot fungus species that can produce manganese peroxidase, lignin peroxidase, and laccase enzymes have been shown in several investigations to be useful in purifying textile wastes tainted with dyestuffs (Asgher et al., 2009). The decolorisation of dyes has been thoroughly investigated in Dichomitus squalens, Daedalea flavida, Irpex flavus, Polyporus sanguineus, Coriolus versicolour, Phanerochaete chrysosoporium and Bjerkandera adusta, Emmia latemargina, and Talaromyces verruculosus (Chadni et al., 2017; Liu et al., 2004).

However, it has also been discovered that certain filamentous fungi, including *Aspergillus quadrilineatus*, *Aspergillus terreus*, *Aspergillus lentulus*, *Aspergillus carbonarius*, *Chaetomium globosum*, and *Penicillium glabrum*, are useful in removing colour (Kaushik & Malik, 2010; Singh & Dwivedi, 2020).

11.5.1.6 Oil Industry

One of the main ingredients in the Mediterranean diet has been olive oil. In the past 60 years, its production has tripled, according to data from the International Olive Council, reaching 3.02 million tons in the 2020–2021 crop year. In the same year, around 74% of the global total came from EU member states. Production for the 2022–2023 crop year is expected to be 2.8 million tons (−17.6%). With an output of almost 3.1 million tons during the last three crop years (2019–2020, 2020–2021, and 2021–2022), the world's consumption of olive oil has grown more quickly than its production (Bungaro, 2022). Nevertheless, this manufacturing is linked to the generation of enormous amounts of toxic and challenging-to-handle waste streams, referred to as olive mill wastes. The two primary waste streams used in each method of extracting olive oil are olive pomace, a solid waste stream, and olive mill wastewater. The so-called traditional discontinuous press extraction method was replaced by a continuous process that uses a two- or three-phase centrifugal extraction system. Importantly, the technology utilised to extract the olive oil determines the kind and quantity of olive mill wastewater generated. Olive mill wastewater contamination of the soil, water, and air is largely caused by olive phenols because of their phytotoxicity and antimicrobial activity. Based on data from the literature, the phenolic compounds found in olive mill wastewaters are both high and low molecular weight substances that fall under the categories of flavonoids, phenolic acids, phenyl alcohols, and secoiridoid derivatives. The polarity of these compounds, the olive cultivar, fruit ripeness, climate, agronomic circumstances, storage conditions before to extraction, and processing procedure all affect olive mill wastewater. The primary phenolic component of olive mill wastewater is hydroxytyrosol.

The oil palm tree, Elaeis guineensis, is the another most important oil crop in the world, accounting for more than half of all traded vegetable fats and oils. Lauric oil, which is derived from the palm kernel, and palm oil, which is derived from the fibrous mesocarp, are the two types of oils that are produced from oil palm fruit. More than 80% of the palm oil produced worldwide comes from

Malaysia and Indonesia. According to González-Diaz and his research group (Gonzalez-Diaz et al., 2021), Indonesia, Malaysia, Thailand, Colombia, and Nigeria are the top five producers of palm oil globally. As the demand for palm oil uses in the production of food (such as margarine, cooking oil, bakery products, dairy products, fried and baked foods), animal feed preparations, and numerous non-food products (such as biodiesel, textiles, and cosmetics) grows, palm oil production has increased steadily over the past few years and is expected to reach 240 million tons by 2050 (Azhar et al., 2017). Despite being praised for its contribution to the economic growth and prosperity of the producing nations, the oil palm industry has also been connected to detrimental environmental effects due to the massive volumes of waste streams generated at both the input and output sides of the oil extraction processes.

The largest volume of any trash is seen in palm oil mill effluent. Approximately half of the water input—roughly 0.7 m^3 per ton of fresh fruit branches processed—ends up as palm oil mill effluent during the milling process, with the remaining 50% being lost to evaporation or to rivers or drains (Gonzalez-Diaz et al., 2021). The palm oil mill effluent is a viscous, colloidal suspension that is primarily composed of water (95–96%), total solids (4–5%), suspended solids (2–4%), oil (0.6–0.7%), and a variety of carbohydrates, including hemicelluloses, simple sugars, organic nitrogen (proteins, amino acids), and organic acids. Additionally, it has significant levels of heavy metals, N, P, K, Mg, Ca, and other plant nutrients.

The ashes from empty fruit bunches are usually used as fertiliser, while the palm kernel shells and palm pressed mesocarp fibre are burned to provide heat and energy for mill operations. Solid wastes are usually burned. Additional strategies to address these wastes include removing cellulosic material from fibres, pyrolysing wastes to produce bio-char or bio-oil, and turning some of these wastes into briquettes to aid in the creation of energy in other sectors. A few of the techniques are costly and labour-intensive, which deters use even if recovering valuable items from solid trash has advantages. Palm oil mill effluent treatment, which is usually accomplished using a number of ponds, still has issues, including releasing greenhouse gases into the atmosphere, occupying a lot of space, and having a long retention period and low treatment efficiency (Poh et al., 2020). As efficient natural decomposition agents for palm oil mill wastes (such as composting), fungi have been suggested because of their ability to accelerate the organic material's breakdown (Dominic & Baidurah, 2022). Solid wastes are usually burned, with the palm kernel shells and palm pressed mesocarp fibre being burned to provide heat and energy for mill operations and the ashes from empty fruit branches being used as fertiliser. Fungi are suggested as efficient natural decomposition agents for palm oil mill wastes (e.g., composting) because of their ability to accelerate the organic material's breakdown (Mohammad et al., 2012). Because palm oil mill effluent contains high levels of proteins, lipids, carbohydrates, and minerals, bioconversion into value-added products through biotechnological techniques has been suggested as an alternate effluent treatment method. Palm oil mill effluent has gained recognition as a great substrate for fungal growth, which leads to its remediation and advanced valorisation.

11.5.1.7 Dairy, Fish and Starch Based Industries

The global increase in demand for dairy products has fuelled the growth of the dairy sector while also contributing to a rise in process wastewater generation. Milk lost in technological cycles, dairy products (skim milk, spoilt milk, and spilt milk), byproducts of processing operations (whey permeate, whey, boiling water, buttermilk, brine), clean-in-place reagents, a variety of cleaning acids, and alkaline detergents used in the cleaning of containers, cans, bottles, equipment tanks, and floors are all considered categories for dairy wastewater based on their source, composition, and process conditions (Sar et al., 2020; Slavov, 2017). Whey, the primary liquid byproduct of the dairy industry, is created during the manufacturing of casein and cheese (Slavov, 2017). Numerous physicochemical and biological techniques can be used to handle dairy waste, particularly whey. Commonly employed physicochemical techniques include advanced oxidation processes, chemical precipitation, and coagulation-flocculation. Nevertheless, these techniques are frequently linked to increased expenses and poorer efficacy in eliminating soluble COD.

Another important global economic activity is the processing of fish. Large amounts of water are unavoidably used during the processing of fish and related byproducts; on average, 11 m^3 tons of fish are processed for sterilisation procedures, canned fish production, fishmeal manufacturing, or raw product washing (Cristóvão et al., 2014). The fish business produces a significant amount of solid waste and wastewater as a result of these activities, which includes soluble nutrients, different types of oil and fat residues, and colloidal and particle forms such as fish residues. When fish industry wastes are released into the receiving water environment directly without any kind of treatment, the dissolved oxygen content drops, causing eutrophication and toxicity. Additionally, the growth and reproduction of aquatic organisms are negatively impacted. Considering the level of contamination while designing an industrial wastewater treatment system is crucial. Colloidal materials, big particles, settleable, and suspended solids are all guaranteed to be removed by pre-treatment techniques like screening, sedimentation, pH adjustment, flocculation, and flotation. The residual organic contaminants are eliminated by aerobic, facultative, and anaerobic bacteria during secondary treatment, also known as biological treatment, which follows pretreatment (Tay et al., 2004).

Food processing facilities, home and commercial kitchens, hotels, restaurants, and other major food production industries are just a few of the many locations that generate food waste. Waste streams from the brewing and starch production industries, as well as waste from starchy root crops including potatoes, cassava, and sweet potatoes, are burned or disposed of, endangering both human health and the environment. However, these wastes may contain minerals, vitamins, or dietary fibre that might be converted into inexpensive media supplements for fungal culture or a valuable green product like methane and ethanol.

Environmental issues arise because potato leftovers, or potato peel waste, are typically thrown away or utilised as animal feed. Rhizopus oryzae has been reported to use potato peel waste to produce lactic acid (298 mL/mg) and ethanol (33 mL/mg). It was demonstrated that *Aspergillus niger* A18 isolates may produce bioflocculants using potato wastewater as a viable substrate. It was advised to use the outflow of potato waste water for irrigation following flocculation (Pu et al., 2018). In order to decolorise textile dyeing wastewater using *Penicillium chrysogenum*, potato starch and wastewater were also employed as promising co-substrates (Lanfranconi et al., 2022).

Processing sweet potatoes also produces a significant amount of waste that might be utilised to make starch. With the aid of *A. niger* extract, sweet potato peel starch was transformed into a high amount of glucose by α-amylase, resulting in the economical generation of ethanol (Pereira et al., 2017). *A.* and *R. oryzae* were cultivated on the wastewater (including suspended particles) of a wheat-starch plant, yielding protein-rich biomass (12 g/L of dry biomass with a protein content of roughly 35%) (Souza Filho et al., 2019).

To ascertain the sustainability of fungal biomass for human nutrition or in feed applications, its nutritional composition must be taken into account. The techno-economic analysis and life cycle assessment of starchy waste valorisation by filamentous edible fungi should also be used to examine the viability of a commercialisation process.

11.6 Concerns, Regulatory and Societal Implications in Using Fungus Biorefineries to Treat Wastewater

In order for the fungal biorefinery to handle wastewater effectively, it is essential to keep fungal species present in the system and maximise the output of already-existing fungus. Thus, environmental elements such as pH and temperature, together with the wastewater's composition, are important factors. Fungal growth may be negatively impacted by any changes in the wastewater's composition or the presence of inhibitors such as acetic acid and furfural produced during wastewater hydrolysis (Sankaran et al., 2010; Szengyel & Zacchi, 2000). Furthermore, fungal interactions within the system may be competitive. Certain fungi may become more prevalent, changing the product yield, even if this competition can benefit fungal development and product production (Jenkins & Grzywacz, 2000). Accordingly, consistent performance depends on preserving the long-term stability of the fungi in the system and putting in place routine fungal population monitoring and control (Kumar et al., 2021). In large-scale reactors, the ideal parameters found in laboratory-scale experiments may not apply, making the reactors inappropriate for producing bioproducts and fungi effectively (Dey et al., 2022). Dey et al. (2022) and Espinosa-Ortiz et al. (2016) with their research groups state that the design of reactors for wastewater treatment necessitates the creation of specialised procedures and optimisation approaches designed especially for fungal production. Fungal biomass may be greatly impacted by difficulties with mass and heat transport in high-volume reactors (Modenbach & Nokes, 2012). Furthermore, even in the best of circumstances, fungal development may be impeded by changes in wastewater composition and fungal output, which can upset the pH balance of the environment.

Therefore, when designing reactors, it is essential to take into consideration factors that affect fungal development. This all-encompassing strategy guarantees that the difficulties associated with mass and heat transfer, pH variations, and scale-up are sufficiently addressed for fungal biorefinery operations in wastewater treatment to be successful. Along with the difficulties outlined above, it is crucial to guarantee economic viability in terms of fungal biomass productivity and product yield; it is not enough to simply succeed in wastewater treatment and product extraction (Chatterjee & Venkata Mohan, 2022). Public concerns must be taken into consideration because there may be difficulties with the acceptability and perception of the fungal species used in wastewater treatment. Addressing concerns about safety and possible environmental effects is crucial for the broad use of fungal biorefinery techniques (Ferreira et al., 2020). Therefore, a thorough strategy that takes into account sociological factors and economic viability in addition to technical difficulties is essential for the effective adoption and deployment of fungal biorefinery techniques.

The use of filamentous fungus for resource recovery from wastewaters offers a viable strategy for the production of sustainable food and feed. When it comes to recycling nutrients, filamentous fungus are effective at absorbing nitrogen and phosphorus from wastewater, which are critical for their development. The creation of fungal biomass that is high in proteins and other nutrients may result from this nutrient recycling (Rousta et al., 2021). Because filamentous fungi are capable of bioconversion, they can successfully turn waste into a resource by converting low-value organic matter in wastewater into high-value proteins, carbohydrates, and byproducts (Negi & Das, 2023).Because the strategy is in line with the concepts of the circular economy and reduces waste and the demand for new resources, using these microorganisms to recover resources from wastewater can therefore support sustainable development (Sankaran et al., 2010). Beyond food and feed, this technique also aids in bioremediation and mediates wastewater's environmental issues while promoting the production of food and feed (Parchami et al., 2021). There are several uses for biomass produced by growing filamentous fungus on wastewater in the food and feed industry. Fungal biomass is ideal for animal feed and maybe for human consumption, particularly in alternative protein products, because of the high protein content of filamentous fungi (Rousta et al., 2021). However, filamentous fungal biomass and its extracts can also be utilised as functional additives in a variety of food and feed products (Rousta et al., 2021).

11.7 Conclusion

Numerous organic compounds and synthetic colours that are resistant to disposal in the environment, particularly in water reservoirs, are found in the effluent water from the textile, paper-pulp, and dyeing industries. Therefore, to lessen their negative effects on the environment, certain types of wastewater should be treated using efficient techniques. The biological methods are less expensive and more ecologically friendly than the currently used physical and chemical methods. However, because of the wide range of pollutants present in industrial effluents, none of these approaches will completely remove them. Therefore, in order to eliminate all currently present hazardous pollutants, a combination of suitable techniques is recommended based on the pollutant type, concentration, and recalcitrance. Filamentous fungal cells and their enzymes have been shown to effectively extract organic components from industrial effluents or convert them into less hazardous substances. It is possible to carefully choose different genera and species of filamentous fungi, or their consortia with other potent microorganisms, to be used in conjunction with other techniques to treat different contaminants found in industrial wastewater.

Further, as the developed and customised systems would be cell-free and more targeted for the removal of contaminants, filamentous fungal enzymes hold great promise for the future of industrial wastewater treatment technology. Although the potential of filamentous fungus to cleanse wastewater has been investigated, their exact function in this process is yet unknown. Prior studies on filamentous fungus in wastewater treatment have mostly been carried out in lab settings; more research is necessary to ascertain whether these techniques are appropriate for use in industrial settings. Therefore, to determine if employing filamentous fungi for large-scale wastewater treatment procedures is feasible, more laboratory testing and pilot scale research are required. Mycoremediation would therefore eventually contribute to lowering the pollution levels in industrial wastewater, protecting the ecosystem and the health of numerous living things on Earth.

References

Abd El-Zaher, E. H. F. (2010). Biodegradation of reactive dyes using soil fungal isolates and Ganoderma resinaceum. *Annals of Microbiology, 60*(2), 269–278. https://doi.org/10.1007/s13213-010-0037-0

Aryal, M. (2021). A comprehensive study on the bacterial biosorption of heavy metals: Materials, performances, mechanisms, and mathematical modellings. *Reviews in Chemical Engineering, 37*(6), 715–754. https://doi.org/10.1515/revce-2019-0016

Asgher, M., Azim, N., & Bhatti, H. N. (2009). Decolorization of practical textile industry effluents by white rot fungus Coriolus versicolor IBL-04. *Biochemical Engineering Journal, 47*(1), 61–65. https://doi.org/10.1016/j.bej.2009.07.003

Azhar, B., Saadun, N., Prideaux, M., & Lindenmayer, D. B. (2017). The global palm oil sector must change to save biodiversity and improve food security in the tropics. *Journal of Environmental Management, 203*(Pt 1), 457–466. https://doi.org/10.1016/j.jenvman.2017.08.021

Braho, V., Sar, T., & Taherzadeh, M. J. (2024). Cultivation of edible filamentous fungi on pomegranate by-products as feedstocks to produce mycoprotein. *Systems Microbiology and Biomanufacturing, 4*(2), 675–686. https://doi.org/10.1007/s43393-023-00212-0

Bungaro, M. (2022). Olive Oil: World Consumption Has Increased Over Last Three Years Available online: https://www.internationaloliveoil.org/rid/?lang=it (accessed on 22 January 2026).

Cabrera, M. N. (2017). *Pulp mill wastewater: Characteristics and treatment*. IntechOpen. https://doi.org/10.5772/67537

Chadni, Z., Rahaman, M. H., Jerin, I., Hoque, K. M. F., & Reza, M. A. (2017). Extraction and optimisation of red pigment production as secondary metabolites from Talaromyces verrucu-

losus and its potential use in textile industries. *Mycology, 8*(1), 48–57. https://doi.org/10.1080/21501203.2017.1302013
Chatterjee, S., & Venkata Mohan, S. (2022). Fungal biorefinery for sustainable resource recovery from waste. *Bioresource Technology, 345*, 126443. https://doi.org/10.1016/j.biortech.2021.126443
Chen, H., Zhang, H., Tian, J., Shi, J., Linhardt, R. J., Ye, T. D. X., & Chen, S. (2019). Recovery of high value-added nutrients from fruit and vegetable industrial wastewater. *Comprehensive Reviews in Food Science and Food Safety, 18*(5), 1388–1402. https://doi.org/10.1111/1541-4337.12477
Chuppa-Tostain, G., Tan, M., Adelard, L., Shum-Cheong-Sing, A., François, J.-M., Caro, Y., & Petit, T. (2020). Evaluation of filamentous fungi and yeasts for the biodegradation of sugarcane distillery wastewater. *Microorganisms, 8*(10), 1588. https://doi.org/10.3390/microorganisms8101588
Citkowska, A., Szekalska, M., & Winnicka, K. (2019). Possibilities of Fucoidan utilization in the development of pharmaceutical dosage forms. *Marine Drugs, 17*(8), 458. https://doi.org/10.3390/md17080458
Cristóvão, R. O., Botelho, C. M., Martins, R. J. E., Loureiro, J. M., & Boaventura, R. A. R. (2014). Primary treatment optimization of a fish canning wastewater from a Portuguese plant. *Water Resources and Industry, 6*, 51–63. https://doi.org/10.1016/j.wri.2014.07.002
Derco, J., & Vrana, B. (2018). Introductory chapter: Biosorption. In J. Derco & B. Vrana (Eds.), *Biosorption*. IntechOpen. https://doi.org/10.5772/intechopen.78961
Dey, P., Rangarajan, V., Singh, J., Nayak, J., & Dilip, K. J. (2022). Current perspective on improved fermentative production and purification of fungal cellulases for successful biorefinery applications: A brief review. *Biomass Conversion and Biorefinery, 12*, 967–995. https://doi.org/10.1007/s13399-020-01227-6
Dhanavade, M. J., & Patil, P. J. (2023). Yeast and fungal mediated degradation of synthetic dyes. In S. P. Govindwar, M. B. Kurade, B.-H. Jeon, & A. Pandey (Eds.), *Current developments in bioengineering and biotechnology* (pp. 371–409). Elsevier. https://doi.org/10.1016/B978-0-323-91235-8.00014-0
Dominic, D., & Baidurah, S. (2022). Recent developments in biological processing technology for palm oil mill effluent treatment—A review. *Biology, 11*(4), 525. https://doi.org/10.3390/biology11040525
Dong, W., Yin, X., Qi, C., Wei, T., Wei, L., Yang, J., Mu, J., Teraoka, H., & Dong, W. (2023). Reduced pigmentation and thyroid hormone disruption in zebrafish embryos caused by industrial sludge near Bohai Bay, China. *Water Biology and Security, 2*(4), 100190. https://doi.org/10.1016/j.watbs.2023.100190
Dusengemungu, L., Kasali, G., Gwanama, C., & Ouma, K. O. (2020). Recent advances in biosorption of copper and cobalt by filamentous fungi. *Frontiers in Microbiology, 11*. https://doi.org/10.3389/fmicb.2020.582016
Einschlag, F. S. G. (2011). *Waste water—Evaluation and management*. IntechOpen. https://doi.org/10.5772/2051
El-Naggar, N. E.-A., Hamouda, R. A., Mousa, I. E., Abdel-Hamid, M. S., & Rabei, N. H. (2018). Biosorption optimization, characterization, immobilization and application of Gelidium amansii biomass for complete Pb2+ removal from aqueous solutions. *Scientific Reports, 8*(1), 13456. https://doi.org/10.1038/s41598-018-31660-7
Espinosa-Ortiz, E. J., Rene, E. R., Pakshirajan, K., Hullebusch, E. D., & Lens, P. N. L. (2016). Fungal pelleted reactors in wastewater treatment: Applications and perspectives. *Chemical Engineering Journal, 283*, 553–571. https://doi.org/10.1016/j.cej.2015.07.068
Esterhuizen, M., Behnam Sani, S., Wang, L., Kim, Y. J., & Pflugmacher, S. (2021). Mycoremediation of acetaminophen: Culture parameter optimization to improve efficacy. *Chemosphere, 263*, 128117. https://doi.org/10.1016/j.chemosphere.2020.128117
Ferreira, J. A., Varjani, S., & Taherzadeh, M. J. (2020). A critical review on the ubiquitous role of filamentous fungi in pollution mitigation. *Current Pollution Reports, 6*(4), 295–309. https://doi.org/10.1007/s40726-020-00156-2

Fishman, C. (2011). *The big thirst: The secret life and turbulent future of water*. Simon and Schuster.

Gao, F., Peng, Y.-Y., Li, C., Yang, G.-J., Deng, Y.-B., Xue, B., & Guo, Y.-M. (2018). Simultaneous nutrient removal and biomass/lipid production by chlorella sp. in seafood processing wastewater. *The Science of the Total Environment, 640–641*, 943–953. https://doi.org/10.1016/j.scitotenv.2018.05.380

Gonzalez-Diaz, A., Pataquiva-Mateus, A., & García-Núñez, J. A. (2021). Recovery of palm phytonutrients as a potential market for the by-products generated by palm oil mills and refineries–A review. *Food Bioscience, 41*, 100916. https://doi.org/10.1016/j.fbio.2021.100916

Goodwin, L., Carra, I., Campo, P., & Soares, A. (2017). Treatment options for reclaiming wastewater produced by the pesticide industry. *International Journal of Water and Wastewater Treatment*. https://doi.org/10.16966/2381-5299.149

Gupta, V. K., & Suhas. (2009). Application of low-cost adsorbents for dye removal—A review. *Journal of Environmental Management, 90*(8), 2313–2342. https://doi.org/10.1016/j.jenvman.2008.11.017

Hallmann, A. (n.d.). *Algal transgenics and biotechnology*. Global Science Book.

Harms, H., Schlosser, D., & Wick, L. Y. (2011). Untapped potential: Exploiting fungi in bioremediation of hazardous chemicals. *Nature Reviews Microbiology, 9*(3), 177–192. https://doi.org/10.1038/nrmicro2519

Helmer, R., & Hespanhol, I. (1997). United Nations environment programme, water supply and sanitation collaborative council. In *Water pollution control: A guide to the use of water quality management principles* (1st ed.). E & FN Spon.

Hollender, J., Singer, H., & McArdell, C. S. (2008). Polar organic micropollutants in the water cycle. In P. Hlavinek, O. Bonacci, J. Marsalek, & I. Mahrikova (Eds.), *Dangerous pollutants (xenobiotics) in urban water cycle* (pp. 103–116). Springer. https://doi.org/10.1007/978-1-4020-6795-2_11

Hubbard, L. E., Kolpin, D. W., Givens, C. E., Blackwell, B. R., Bradley, P. M., Gray, J. L., Lane, R. F., Masoner, J. R., McCleskey, R. B., Romanok, K. M., Sandstrom, M. W., Smalling, K. L., & Villeneuve, D. L. (2021). Food, beverage, and feedstock processing facility wastewater: A unique and underappreciated source of contaminants to U.S. streams. *Environmental Science & Technology, 56*(2), 1028. https://doi.org/10.1021/acs.est.1c06821

Hussein, H., Ibrahim, S. F., Kandeel, K., & Moawad, H. (2004). Biosorption of heavy metals from waste water using pseudomonas sp. *Electronic Journal of Biotechnology, 7*(1), 30–37. http://www.scielo.cl/scielo.php?script=sci_abstract&pid=S0717-34582004000100006&lng=es&nrm=iso&tlng=en

Jenkins, N. E., & Grzywacz, D. (2000). Quality control of fungal and viral biocontrol agents—Assurance of product performance. *Biocontrol Science and Technology, 10*(6), 753–777. https://doi.org/10.1080/09583150020011717

Kamali, M., Gameiro, T., Costa, M. E. V., & Capela, I. (2016). Anaerobic digestion of pulp and paper mill wastes—An overview of the developments and improvement opportunities. *Chemical Engineering Journal, 298*, 162–182. https://doi.org/10.1016/j.cej.2016.03.119

Kanamarlapudi, S. L. R. K., KumarChintalpudi, V., Muddada, S., Kanamarlapudi, S. L. R. K., KumarChintalpudi, V., & Muddada, S. (2018). Application of biosorption for removal of heavy metals from wastewater. In *Biosorption*. IntechOpen. https://doi.org/10.5772/intechopen.77315

Karimi, S., Mahboobi Soofiani, N., Lundh, T., Mahboubi, A., Kiessling, A., & Taherzadeh, M. J. (2019). Evaluation of filamentous fungal biomass cultivated on vinasse as an alternative nutrient source of fish feed: Protein, lipid, and mineral composition. *Fermentation, 5*(4), 99. https://doi.org/10.3390/fermentation5040099

Kaudare, S., Mishra, H., Student, M., & Co-Ordinator, M. (2019). Review on wastewater treatment technologies. *International Journal of Engineering Research & Technology, 6*(2), 1.

Kaushik, P., & Malik, A. (2010). Alkali, thermo and halo tolerant fungal isolate for the removal of textile dyes. *Colloids and Surfaces B: Biointerfaces, 81*(1), 321–328. https://doi.org/10.1016/j.colsurfb.2010.07.034

Kc, S., Upadhyaya, J., Joshi, D. R., Lekhak, B., Kumar Chaudhary, D., Raj Pant, B., Raj Bajgai, T., Dhital, R., Khanal, S., Koirala, N., & Raghavan, V. (2020). Production, characterization, and industrial application of pectinase enzyme isolated from fungal strains. *Fermentation, 6*(2), 59. https://doi.org/10.3390/fermentation6020059

Khan, N. A., Ahmed, S., Farooqi, I. H., Ali, I., Vambol, V., Changani, F., Yousefi, M., Vambol, S., Khan, S. U., & Khan, A. H. (2020). Occurrence, sources and conventional treatment techniques for various antibiotics present in hospital wastewaters: A critical review. *TrAC - Trends in Analytical Chemistry, 129*, 115921. https://doi.org/10.1016/j.trac.2020.115921

Koller, M., Saleh, H. M., Koller, M., & Saleh, H. M. (2018). Introductory chapter: Introducing heavy metals. In *Heavy metals*. IntechOpen. https://doi.org/10.5772/intechopen.74783

Kumar, A., Yadav, A. N., Mondal, R., Kour, D., Subrahmanyam, G., Shabnam, A. A., Khan, S. A., Yadav, K. K., Sharma, G. K., Cabral-Pinto, M., Fagodiya, R. K., Gupta, D. K., Hota, S., & Malyan, S. K. (2021). Myco-remediation: A mechanistic understanding of contaminants alleviation from natural environment and future prospect. *Chemosphere, 284*, 131325. https://doi.org/10.1016/j.chemosphere.2021.131325

Kunjadia, P. D., Sanghvi, G. V., Kunjadia, A. P., Mukhopadhyay, P. N., & Dave, G. S. (2016). Role of ligninolytic enzymes of white rot fungi (Pleurotus spp.) grown with azo dyes. *Springerplus, 5*(1), 1487. https://doi.org/10.1186/s40064-016-3156-7

Kurade, M. B., Waghmode, T. R., Chakankar, M., Awasthi, M. K., Dawkar, V., & Jeon, B.-H. (2023). Microbial consortium as an effective biocatalyst for the decolorization of textile dyes wastewater. In S. P. Govindwar, M. B. Kurade, B.-H. Jeon, & A. Pandey (Eds.), *Current developments in bioengineering and biotechnology* (pp. 473–511). Elsevier. https://doi.org/10.1016/B978-0-323-91235-8.00007-3

Lanfranconi, I., Ceretta, M. B., Bertola, N., Wolski, E. A., & Durruty, I. (2022). Textile dyeing wastewater treatment by Penicillium chrysogenum: Design of a sustainable process. *Water Science and Technology: A Journal of the International Association on Water Pollution Research, 86*(2), 292–301. https://doi.org/10.2166/wst.2022.204

Lawson, H. D., Walton, S. P., & Chan, C. (2021). Metal-organic frameworks for drug delivery: A design perspective. *ACS Applied Materials & Interfaces, 13*(6), 7004–7020. https://doi.org/10.1021/acsami.1c01089

Legorreta-Castañeda, A. J., Lucho-Constantino, C. A., Beltrán-Hernández, R. I., Coronel-Olivares, C., & Vázquez-Rodríguez, G. A. (2020). Biosorption of water pollutants by fungal pellets. *Water, 12*(4), 4. https://doi.org/10.3390/w12041155

Li, L., Liang, T., Liu, W., Liu, Y., & Ma, F. (2020). A comprehensive review of the mycelial pellet: Research status, applications, and future prospects. *Industrial & Engineering Chemistry Research, 59*(39), 16911–16922. https://doi.org/10.1021/acs.iecr.0c01325

Liu, W., Chao, Y., Yang, X., Bao, H., & Qian, S. (2004). Biodecolorization of azo, anthraquinonic and triphenylmethane dyes by white-rot fungi and a laccase-secreting engineered strain. *Journal of Industrial Microbiology and Biotechnology, 31*(3), 127–132. https://doi.org/10.1007/s10295-004-0123-z

Madadi, R., & Bester, K. (2021). Fungi and biochar applications in bioremediation of organic micropollutants from aquatic media. *Marine Pollution Bulletin, 166*, 112247. https://doi.org/10.1016/j.marpolbul.2021.112247

Manekar, P., Patkar, G., Aswale, P., Mahure, M., & Nandy, T. (2014). Detoxifying of high strength textile effluent through chemical and bio-oxidation processes. *Bioresource Technology, 157*, 44–51. https://doi.org/10.1016/j.biortech.2014.01.046

Marchlewicz, A., Guzik, U., Hupert-Kocurek, K., & Wojcieszyńska, D. (2023). Evaluation of the defined bacterial consortium efficacy in the biodegradation of NSAIDs. *Molecules, 28*(5), 2185. https://doi.org/10.3390/molecules28052185

Michalska, K., Goszkiewicz, A., Skalska, K., Kołodziejczyk, E., Markiewicz, J., Majzer, R., & Siedlecki, M. (2021). Treatment of industrial wastewaters and liquid waste by fungi. In *Encyclopedia of mycology* (pp. 662–682). Elsevier. https://doi.org/10.1016/B978-0-12-819990-9.00055-X

Modenbach, A. A., & Nokes, S. E. (2012). The use of high-solids loadings in biomass pretreatment—A review. *Biotechnology and Bioengineering, 109*(6), 1430–1442. https://doi.org/10.1002/bit.24464

Mohammad, N., Alam, M. Z., Kabbashi, N. A., & Ahsan, A. (2012). Effective composting of oil palm industrial waste by filamentous fungi: A review. *Resources, Conservation and Recycling, 58*(C), 69–78. https://ideas.repec.org//a/eee/recore/v58y2012icp69-78.html

More, T. T., Yan, S., Tyagi, R. D., & Surampalli, R. Y. (2010). Potential use of filamentous fungi for wastewater sludge treatment. *Bioresource Technology, 101*(20), 7691–7700. https://doi.org/10.1016/j.biortech.2010.05.033

Morya, R., Kumar, M., Tyagi, I., Kumar Pandey, A., Park, J., Raj, T., Sirohi, R., Kumar, V., & Kim, S.-H. (2022). Recent advances in black liquor valorization. *Bioresource Technology, 350*, 126916. https://doi.org/10.1016/j.biortech.2022.126916

Münze, R., Hannemann, C., Orlinskiy, P., Gunold, R., Paschke, A., Foit, K., Becker, J., Kaske, O., Paulsson, E., Peterson, M., Jernstedt, H., Kreuger, J., Schüürmann, G., & Liess, M. (2017). Pesticides from wastewater treatment plant effluents affect invertebrate communities. *Science of the Total Environment, 599*, 387–399. https://doi.org/10.1016/j.scitotenv.2017.03.008

Nair, R. B., & Taherzadeh, M. J. (2016). Valorization of sugar-to-ethanol process waste vinasse: A novel biorefinery approach using edible ascomycetes filamentous fungi. *Bioresource Technology, 221*, 469–476. https://doi.org/10.1016/j.biortech.2016.09.074

Nasirabadi, P. S., Saljoughi, E., & Mousavi, S. M. (2016). Membrane processes used for removal of pharmaceuticals, hormones, endocrine disruptors and their metabolites from wastewaters: A review. *Desalination and Water Treatment, 57*(51), 24146–24175. https://doi.org/10.1080/19443994.2016.1140081

Negi, B. B., & Das, C. (2023). Mycoremediation of wastewater, challenges, and current status: A review. *Bioresource Technology Reports, 22*, 101409. https://doi.org/10.1016/j.biteb.2023.101409

Nitayavardhana, S., Issarapayup, K., Pavasant, P., & Khanal, S. K. (2013). Production of protein-rich fungal biomass in an airlift bioreactor using vinasse as substrate. *Bioresource Technology, 133*, 301–306. https://doi.org/10.1016/j.biortech.2013.01.073

Ojima, Y., Kosako, S., Kihara, M., Miyoshi, N., Igarashi, K., & Azuma, M. (2019). Recovering metals from aqueous solutions by biosorption onto phosphorylated dry baker's yeast. *Scientific Reports, 9*(1), 225. https://doi.org/10.1038/s41598-018-36306-2

Olicón-Hernández, D. R., González-López, J., & Aranda, E. (2017). Overview on the biochemical potential of filamentous fungi to degrade pharmaceutical compounds. *Frontiers in Microbiology, 8*. https://doi.org/10.3389/fmicb.2017.01792

Parchami, M., Ferreira, J. A., & Taherzadeh, M. J. (2021). Starch and protein recovery from brewer's spent grain using hydrothermal pretreatment and their conversion to edible filamentous fungi—A brewery biorefinery concept. *Bioresource Technology, 337*, 125409. https://doi.org/10.1016/j.biortech.2021.125409

Pearce, G. K. (2008). UF/MF pre-treatment to RO in seawater and wastewater reuse applications: A comparison of energy costs. *Desalination, 222*(1), 66–73. https://doi.org/10.1016/j.desal.2007.05.029

Pereira, C. R., Resende, J. T. V., Guerra, E. P., Lima, V. A., Martins, M. D., & Knob, A. (2017). Enzymatic conversion of sweet potato granular starch into fermentable sugars: Feasibility of sweet potato peel as alternative substrate for α-amylase production. *Biocatalysis and Agricultural Biotechnology, 11*, 231–238. https://doi.org/10.1016/j.bcab.2017.07.011

Poh, P. E., Wu, T. Y., Lam, W. H., Poon, W. C., & Lim, C. S. (2020). Palm oil milling wastes. In *Waste management in the palm oil industry: Plantation and milling processes* (pp. 21–44). Springer. https://doi.org/10.1007/978-3-030-39550-6_3

Premaratne, M., Nishshanka, G., Liyanaarachchi, V., Nimarshana, P., & Ariyadasa, T. U. (2021). Bioremediation of textile dye wastewater using microalgae: Current trends and future perspectives. *Journal of Chemical Technology & Biotechnology, 96*(12), 3249–3258. https://doi.org/10.1002/jctb.6845

Pu, S., Ma, H., Deng, D., Xue, S., Zhu, R., Zhou, Y., & Xiong, X. (2018). Isolation, identification, and characterization of an Aspergillus niger bioflocculant-producing strain using potato starch wastewater as nutrilite and its application. *PLoS One, 13*(1), e0190236. https://doi.org/10.1371/journal.pone.0190236

Rezaei, R., Aghapour, A. A., & Khorsandi, H. (2022). Investigating the biological degradation of the drug β-blocker atenolol from wastewater using the SBR. *Biodegradation, 33*(3), 267–281. https://doi.org/10.1007/s10532-022-09979-w

Rizzo, L., Manaia, C., Merlin, C., Schwartz, T., Dagot, C., Ploy, M. C., Michael, I., & Fatta-Kassinos, D. (2013). Urban wastewater treatment plants as hotspots for antibiotic resistant bacteria and genes spread into the environment: A review. *Science of the Total Environment, 447*, 345–360. https://doi.org/10.1016/j.scitotenv.2013.01.032

Rodriguez-Narvaez, O. M., Peralta-Hernandez, J. M., Goonetilleke, A., & Bandala, E. R. (2017). Treatment technologies for emerging contaminants in water: A review. *Chemical Engineering Journal, 323*, 361–380. https://doi.org/10.1016/j.cej.2017.04.106

Romera, E., González, F., Ballester, A., Blázquez, M. L., & Muñoz, J. A. (2006). Biosorption with algae: A statistical review. *Critical Reviews in Biotechnology, 26*(4), 223–235. https://doi.org/10.1080/07388550600972153

Rousis, N. I., Zuccato, E., & Castiglioni, S. (2017). Wastewater-based epidemiology to assess human exposure to pyrethroid pesticides. *Environment International, 99*, 213–220. https://doi.org/10.1016/j.envint.2016.11.020

Rousta, N., Hellwig, C., Wainaina, S., Lukitawesa, L., Agnihotri, S., Rousta, K., & Taherzadeh, M. J. (2021). Filamentous fungus aspergillus oryzae for food: From submerged cultivation to fungal burgers and their sensory evaluation-A pilot study. *Foods, 10*(11), 2774. https://doi.org/10.3390/foods10112774

Ryu, H., Ryu, H. W., & Cho, K. S. (2017). Characterization of dye decolorization in cell-free culture broth of Trametes versicolor CBR43. *Journal of Microbiology and Biotechnology, 27*(1), 155–160. https://doi.org/10.4014/jmb.1608.08008

Saleh, I. A., Zouari, N., & Al-Ghouti, M. A. (2020). Removal of pesticides from water and wastewater: Chemical, physical and biological treatment approaches. *Environmental Technology & Innovation, 19*, 101026. https://doi.org/10.1016/j.eti.2020.101026

Sankaran, S., Khanal, S. K., Jasti, N., Jin, B., III, & Leeuwen, J. H. V. (2010). Use of filamentous fungi for wastewater treatment and production of high value fungal Byproducts: A review. *Critical Reviews in Environmental Science and Technology*. https://doi.org/10.1080/10643380802278943

Sar, T., Ferreira, J. A., & Taherzadeh, M. J. (2020). Bioprocessing strategies to increase the protein fraction of Rhizopus oryzae biomass using fish industry sidestreams. *Waste Management, 113*, 261–269. https://doi.org/10.1016/j.wasman.2020.06.005

Satari, B., Karimi, K., Taherzadeh, M. J., & Zamani, A. (2016). Co-production of fungal biomass derived constituents and ethanol from citrus wastes free sugars without auxiliary nutrients in airlift bioreactor. *International Journal of Molecular Sciences, 17*(3), 302. https://doi.org/10.3390/ijms17030302

Shamim, S. (2018). Biosorption of heavy metals. In *Biosorption*. IntechOpen. https://doi.org/10.5772/intechopen.72099

Sharma, K. R., Giri, R., & Sharma, R. K. (2020). Lead, cadmium and nickel removal efficiency of white-rot fungus Phlebia brevispora. *Letters in Applied Microbiology, 71*(6), 637–644. https://doi.org/10.1111/lam.13372

Sharma, P., Iqbal, H. M. N., & Chandra, R. (2022). Evaluation of pollution parameters and toxic elements in wastewater of pulp and paper industries in India: A case study. *Case Studies in Chemical and Environmental Engineering, 5*, 100163. https://doi.org/10.1016/j.cscee.2021.100163

Siddiquee, S., Rovina, K., & Azad, S. A. (2015). Heavy metal contaminants removal from wastewater using the potential filamentous fungi biomass: A review. *Journal of Microbial & Biochemical Technology, 7*(6), 1000243. https://doi.org/10.4172/1948-5948.1000243

Singh, C., Vyas, D., Singh, C., & Vyas, D. (2021). Biodegradation by fungi for humans and plants nutrition. In *Biodegradation technology of organic and inorganic pollutants*. IntechOpen. https://doi.org/10.5772/intechopen.99002

Singh, G., & Dwivedi, S. K. (2020). Decolorization and degradation of direct Blue-1 (azo dye) by newly isolated fungus aspergillus terreus GS28, from sludge of carpet industry. *Environmental Technology & Innovation, 18*, 100751. https://doi.org/10.1016/j.eti.2020.100751

Slavov, A. K. (2017). General characteristics and treatment possibilities of dairy wastewater—A review. *Food Technology and Biotechnology, 55*(1), 14–28. https://doi.org/10.17113/ftb.55.01.17.4520

Souza Filho, P. F., Zamani, A., & Taherzadeh, M. J. (2019). Edible protein production by filamentous fungi using starch plant wastewater. *Waste and Biomass Valorization, 10*(9), 2487–2496. https://doi.org/10.1007/s12649-018-0265-2

Srivastava, A., Singh, P., Srivastava, S., & Singh, S. (2023). Reclamation and characterization of value-added products from pulp and paper mill effluent using microbial fuel cell. *Materials Today Proceedings*. https://doi.org/10.1016/j.matpr.2023.02.415

Strong, P. J., & Burgess, J. E. (2008). Treatment methods for wine-related and distillery wastewaters: A review. *Bioremediation Journal, 12*(2), 70–87. https://eprints.qut.edu.au/203986/

Szengyel, Z., & Zacchi, G. (2000). Effect of acetic acid and furfural on cellulase production of Trichoderma reesei RUT C30. *Applied Biochemistry and Biotechnology, 89*(1), 31–42. https://doi.org/10.1385/abab:89:1:31

Tamjidi, S., & Ameri, A. (2020). A review of the application of sea material shells as low cost and effective bio-adsorbent for removal of heavy metals from wastewater. *Environmental Science and Pollution Research International, 27*(25), 31105–31119. https://doi.org/10.1007/s11356-020-09655-7

Tay, J.-H., Show, K.-Y., & Hung, Y.-T. (2004). Seafood processing wastewater treatment. In *Handbook of industrial and hazardous wastes treatment* (2nd ed.). CRC Press.

Telke, A. A., Kadam, A. A., Jagtap, S. S., Jadhav, J. P., & Govindwar, S. P. (2010). Biochemical characterization and potential for textile dye degradation of blue laccase from aspergillus ochraceus NCIM-1146. *Biotechnology and Bioprocess Engineering, 15*(4), 696–703. https://doi.org/10.1007/s12257-009-3126-9

The Challenge of Micropollutants in Aquatic Systems. (n.d.). Retrieved November 19, 2024, from https://www.science.org/doi/10.1126/science.1127291

Toczyłowska-Mamińska, R. (2017). Limits and perspectives of pulp and paper industry wastewater treatment—A review. *Renewable and Sustainable Energy Reviews, 78*, 764–772. https://doi.org/10.1016/j.rser.2017.05.021

Topare, N. S., Attar, S. J., & Manfe, M. M. (2011). Sewage/wastewater treatment technologies: A review. *Scientific Reviews and Chemical Communications, 1*(1). https://www.tsijournals.com/abstract/sewagewastewater-treatment-technologies-a-review-11194.html

Tracking Clean Energy Progress 2023 – Analysis Available online: https://www.iea.org/reports/tracking-clean-energy-progress-2023 (accessed on 22 January 2026).

Trellu, C., Olvera Vargas, H., Mousset, E., Oturan, N., & Oturan, M. (2021). Electrochemical technologies for the treatment of pesticides. *Current Opinion in Electrochemistry, 26*, 100677. https://doi.org/10.1016/j.coelec.2020.100677

Veiter, L., Rajamanickam, V., & Herwig, C. (2018). The filamentous fungal pellet-relationship between morphology and productivity. *Applied Microbiology and Biotechnology, 102*(7), 2997–3006. https://doi.org/10.1007/s00253-018-8818-7

Viuda-Martos, M., Fernandez-Lopez, J., Sayas-Barbera, E., Sendra, E., & Perez-Alvarez, J. A. (2011). Physicochemical characterization of the orange juice waste water of a citrus by-product. *Journal of Food Processing and Preservation, 35*(2), 264–271. https://doi.org/10.1111/j.1745-4549.2009.00450.x

Westlund, P., & Yargeau, V. (2017). Investigation of the presence and endocrine activities of pesticides found in wastewater effluent using yeast-based bioassays. *The Science of the Total Environment, 607*, 744–751. https://doi.org/10.1016/j.scitotenv.2017.07.032

White, C., Wilkinson, S. C., & Gadd, G. M. (1995). The role of microorganisms in biosorption of toxic metals and radionuclides. *International Biodeterioration and Biodegradation, 35*(1-3), 17–40. https://doi.org/10.1016/0964-8305(95)00036-5

Winker, M., Faika, D., Gulyas, H., & Otterpohl, R. (2008). A comparison of human pharmaceutical concentrations in raw municipal wastewater and yellowwater. *The Science of the Total Environment, 399*(1–3), 96–104. https://doi.org/10.1016/j.scitotenv.2008.03.027

Yadav, M., Gupta, R., & Sharma, R. K. (2019). Green and sustainable pathways for wastewater purification. In *Advances in water purification techniques* (pp. 355–383). Elsevier. https://doi.org/10.1016/B978-0-12-814790-0.00014-4

Yaseen, D. A., & Scholz, M. (2019). Textile dye wastewater characteristics and constituents of synthetic effluents: A critical review. *International Journal of Environmental Science and Technology, 16*(2), 1193–1226. https://doi.org/10.1007/s13762-018-2130-z

Zhang, Z. Y., Jin, B., Bai, Z. H., & Wang, X. Y. (2008). Production of fungal biomass protein using microfungi from winery wastewater treatment. *Bioresource Technology, 99*(9), 3871–3876. https://doi.org/10.1016/j.biortech.2006.10.047

Chapter 12
Fungal-Based Nanosystem: Futuristic Tools for Controlling Fungal Infections

Ankita Yadav, Divya Chauhan, Neetu Talreja, Shuphiya Khan, and Mohammad Ashfaq

Abstract Fungal infections pose a serious risk to human health, especially to those with weakened immune systems. Although fungi are common and difficult to treat, it might be challenging to manage these infections because there aren't many antifungal medication options. Furthermore, these medications can cause fungus to develop drug resistance, making treatments less effective. Fungal nanosystems can be considered one of the most prospective research domains. These nanosystems employ nano-scaled particles extracted from or based on fungi for enhancing fungi for combating fungal infections due to its unique properties of fungal-based nanomaterials. These devices would be able to easily identify fungal cells, deliver antifungal medications to the scene, and enhance the host's immune response. Furthermore, fungal-based nanosystems should eventually be able to help with the faster and more accurate identification of fungal infections. This book chapter aims to describe studies in which nanomaterials/nanosystems were synthesized using fungi as reducing agents, discussing the mechanisms and optimization of the synthesis, as well as the applications. The literature shows that various species of fun-

A. Yadav
Department of Bioscience and Biotechnology, Banasthali Vidyapith, Newai, Rajasthan, India

D. Chauhan
Department of Drinking Water and Sanitation, Ministry of Jal Shakti, New Dehli, India

N. Talreja (✉)
Faculty of Science and Technology, Department of Science, Alliance University, Bengaluru, Karnataka, India

S. Khan
Department of Bioscience and Biotechnology, Banasthali Vidyapith, Newai, Rajasthan, India

Shriram Institute for Industrial Research (SIFIR), Gurugram, Haryana, India

M. Ashfaq (✉)
Department of Biotechnology, School of Sciences, Woxsen University, Hyderabad, Telangana, India

Department of Mechanical Engineering, Faculty of Engineering, Chulalongkorn University, Pathumwan, Bangkok, Thailand

C. M. Hussain et al. (eds.), *Fungi in Waste Water Treatment*, Springer Water,
https://doi.org/10.1007/978-3-032-12693-1_12

gus have potential for use in biogenic synthesis, enabling the production of nanomaterials/nanosystems and this nanosystem is used for controlling fungal infection. Nanosystems/nanomaterials synthesized using fungi enable the control of fungal infection, with low toxicity and good biocompatibility. The potential applications of fungal-based nanosystems as fungal infection control solutions in the future. Fungal-based nanosystems, which build on current research trends, have the potential to become more effective in the fight against fungal infections in the future.

Keywords Nanosystem · Infection · Antifungal · Nanomaterials · Fungi

12.1 Introduction

Fungal pathogens threaten the health and food of the three most prosperous domains of the animal kingdom and humanity (Garibotto et al., 2010; Leyva Salas et al., 2017). Concerning human health, fungi may be responsible for life-threatening infections (mycosis), poisoning from toxins (like botulism) and allergic reactions to proteins found in fungi (Kathiravan et al., 2012). Pathogenic species that are exogenous, endogenous, or opportunistic can cause fungal infections (Garibotto et al., 2010). Invasive fungal infections are more likely to occur in individuals with compromised, impaired, or underdeveloped immune systems, such as infants, cancer patients undergoing chemotherapy, organ transplant recipients, burn patients, and individuals suffering from acquired immune deficiency syndrome (AIDs) (Kathiravan et al., 2012). Invasive fungal infections have become much more common over the last few decades. Fungal infections affect more than one billion people yearly, causing ~1.6 million deaths worldwide (Brown et al., 2012; Tufa and Denning, 2019). Four genera of fungi (*Aspergillus* species, *Pneumocystis* species, *Candida* species, and *Cryptococcus* species) account for about 90% of deaths due to fungi (Kauffman and Hedderwick, 2006; Denning and Hope, 2010), specifically for *Candida* spp., invasive fungal infections, and *Aspergillus* spp. Mortality rates if an infection occurs can reach 30–50% (Denning and Hope, 2010). *Candida* spp. also raises the concern that they contribute to ~40% of the global mortality from all forms of fungal disease (Patterson, 2007; Perfect, 2012). In addition, Tinea pedis, a superficial infection in patients, is associated with species like *Trichophyton* spp., *Epidermophyton* spp., and *Microsporum* spp. (Tampieri, 2004). Therefore, effectively controlling infections due to these fungi is critical for human health.

The global rise of antimicrobial resistance represents a significant threat to human health. Opportunistic fungal pathogens for which antifungals have been developed have shown natural resistance to many of these drugs (Fisher et al., 2018). In an era where the consecutive rise of invasive fungal infections is adding to the burden of contemporary medicine, a global failure in the ability to manage these organisms is looming, and this can only be achieved by improving the management of current antifungals, captivating further discoveries of new antifungals, and further expanding our search for alternative antifungal options utilising emerging technologies (Kathiravan et al., 2012).

Nanobiotechnology is a promising strategy in disease management for quick detection of diseases, potential fungicides (nano fungicides), and effective delivery systems for fungicides to plant systems (Thakkar et al., 2010). Nanosystems are usually on the submicrometer scale by definition, and their behaviours are dependent on various factors, including the administration route and the exposure period in the blood and human immune response (Seaberg et al., 2021). In addition to the nature of the nano-drug delivery systems, the physicochemical properties also seem to be of great importance for the drug release behaviour (Ruenraroengsak et al., 2010). A plethora of studies were done on the matrix of antifungal drug delivery nanosystems, algal-derived dendrimers, micelles, liposomes, ferritin, and dendrimer-micelle nanosystems to establish a better mechanism of action against fungal pathogens from both animal and plant origins. The hollow core of these devices is susceptible to heat and electromagnetic radiation (a nanocapsule) (Nami et al., 2021). They are efficient in biomedical systems; thus, they are used in some delivery systems (Tiwari et al., 2008). Inorganic nanoparticles (NPs) are generally classified as metal and metal oxide–based NPs. Metallic NPs can be made both top-down and bottom-up. Lead (Pb), copper (Cu), silver (Ag), iron (Fe), gold (Au), Cadmium (Cd), Cobalt (Co), Zinc (Zn), and Aluminum (Al) are the metallic NPs that are utilized. Small quantities of NPs between 10 and 100 nm, the crystalline and amorphous structure, surface charge, surface charge density, pore size, the spherical and tube shape, and environmental dependence are a few examples (Salavati-Niasari et al., 2008; Chauhan et al., 2024; Omar et al. 2019). Thin-layer NPs are made up of several carbon layers. These substances can include graphene, fullerenes, carbon nanotubes (CNTs), carbon nanofibre (CNFs), carbon black, and nanoscale-activated carbon in certain situations. This application-based spectrum is enormous, including antifungal drugs and diagnostic apparatus.

The control of fungal infections is the most significant advantage of fungal-based nanosystems. These nanosystems can be engineered to combat fungal pathogens based on various mechanisms such as cell wall and membrane disruption, reactive oxygen species (ROS), DNA degradation, and radical generation. Besides antifungal response, fungal-based nanosystems may act on pathogenic mechanisms to confer protection against diseases, such as vaccination and biofilm inhibition (Du et al., 2021; Xin et al., 2023; Ashfaq et al., 2017; Ashfaq et al., 2014). Innovations in nanotechnology will also open up better delivery systems for antifungal drugs, thus allowing continuous discontinuation and targeting (Mishra et al., 2021). Consequently, developing efforts and consequences are therapeutically decreasing. Increasing attention is being paid to these developments in the fight against drug-resistant fungi, a growing threat to global health. Fungal-based nanosystems have attracted much attention because of their enormous potential for widespread medicinal use, vaccine development, and agricultural disease control (Abbas and Krishnan, 2020; Liu et al., 2024). This chapter intends to give a background of nanosystems based on fungi. First, fungal-based nanosystems are either synthesized intracellularly or extracellularly, so the synthesis of fungal-based nanosystems is categorized based on intracellular and extracellular. The second one involves application on different NPs such as metal NPs, carbon NPs, metal-organic frameworks, polymeric NPs, and other types of NPs. Finally, it talks about the various ways that antifungals work and the evolution of nanomaterials for treating and vaccinating against infection.

12.2 Synthesis of Fungal-Based Nanosystem

There are practical advantages for fungi, such as nanoparticle (NP) synthesis. Because fungi are not generally exposed to high concentrations of toxic metals, their natural ability to synthesize more proteins enables them to evolve to a state resistant/reductive to high concentrations of metal ions by the mechanism of reduction from multivalent into low toxicity forms (Sachin and Karn, 2021).

Since fungi are easier to work with in the lab and secrete large numbers of enzymes, they may also be interesting for NP production. According to research, the species *Penicillium*, *Aspergillus*, and *Fusarium* show some promise for the synthesis of various metal NPs (MNPs) outside of cells. Possible intracellular production of NPs by *Verticillium* Sp. (Araujo et al., 2020, Tsivileva et al., 2021). Table 12.1 summarizes the synthesis of NPs from varied strains of fungi.

Table 12.1 Utilizing several fungal strains as biological sources to produce NPs

'Biological source	Production site	NPs	Comments	References
Aspergillus clavatus	Extracellular	AgNPs	Antibacterial activity against mycobacterium aureus (MRSA) and mycobacterium epidermidis (MRSE) that are resistant to antibiotics	Saravanan and Nanda (2010)
Aspergillus flavus	Extracellular	Ag		Vigneshwaran et al. (2007)
Aspergillus Niger	Extracellular	Ag	Outstanding efficacy against both gram-negative and gram-positive bacteria	Gade et al. (2008)
Fusarium oxysporum	Extracellular	Zr	Can be useful for space applications and quantum dots.	Bansal et al., (2004)
Fusarium oxysporum	Extracellular	Tetragonal barium titanate	In the electronics sector to create ultrahigh density non-volatile ferromagnetic memory and minuscule capacitors.	Bansal et al. (2006)

Table 12.1 (continued)

'Biological source	Production site	NPs	Comments	References
Fusarium oxysporum	Extracellular from aqueous anionic complexes	TiO_2 and SiO_2	They have uses in the cosmetics sector and can serve as inorganic building blocks for the production of several layers on polystyrene (PS) sphere templates.	Bansal et al., (2005)
Fusarium moniliforme	Extracellular	Bimetallic Au-Ag alloy NPs		Bansal et al. (2005)
Fusarium semitectum	Extracellular	Ag	Possible uses in biomedicine	Basavaraja et al., (2008)
Cladosporium cladosporioides	Extracellular	Ag		Balaji et al. (2009)
Fusarium oxysporum and *Verticillium*	Intracellular and extracellular	Ag	Can create NPs that are trapped as a film in the biomass.	Senapati et al. (2004)
Fusarium solani	Extracellular	Ag		Ingle et al. (2009)
Pennicillum brevicompactum	Extracellular	Ag		Shaligram et al., (2009)
Phaenerochaete chrysosporium	Extracellular	Ag		Vigneshwaran et al. (2006)
Phoma glomerata	Extracellular	Ag and au	Thorough bactericidal action against the majority of gram-negative bacteria	Birla et al. (2009)
Streptomyces hygroscopicus	Extracellular	Ag	Greatly inhibited the growth of medically important pathogenic bacteria, gram-positive and gram-negative bacteria, and yeast.	Sadhasivam et al. (2010)
Trichoderma dresses	Extracellular	Ag	Antimicrobial effects	Mansoori (2013)
Tricoderma viride	Extracellular	Ag	Particle size decreases with an increase in reaction temperature, while a reduction in reaction temperature leads to	Fayaz et al. (2009)
Tricoderma viride	Extracellular	Ag	Emissions of blue-orange light have use in biosensors and bioimaging	Fayaz et al. (2010)
Verticillium lute album	Intracellular	Au		Gericke and Pinches (2006)
Verticillium sp	Intracellular	Au and Ag		Mukherjee et al. (2001), Sastry et al. (2003)

12.2.1 Intracellular Versus Extracellular Biosynthesis

Intracellular or extracellular (extracellular secreted to surrounding) outside their plasma membrane, fungi NP production. If we are talking about intracellular synthesis, instead of the solution, it forms on the fungal cell's inner surface. The ions are held on the fungal cell surface after their reduction via enzymes (reductases) inside the cytoplasmic membrane or cell wall. Combined with intracellular biosynthesis, it allows more control of the NP's size and shape distributions. Harvesting and recovering products, however, is less practical and economical (Macdonald and Smith, 1996). However, extracellular synthesis can synthesize a wider variety of NP systems because it is more adaptable. Additionally, as seen in Fig. 12.1, the proteins and reducing agents produced by the fungus maintained the extracellularly formed NPs, providing them an edge over intracellular production (Macdonald and Smith, 1996; Kumar and McLendon, 1997; Dhillon et al., 2012). The researchers found that the fungal biomass produced at least four high molecular weight proteins that were non-patholytic NP. Among these proteins was an NADH-dependent reductase enzyme specific to the tested strain. Fluorescence spectra, on the other hand, revealed emission bands that matched these proteins' natural forms in solution or affixed to NPs (Durán et al., 2007).

Additionally, neither the proteins' attachment to the NPs nor the decrease in metal ions had an impact on the tertiary structure of the proteins. A certain amount of clarification has already been provided regarding the reduction of metal ions and

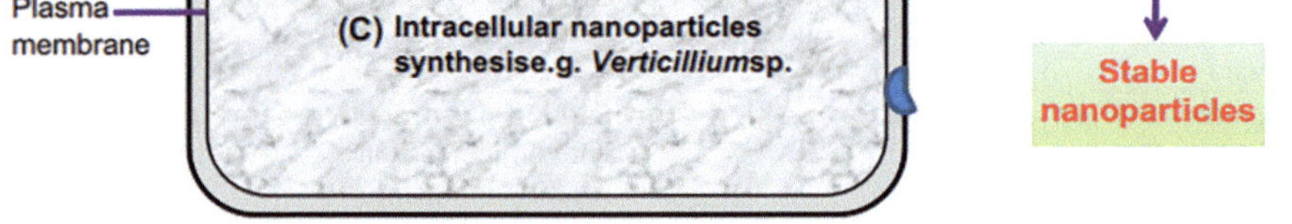

Fig. 12.1 Hypothetical model of NPs biosynthesis by Fungus: (**a**) electrostatic interaction between the metal ion and the enzyme present in the cell wall; (**b**) reduction of M^+ to Mo state by the enzyme; (**c**) intracellular biosynthesis of NPs, and (**d**) extracellular biosynthesis of NPs by fungi. The image was reproduced with the permission (Dhillon et al., 2012)

protein surface interaction with the NPs (Anil Kumar et al., 2007). Additional research verified the possibility of extracellular production of several metal NPs by genera like *Fusarium*, *Aspergillus*, and *Penicillium*. Simultaneously, the filtrate of cells obtained yielded extracellular NPs of Au and Ag. The sizes of these NPs were also strongly stable (ranging from 2 to 50 nm) (Mukherjee et al., 2002; Ahmad et al., 2003). Senapati et al. studied the first biological synthesis within bacteria and fungi was carried out by another pioneering study, wherein the fungus *Verticillium* and *Fusarium oxysporum* were initially employed to generate intracellular and extracellular AgNPs, respectively (Senapati et al., 2004). We have exciting potential here, as this outcome now allows the NPs to become trapped as a film within the biomass or produced in solution, with commercial opportunity associated with both. The study by Ahmad et al. additional information also corroborated the findings, which used the fungi cultures, which illustrated the critical aspect of the growth conditions at the time of producing the NPs (Ahmad et al., 2005). The result showed that Au ions, while incubated with *Trichothecium* sp. NPs were generated while biomass was growing in stationary phase conditions. NPs within cells were formed when shaking. This may be explained by releasing some enzymes or proteins synthesizing NPs into their medium when growth is stationary, but not when shaking occurs.

12.2.2 Verticillium

A species of the acidophilic fungus, *Verticillium* sp., can generate Au and Ag NPs when examined with $AuCl^{4-}$ and Ag^+ ions (Sastry et al., 2003). AgNPs have been in the spotlight in recent years as they possess broad-spectrum antibacterial activity as well as other biomedical applications. Mukherjee et al., when exposed to Ag ions, *Verticillium* sp. demonstrated the ability to generate AgNPs intracellularly. However, a higher type of oxidative stress was presented by *Verticillium luteoalbum* when exposed to Au-ions at different pHs (Mukherjee et al., 2001). In contrast, the ability to achieve varying pH levels was not significantly different, showing that the typical factors, such as the number, form, and size of particles produced per unit cell, were impacted (Mukherjee et al., 2001, Gericke and Pinches, 2006).

12.2.3 Fusarium

Fusarium oxysporum has also been picked for its characteristics, which result in the development of an extremely durable Ag hydrosol (Ahmad et al., 2003). There is a great deal of room for expansion, particularly in the large-scale production of semiconductor materials made of metal oxide, given the capacity of biological systems to regenerate and the ability of fungi like *F. oxysporum* to hydrolyze metallic complexes that are never present throughout their life cycle. Just last June, Ingle et al. later, Bansal et al. challenged a filamentous phytopathogen, *F. solani*, with $AgNO_3$ (1 mM), producing spherical polydispersed NPs having a 16.23 nm standard diameter. Furthermore, *F. oxysporum* produces TiO_2 and SiO_2 NPs from the aqueous anionic complex SiF_6^{2-} and TiF_6^{2} (Ingle et al., 2008; Bansal et al., 2006). Research has shown that the fungus secretes proteins and reducing agents that extracellularly stabilize the NPs (Durán et al., 2005).

12.2.4 Penicillium

To create AgNPs, the genus *Penicillium* has emerged as a strong contender. Highly uniform NPs (high zeta potential) are formed due to electrostatic repulsion and remain stable at pH above 8. Sadowaski et al. and Kathiresan et al. Spectrophotometric confirmation of *P. fellutanum's* great efficacy in biosynthesizing AgNPs has been established from coastal mangrove sediments, as evidenced by a 430 nm UV-Vis absorption peak (Sadowski et al., 2008; Kathiresan et al., 2009). The NPs generation dose-dependent optimization studies were conducted at pH 6, 0.35% NaCl, 5 °C, and 1 mM $AgNO_3$ challenge for 24 h. A protein band with an average molecular weight of 70 kDa has been identified in the purified fractions using denatured polyacrylamide gel electrophoresis (PAGE) analysis. The fungus *Penicillium brevicompactum* WA 2315's supernatant was used to create AgNPs. The produced NPs' capacity to stay stable and scatter in aqueous solution with no clumping together is credited as suggesting possible biocompatibility (Shaligram et al., 2009). Through the FTIR spectrum, we supported the idea that proteins are involved in synthesizing AgNPs. It would facilitate cheaper fabrication of a variety of nanostructures based on NPs.

12.2.5 Trichoderma

Recently, in a US patent, Mansoori et al. described the method for the extracellular synthesis of NPs by *Trichoderma reesei*. UV–vis spectrophotometry, FTIR, and TEM achieved confirmation of NPs biosynthesis of 5–50 nm in diameter (Mansoori, 2013). *Trichoderma asperellum* was utilized to create 13–18 nm-sized extracellular AgNPs. Likewise, Mohammed-Fayaz et al. under ambient conditions, *Trichoderma viride* synthesized AgNPs of extracellular and Ag hydrosol of 2–4 nm diameter. *Trichoderma reesei* is known for producing more specific extracellular enzymes than other fungi (Fayaz et al., 2010).

12.2.6 Other Fungal Strains

Phoma glomerata is used to produce extracellular AgNPs, and its antibacterial properties against *Pseudomonas aeruginosa*, *Staphylococcus aureus*, and *Escherichia coli* are examined in vitro (Birla et al., 2009). In a study conducted by Sadhasivam and colleagues, AgNPs were biosynthesized successfully by the fungus *Streptomyces hygroscopicus* (Sadhasivam et al., 2010). Balaji et al. *Cladosporium cladosporioides* process litter in marsh areas and were to biosynthesize AgNPs. In liquid media (g/L) containing 0.02 M, $FeCl_3$; 0.001 M, $MnSO_4$ H_2O, $CuSO_4$ $5H_2O$, 0.003 M; $ZnSO_4$ $7H_2O$, 0.01 M; Na_2MoO_3 $2H_2O$ 0.0015 M; K_2 HPO_4, 2.5 M, KNO_3, $MgSO_4$ $7H_2O$ and glucose, fungi were cultivated for 1 week. The broth underwent filtering and distilled water washing until 100 ml of double-distilled water and 0.01 ml of Ag^+ metal ion solution was combined in a 10 ml clear supernatant. The preparation was shaken and then maintained in a shaker at a temperature of 27 °C for 78 h. Using TEM and FTIR, NPs' dimensions and form were determined (Balaji et al. 2009). Vigneshwaran et al. and Sanghi and Verma *Phaenerochaete chrysosporium* and *Coriolus versicolor* have both created polydispersed AgNPs with diameters ranging from 15 to 60 and 25 to 75 nm, respectively (Vigneshwaran et al. 2006; Sanghi and Verma, 2009). There are numerous industrial uses for the fungal strains mentioned above. For instance, *Coriolus versicolor* has been used as a mushroom for traditional Asian medicinal treatments, *Streptomyces* is employed in pharmaceutical sectors, and *Phaenerochaete chrysosporium* is utilized for the synthesis of lignolytic enzymes. They can also use their waste biomass to manufacture NPs. The function of AgNP biogenesis by fungi is to establish a safe biological route for developing metal NPs for use in drug discovery programs (Wright, 2005; Wright and Sutherland, 2007). Thus, fungi can rapidly secrete proteins, making this biosynthesis method much more productive than bacteria. The method of fungal-mediated synthesis of NPs is economical, environment-friendly, and highly efficient since fungi are known to secrete enormous amounts of enzymes and proteins. Synthesis both intracellular and extracellularly is feasible and extracellular routes provide easier recovery and stabilization through fungal proteins.

12.3 Application of Fungal Nanosystem

12.3.1 M-NPs

Michael Faraday began researching the application of metals for the creation of NPs in the nineteenth century and was the first to show how a colloid metallic containing metallic particles in the solution '(Brust et al., 1994). The sizes of M-NPs range from 1 to 100 nm.' M-NPs offer benefits in the form of easy breakdown, UV photoactivity, chemical stability, low toxicity, minimal side effects, a modulus of pore size, the usefulness of antifungal, antimicrobial activity, and high anti-pathogenic resistance ability (Kamat, 2002; Ghosh et al., 2008). Due to the prospective antimicrobial efficacy, different kinds of nanomaterials and their properties and metabolites are being explored. Antifungal solid activity of different MNPs, ZnO, CuO, TiO_2, Si, Au, CaO, MgO, Ag, and Ag_2O (Abd-Elsalam et al., 2017). Metal & MO-NPs, unfortunately, were absorbed because they were effective against recalcitrant plant pathogens leading our agricultural systems to suffer significant financial losses (Pandey et al., 2018).

12.3.2 Ag-NPs

Ag-NPs, the most potential M-NPs, have left no field untouched due to their precise role. Thus, concerning Ag-NPs as antimicrobial agents, it is encouraging and exciting that, in most cases, the purpose of their application was to control plant pathogenic diseases (Jo et al., 2009; Park et al., 2006; Kim et al., 2007, Rubina et al., 2017),' and in some instances even both directly and indirectly. These scientists have demonstrated Ag-NPs' antibacterial properties against a variety of plant diseases. Ag-NPs of 20–30 nm size are better penetrators and colonizers into plant tissues. Compared to synthetic fungicides, these NPs are far less toxic and typically have a high potential for lowering fungal plant illnesses that are caused by spores (Jo et al., 2009). According to Kim et al. (2007), the process of eliminating Ag metal inactivates the enzyme system, which kills unicellular microbes (Kim et al., 2007). Ag-NPs have been shown to stimulate plant growth (Sharon et al., 2010). According to Oh et al. (2006), they have shown antifungal action against *Botrytis cinerea.* In both field and greenhouse studies, powdery mildew on pumpkin at 0.3 ppm was successfully controlled by nanosized silica-silver (1–5 nm), which is created when $AgNO_3$ reacts with sodium silicate (Na_2SiO_3) and a water-soluble polymer (Oh et al., 2006, Park et al., 2006). The fungicidal effect of Ag-NPs against the *Aspergillus*, *Penicillium*, and *Trichoderma* species was evaluated (Petica et al., 2008). *Magnaporthe oryzae* and *Bipolaris sorokiniana* colony formation is influenced by Ag ions and Ag-NPs (20–30 nm). Apart from their impact on *Raffaelea* sp. and *Golovinomyces cichoracearum* or *Sphaerotheca fusca* (Lamsal et al., 2011), each ionic and Ag-NPs also markedly decreased disease severity on perennial ryegrass (Lolium perenne) 3 h before spore inoculation (Kim et al., 2009). Additionally, Min et al. have employed Ag-NPs (4–8 nm) as an alternative to synthetic fungicides in the dose-dependent management of sclerotia-forming plant pathogenic fungi, including *Rhizoctonia solani*, *Sclerotinia sclerotiorum*. In these cases, hyphae damage after Ag-NPs exposure was critical, leading to hyphae plasmolysis and crumbling (Min et al., 2009). The results above show how Ag-NPs affect certain plant pathogens, which can contribute to severe crop losses.

12.3.3 Cu-NPs

Cu compounds have been in the category of antimicrobials for numerous years, since the eighteenth century. Developing new biocide classes is crucial for better fungicidal activity and lower environmental toxicity (Abd-Elsalam and Alghuthaymi, 2015). The antifungal assessment and advancements related to Cu-based NPs have substantially contributed to their development as antifungal agents with promising attributes desirable for fungicide application. The activity of an antifungal nature was registered for the produced Cu-NPs against filamentous fungi such as *Penicillium chrysogenum, Fusarium solani, Alternaria alternata, and Aspergillus flavus* (Ghasemian et al., 2012). Saharan et al. also reported that, with the help of the 0.1% level, CuNPs (196.4 ± 2.2 nm) discharged 89.5%, 63.0%, and 60.1% of their inhibitory effects on the growth of *A. alternata*, *M. phaseolina*, and *R. solani*, respectively (Saharan et al., 2013). The study found that three specially shaped Cu-based NPs were well effective as a treatment for protecting tomato plants in Mediterranean regions against Phytophthora infestans, unlike certain synthetic fungicides, which induce late blight tomato and potato disease (Giannousi et al., 2013). According to Giannousi et al., Cu_2O, CuO, and Cu/Cu_2O were all synthesized with the same diameters (11–14 nm) and forms (almost spherical). They were between 40% effective 3 days after the treatment and 6 days after treatment, 61% were effective. This suggested that improving treatment activity was influenced by time. Cu-NPs (3–10 nm) were recently reported to have a better in vitro antifungal effect than the widely used commercial fungicide bavistin in opposition to Phoma destructive *F. oxysporum*, *A. alternata*, and *C. lunata* and confirmed to have comparable antifungal efficacy. At 0.12% and 0.10% of CuNPs, respectively, these studies showed a noteworthy influence on the mycelial development and spore germination of *A. solani* and *F. oxysporum* (Giannousi et al., 2013).

12.3.4 Zn-NPs

Because of its crystalline form, huge surface area, and superior mechanical qualities, ZnO NPs can be fixed in polymer matrix materials to enhance the other materials. Nonetheless, ZnO requires more research as a fungicide for other plant fungal pathogens. For certain species, such as plant and food-borne diseases, ZnO may have antibacterial properties. ZnNPs decreased growth in *A. flavus*, *B. cinerea*, *A. niger*, and *Penicillium expansum* (He et al., 2011; Jayaseelan et al., 2012). *B. cinerea* at 70 ± 15 nm, ZnO NPs dramatically suppressed, causing mat deformation and altered cell shape. On the other hand, fungal hyphae die when the growth of *P. expansum* conidiophores and conidia is inhibited (He et al., 2011). The newly formed nano ZnO derived from zinc nitrate inhibited the growth of a major fungal pathogen.

12.3.5 Carbon Nanomaterials (CNMs)

CNMs possess multiple antifungal properties and are considered potential green materials for nanoformulations targeting fungal pathogens. Gorczyca et al. looked at how commercial multi-walled carbon nanotubes (MWCNTs) affected the entomopathogenic fungus *Paecilomyces fumosoroseus's* conidia (Gorczyca et al., 2009). The CNMs strictly regulated spore formation and mycelial production. The polyethylene (PE) films were made by solution-casting the CNTs nanocomposite substrate from boiling xylene. This nanocomposite has potent antifungal activity in packaged foods (Asgari et al., 2014). Against the pathogenic phytopathogenic fungus, *F. graminearum* and *F. poae*, the antifungal effects of six types of CNMs, including C60, GO, rGO, SWCNTs, MWCNTs, and AC, were tested (Wang et al., 2014). SWCNTs > MWCNTs > GO > rGO according to the antifungal activity time course data from left to right. Neither C60 nor AC showed any evident antifungal growth suppression. According to Wang et al., MWCNTs with particular surface groups have antifungal spots against *F. graminearum* their findings showed that the surface-modified MWCNTs had a germination rate of just 18%, which was three times lower than that of the comparable pure MWCNTs, and prevented spore germination (range of 0–108 spores) (Wang et al., 2017).

12.3.6 Metal-Organic Frameworks (MOFs)

MOFs are a novel subclass of porosity crystalline materials with a high degree of order consisting of metal nodes and organic ligands that absorb tremendous interest due to a unique structure to adjust for specific biomedical uses such as sensing, imaging, and controlled drug release (Abednatanzi et al., 2019; Lin et al., 2023). Currently, there are about 90,000 types of MOFs. For instance, Huang et al. described a fast preparation of CAU-17 of Bi-MOFs aligned crystallinity and proved its robust antimicrobial activity (Huang et al., 2023). Century in controlling diseased states caused by *P. gingivalis* and the action mechanisms involving the removal of H_2S and the interactions having functional groups S-, N-, and O-found in biomacromolecules (Huang et al., 2023). As such, Fig. 12.2 explains the Bi-MOF reactivity towards H_2S can make it possible to find antibiotic-tolerant *S. aureus* that is H_2S enriched, and the endogenously generated Bi_2S_3 product makes PA imaging easier for in situ and accurate theragnostic control of infections (Yuan et al., 2022).

The porphyrin derivatives (Zn–TCPP) combined over the Zn ion have been synthesized to form MOFs nanorods that turn luminous when exposed to the wound site's acidic environment. Interestingly, these Zn-TCPP nanorods could be utilized for the first time to provide highly effective fluorescence-imaging-guided PDT inhibition of microorganisms (Zhang et al., 2021).

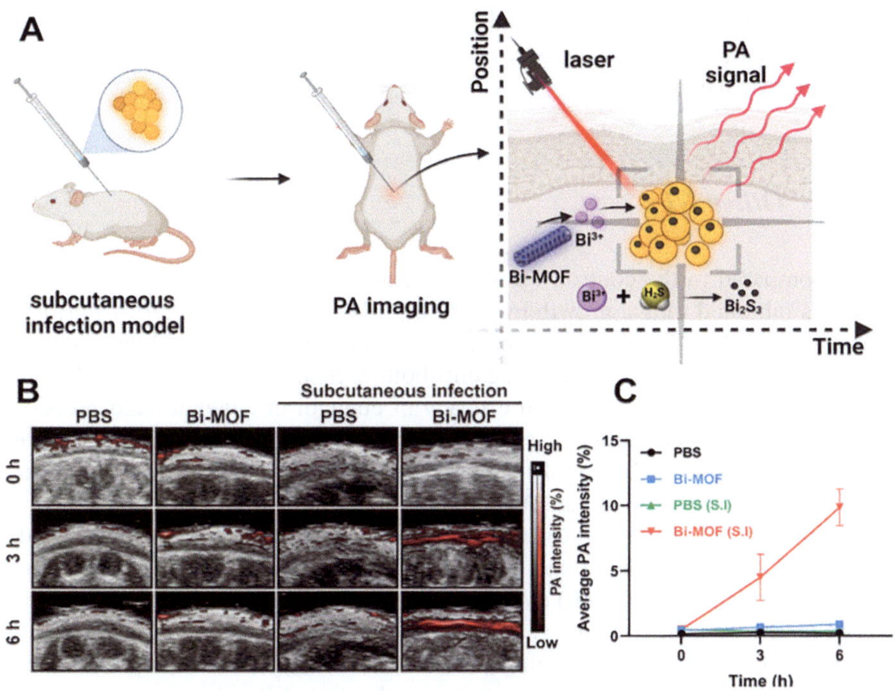

Fig. 12.2 Bi-MOF uses laser-triggered photoacoustic (PA) imaging to provide real-time in situ infection diagnosis. (**a**) Diagrammatic representation of the use of Bi-MOF for diagnosis in a model of subcutaneous infection in mice. (**b**) Typical in vivo photoacoustic pictures. PBS and 200 μg/ml Bi-MOF were injected into the subcutaneous cavities of either normal or infected animals. Following injection, photoacoustic imaging was observed at 0, 3, and 6 h. (**c**) The average PA intensity in the mice's subcutaneous cavity in (**b**) is quantitatively analysed. The values ($n = 3$) are means ± SDs. Figure was reproduced with the permission (Yuan et al., 2022)

12.3.7 Polymeric NPs

Natural and synthetic polymers possess extraordinary chemical, physical, mechanical, and structural properties. Utilizing these merits, they cannot only be self-assembled to form nanomaterials themselves but also play a role as additive components after further modifications on other nanomaterials to actively participate as biomedical application carriers. According to Spirescu et al. polymeric NPs have been developed as nanotheranostics in degradable and stimuli-responsive platforms. Proteins, peptides, nucleic acids, and other tiny or even biomacromolecules can be delivered via these particles to positively direct against pathogenic microbes and their accumulations (Spirescu et al., 2021). Polymeric NPs can be imparted with antimicrobial properties conventionally, (1) synthesized by antimicrobial cationic/anionic polymers or (2) metal/MONPs and antibiotics are examples of encapsulated antimicrobial agents. By combining and applying different strategies, they can work as a nanosystem to eliminate microorganisms and their biofilms. More significantly, the NPs' combined cationic chemotherapy and photodynamic treatments (irradiation with 660 nm light) may also inhibit the growth of *S. aureus* and *E. coli* (Spirescu et al., 2021).

12.3.8 Other NPs

When compared to the control, the thiamine di-lauryl sulfate (TDS) nanoparticle solutions inhibited fungal growth by 80% towards *Colletotrichum gloeosporioides* (Seo et al., 2011). Similarly, amphotericin B-nanoemulsion gel showed strong, dose-dependent antifungal efficacy against both *A. fumigatus* and *A. niger*. Method of preparation and mode of action efficacy of eugenol or eugenol nanoemulsions against *Fusarium oxysporum*. The antifungal activity of eugenol has been studied after the preparation of eugenol nanoemulsions and may have supported against *F. oxysporum* (Abd-Elsalam and Alghuthaymi, 2015). 10% eugenol was shown to have the highest antifungal growth both in vitro and in vivo. In a recent study, Priyanka et al. reported the impact of TiNPs on the mycorrhizal behaviour of arbuscular mycorrhizal fungi (AMF) (Mycorrhizal rice plant (MR plants)) seedling cuts set out for 25, 50, and 100 mg per plant. In the case of plants, the presence of TiO_2 NPs prevented roots from colonizing AMF. Additionally, the wood may be shielded from fungi that cause brown and white rot by the TiO_2 NPs (Priyanka et al., 2017, De Filpo et al., 2013).

Metallic and carbon-based NPs exhibit significant antifungal and antimicrobial properties, making them promising alternatives to synthetic fungicides. Ag, Cu, and Zn NPs are particularly effective against plant pathogens, enhancing crop protection and reducing environmental impact. Carbon nanomaterials (e.g., CNTs, graphene oxide) and MOF offer eco-friendly, multifunctional solutions for infection control. Additionally, polymeric and other specialized NPs demonstrate versatile applications, including targeted drug delivery and enhanced microbial eradication.

12.4 Antifungal Mechanisms

According to the examined microorganisms' rhizosphere competency, stimulating factors, parasitism, antibiosis, altered metabolism, and numerous other factors, the biofungicide's antifungal modes of action differ (Saraf et al., 2014; Shrivastava et al., 2014). Despite the lack of sufficient research to properly understand the mechanisms, they seem to be similar to the transmembrane disruption-based antibacterial effects of NMs. The following outcomes result from NMs to the fungal innermost membrane: (1) damage to peptidoglycan, (2) harmful metal ions in solution, (3) change of cellular pH via electro-plumb ion pumps, (4) Production of ROS and damage to nuclear material (5). Ag has a far more dramatic antifungal potential than other metals because its ions deactivate the action of fungal cell wall thiol clusters, which disrupts transmembranes, energy metabolism, and the electron transport chain. Additional processes include cell lysis, reduced membrane permeability, dissociation of the respiratory cycle-related enzyme complexes, and abnormalities in fungal DNA mutations (Velmurugan et al., 2009). AgNPs have been demonstrated to impact several biological systems positively. However, their efficacy is factor-dependent, and as particle size increases, activity decreases. Studies showed that compared to spherical and rod morphologies, the truncated triangular particle structure had a greater "tidal" effect (Pal et al., 2007; Panáček et al., 2009). The findings on fungal hyphae and their deformation, along with the wider effects on resting stages (spores) and fruiting bodies (sclerotia), thus call for further investigation into the antifungal mechanisms of NMs (Ul Haq and Ijaz, 2019).

12.4.1 Membrane Disruption

AgNPs have antifungal action through adsorption on the cell surfaces of bacteria and fungi (Karimi et al., 2012). Even though the surfactant agent sodium dodecyl sulfate (SDS) for NPs enhanced the integration of AgNPs into the fungal cells by binding to the proteins and lipids of cell membranes, toxicity related to membrane adhesion was also observed for composite substance Ag-SiO_2 NPs (Oh et al., 2006; Panáček et al., 2009). These NPs stick to the cells to physically cut them, block water channels, increase membrane permeability, and possibly even enter fungal cells to cause cell death. Since chitin and cellulose, the primary constituents of the cell wall, are more robust than peptide polysaccharides, fungal spores are resistant to mechanical disintegration. Accordingly, CNTs are unlikely to harm spores' cytoplasmic membranes (Wang et al., 2017). The spore's surface absorbs the MWCNT aggregates, which obstruct the water route and prevent germination. The effect of MWCNT on spore survival might also be explained by the suggested mechanism (Wang et al., 2017; Wang et al., 2014). As previously documented by chitosan NC treatment of seeds resulted in a more thorough development of plant defence mechanisms. Chitosan can inhibit the growth of microorganisms by either destabilizing membranes (secondary mode) or disrupting protein synthesis (primary mode) (Márquez et al., 2013). Furthermore, ZnO NPs liquefy cytoplasmic components, reduce the cytoplasm's electron-dense content, and offer superior fungal cell wall separation (Sirelkhatim et al., 2015; Arciniegas-Grijalba et al., 2017).'

12.4.2 Reactive Oxygen Species (ROS)

By producing an accumulation of ROS, species including ZnO, TiO_2, and SiO_2 NPs damage the cell membranes of *A. niger* and *F. oxysporum* (Gondal et al., 2012; Patra et al., 2012). The action modes of Cu-NPs, NCs made of Cu and Zn chitosan are almost identical as they all can generate free radicals and damage the cell membrane

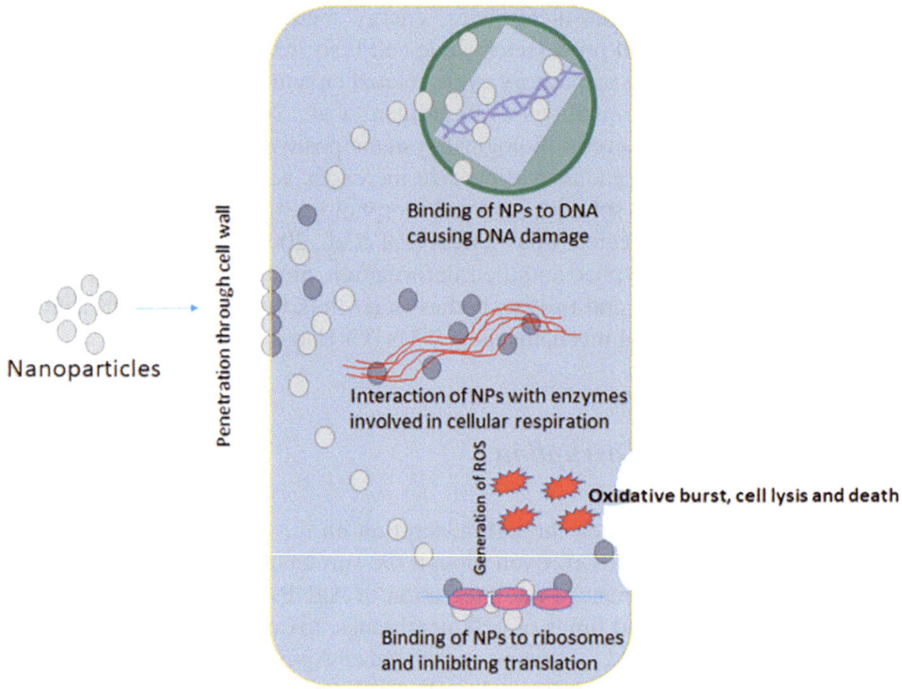

Fig. 12.3 By interacting with the cellular machinery, NPs have antifungal effects on fungal cells and their functioning. This results in cellular pathway arrest and ROS production, which causes an oxidative burst and fungal death. Figure was reproduced with the permission (Ul Haq and Ijaz, 2019)

(Xie et al., 2011). In addition, zinc is an essential plant micronutrient, absorbed through diffusion in plants and transported with high specificity as a divalent ion. When *A. niger*, *A. oryzae*, and *F. oxysporum* vegetative mycelia came into physical contact with reduced rGO nanosheets, the organic functional molecules of chitin and the polysaccharides of the fungal cell wall underwent chemical modification, resulting in the production of ROS (Sawangphruk et al., 2012). However, NPs disrupted the fungus's ROS-scavenging pathway, which led to the buildup of ROS (Liu et al., 2016). Intracellular materials leak out of *A. niger*, *A. oryzae*, and *F. oxysporum* due to Cu-BTC-induced cell membrane degradation which may subsequently disrupt the cell entirely and lead to death. Cu-BTC's antifungal activity is seemingly linked to ROS produced by the Cu-BTC which acts by reducing oxygen (Bouson et al., 2017). Figure 12.3 shows the cellular machinery NP antifungal effects on fungal cells and cellular pathway arrest and ROS production causes an oxidative burst and fungal death.

12.4.3 DNA Degradation

According to Borkow and Gabbay, by generating hydroxyl radicals, Cu fungicides can harm proteins, lipids, DNA, and other macromolecules, which can suppress plant infections and treat a variety of fungal illnesses (Borkow and Gabbay, 2005). Although significantly more Cu is liberated from these micro-sized particulates (Midander et al., 2009), CuNPs appear far more cytotoxic and genotoxic in vitro. Nucleic acids and carbohydrates were lumped after the exposure of *B. cinerea* stems on the fungal hyphae to nano-ZnO, leading to abnormal heifer cell systems (He et al., 2011). According to Hernández-Lauzardo et al. the introduction of the chitosan oligomer into the microorganism's cells, which prevents DNA transcription to mRNA and hence limits cell proliferation, is the other primary process. Cu-chitosan nanocomposites should be able to attach to DNA or mRNA and pass through fungal mobile partitions. By interfering with fungal metabolism and replicate, it may also eliminate the infections (Hernández-Lauzardo et al., 2011).

12.4.4 Radical Formation

Another method that MNPs are harmful to pathogenic fungi is by ion release (Hao et al., 2017). However, the liberated Cu ions suppress fungus by producing reactive hydroxyl radicals (Borkow and Gabbay, 2005). Because of the mechanism of ionic and chelating interactions, Cu-chitosan nanocomposites showed notably greater Cu deposition in porous regions which in turn inhibited the activities of fungal pathogen enzymes during pathogenesis. Along with Cu, CuO, and Cu_2O NPs, the most significant toxicity mechanism of Cu-based nanomaterials is the release of Cu^{2+}, which is typically responsible for the improved antifungal activities of CuO NPs (Kanhed et al., 2014, Zarzuela et al., 2017).

12.5 The Current Scenario of Fungal Infection Treatment and Vaccination Using Nanotechnology

Currently, available antifungals are limited by their significant side effects, lack of selectivity, poor biodistribution, ineffectiveness as a treatment, and negative effects in systemic fungal infections. As a controlled and site-specific drug delivery system pattern, NPs may enhance mycosis treatment without impairing the patient's quality of life. This proves this as an alternative to the dataset problem within nanotechnology. This covers different genera based on the state of NPs in the treatment of fungal diseases and immunization.

12.5.1 Candidiasis

Candida species are commensal inhabitants of the human microbiota. These microorganisms may also infect humans but are well-controlled when the biological barriers are intact and the immune system functions optimally. However, they cause serious infections when the biological barriers are lowered or there is a state of immunosuppression. *Candida* species are among the most prevalent fungi that infect humans. With *Candida albican* species being the most pathogenic, resulting in both systemic and superficial diseases (Pfaller and Diekema, 2010). Regarding the other species, the most menacing is *C. auris*, owing to its unprecedented propensity for intrinsic multidrug resistance. Three classes of antifungals are available for treating invasive *candidiasis*: polyenes, azoles and echinocandins'(Tumbarello et al., 2007, Sawant and Khan, 2017). The anticandida properties of certain NPs have been used to their advantage. For example, the antifungal activity of AuNPs in *C. albicans* biofilms has been studied because, when combined with a photosensitizer, they may improve photodynamic therapy. The pathogen's cell membrane may be disrupted by AuNPs' direct interactions with proteins and lipids. Additionally, there may be a lower chance of pathogens developing resistance to photodynamic treatment when photosensitizers are coupled with MNPs (Maliszewska et al., 2017). The nontoxicity of curcumin is supported by its wide range of antibiotic activity. Although curcumin extracts exhibit high anti-cancer activity, their therapeutic applications are limited by their low blood bioavailability due to degradation and rapid elimination from the systemic circulation. AgNPs are designed to improve the delivery of curcumin. AgNPs and curcumin inhibit fluconazole-resistant *Candida* glabrata and *C. albicans* in a concentration-related manner (Paul et al., 2018).

12.5.2 Cryptococcosis

Patients with cryptococcosis, a common and potentially fatal systemic mycosis brought on by *C. neoformans/C. gattii* species complexes have significant rates of morbidity and mortality and in particular, in immunocompromised individuals and third-world countries. Inhalation of fungal propagules initiates the disease that is asymptomatic in otherwise healthy individuals and can be resolved spontaneously. Transmission from asymptomatic carriers happens often, and in immunosuppressed cases (e.g., those with HIV), relapse can occur, leading to disease. The primary infection in immunocompromised hosts is less likely to occur. In either case, the mycosis commonly manifests as meningoencephalitis and has an impact on the brain (Kwon-Chung et al., 2014). While the most affected risk group by *C. neoformans* are individuals infected by HIV, patients in chemotherapy or undergoing immunosuppression are other at-risk groups. Significantly, although most *C. gattii* cases occur in immunocompetent individuals, other risk factors may affect disease development (Marr et al., 2012; Chen et al., 2014). The best treatment for

Cryptococcosis, whether it presents as severe pulmonary cryptococcosis or cryptococcal meningitis, is combination therapy with AmB + 5-fluorocytosine, coupled with fluconazole as a maintenance drug for weeks to a lifetime (Perfect et al., 2010). Although nations are not able to procure 5-fluorocytosine fluconazole, it can be utilized instead (in combination with AmB) (Perfect et al., 2010). Fluconazole high-dose may be used as primary therapy in little-resourced settings. Amphotericin B deoxycholate is an old but indispensable drug for treating invasive fungal disease.

12.5.3 Paracoccidioidomycosis (PCM)

Species of thermo-dimorphic fungus belonging to the genus *Paracoccidioides* are responsible for the systemic mycosis known as PCM. PCM has an acute and a chronic clinical form (Shikanai-Yasuda et al., 2017), It will be called the chronic form (adult) and the acute/subacute form (juvenile) in this study. Chemotherapy is used to treat PCM; the primary classes are sulfanilamides like Bactrim®, polyenes like AmB, and azole drugs like fluconazole (Shikanai-Yasuda et al., 2017; Souza and Amaral, 2017). One approach to this is conjugating antibodies to the surface of the NPs, which commonly target them. Other types of molecules can also address NPs against specific tissues. While polysorbate helps medications that target the brain, the use of nanocarrier devices to deliver dimercaptosuccinic acid (DMSA) has mostly been found to direct its NPs to the lung. Still, the pathways through which this tropism takes place have been incompletely defined (Amaral et al., 2009). AmB-loaded PLGA and DMSA NPs (Nano-D-AMB) were employed to treat chronic PCM brought on by *P. brasiliensis*. Nano-D-AMB (6 mg/kg) was given to BALB/c mice every 72 h after a 30-day infection period. The treated mice showed no signs of stress (hypotrichosis and piloerection), no abnormalities of the kidneys or liver, and less weight loss than the AmB deoxycholate-treated group. Furthermore, the formulation did not increase the cytotoxic or genotoxic effects (Amaral et al., 2009). A recent study revealed that mice's lungs, liver, and spleen maintained a significant amount of nano-D-AMB. When tested against *P. brasiliensis*, ITZ-loaded DMSA-PLGA NPs demonstrated a PLGA-ITZ-MIC that was lower than ITZ alone and a lower cytotoxicity than the free drug (Cunha-Azevedo et al., 2011).

12.5.4 Histoplasmosis

Due to the theriomorphic fungus *H. capsulatum*, histoplasmosis is an endemic, systemic mycosis (Kauffman, 2007). According to the authors, histoplasmosis can range from a latent infection that occurs worldwide to deep pulmonary mycosis and/or systemic disease, depending on the infectious dosage and the host's immune state. *H. capsulatum* grows as a yeast, which has means by which it avoids death within phagocytic cells. It is also a potential reservoir of infection since it is not easily reachable by treating the immune response. Still, more importantly, most antifungal therapies are affected by a lack of interaction with their target in the cell, which can happen if the drug is not allowed to cross the macrophage membrane, where the fungus resides (Edwards et al., 2013). The only histoplasmosis treatment that has been investigated is AmBisome®. In this trial, 81 AIDS patients with moderate to severe histoplasmosis were treated with AmB deoxycholate and AmBisome®. 4 of 22 patients treated with AmB deoxycholate, 14 (64%) were successfully treated (3 died). Adverse events were observed in 63%, and nephrotoxicity occurred in 37%. In a study of AmBisome®, 88% of the 51 patients were successful. One patient did, however, pass away, and 25% of cases experienced adverse medication reactions, with 9% developing nephrotoxicity (Edwards et al., 2013).

The nanosystem is a well-studied approach for antifungals and more effective vaccines. It was a creative and economical approach to creating novel antifungals and therapies. This nanotechnology platform has drawn the eye of researchers since nanotherapeutics provide an excellent platform that combines low-toxicity materials capable of slow-direct medication and antigen targeting. The current status of NPs in the creation of innovative antifungal medications and vaccines that target systemic mycoses, with an emphasis on *Aspergillus* sp., *Candida* sp., *Cryptococcus* sp., *Paracoccidioides* sp., *Histoplasma* sp., and *Coccidioides* sp. (Edwards et al., 2013).

12.6 Conclusion

Advancements in nanotechnology have created increasingly new opportunities for health and productivity improvements over the past few decades. Fungal infectious diseases are an emerging and highly relevant issue in our current landscape, as they are one of the most common indications of contagious diseases annually. Also, the mortality rate caused by these diseases is increasing. Nanotechnology applied to identify novel targets and search for fungicides is still at the concept stage. The antifungal use of plant and MONPs, MNPs, natural products in combination with MNPs, and hybrid NPs composed of biodegradable materials like chitosans combined with other antifungal agents to provide an alternative to synthetic fungicides against pathogenic fungi. Although many antifungal agents are effective against these fungal infectious diseases, the low bioavailability, poor water solubility, side effects, inadequate tissue penetration, drug resistance and neutralization, and insufficient pharmacological dynamics of these agents have prompted researchers to find ways to improve existing agents. Beyond these nanostructures are further disease detection, diagnostics, and therapeutic applications.

References

Abbas, H. S., & Krishnan, A. (2020). Magnetic nanosystems as a therapeutic tool to combat pathogenic fungi. *Advanced Pharmaceutical Bulletin, 10*, 512.

Abd-Elsalam, K., & Alghuthaymi, M. (2015). Nanobiofungicides: Are they the next-generation of fungicides. *Journal of Material Science and Nanotechnology, 2*, 38–40.

Abd-Elsalam, K. A., Hashim, A. F., Alghuthaymi, M. A., & Said-Galiev, E. (2017). Nanobiotechnological strategies for toxigenic fungi and mycotoxin control. In A. M. Grumezescu (Ed.), *Food preservation*. Academic.

Abednatanzi, S., Derakhshandeh, P. G., Depauw, H., Coudert, F.-X., Vrielinck, H., Van Der Voort, P., & Leus, K. (2019). Mixed-metal metal–organic frameworks. *Chemical Society Reviews, 48*, 2535–2565.

Ahmad, A., Mukherjee, P., Senapati, S., Mandal, D., Khan, M. I., Kumar, R., & Sastry, M. (2003). Extracellular biosynthesis of silver nanoparticles using the fungus fusarium oxysporum. *Colloids and Surfaces B: Biointerfaces, 28*, 313–318.

Ahmad, A., Senapati, S., Khan, M. I., Kumar, R., & Sastry, M. (2005). Extra-/intracellular biosynthesis of gold nanoparticles by an alkalotolerant fungus, Trichothecium sp. *Journal of Biomedical Nanotechnology, 1*, 47–53.

Amaral, A. C., Bocca, A. L., Ribeiro, A. M., Nunes, J., Peixoto, D. L., Simioni, A. R., Primo, F. L., Lacava, Z. G., Bentes, R., & Titze-De-Almeida, R. (2009). Amphotericin B in poly (lactic-co-glycolic acid)(PLGA) and dimercaptosuccinic acid (DMSA) nanoparticles against paracoccidioidomycosis. *Journal of Antimicrobial Chemotherapy, 63*, 526–533.

Anil Kumar, S., Abyaneh, M. K., Gosavi, S., Kulkarni, S. K., Pasricha, R., Ahmad, A., & Khan, M. (2007). Nitrate reductase-mediated synthesis of silver nanoparticles from AgNO3. *Biotechnology Letters, 29*, 439–445.

Araujo, V. H. S., Duarte, J. L., Carvalho, G. C., Silvestre, A. L. P., Fonseca-Santos, B., Marena, G. D., Ribeiro, T. D. C., Dos Santos Ramos, M. A., Bauab, T. M., & Chorilli, M. (2020). Nanosystems against candidiasis: A review of studies performed over the last two decades. *Critical Reviews in Microbiology, 46*, 508–547.

Arciniegas-Grijalba, P., Patiño-Portela, M., Mosquera-Sánchez, L., Guerrero-Vargas, J., & Rodríguez-Páez, J. (2017). ZnO nanoparticles (ZnO-NPs) and their antifungal activity against coffee fungus Erythricium salmonicolor. *Applied Nanoscience, 7*, 225–241.

Asgari, P., Moradi, O., & Tajeddin, B. (2014). The effect of nanocomposite packaging carbon nanotube base on organoleptic and fungal growth of Mazafati brand dates. *International Nano Letters, 4*, 1–5.

Ashfaq, M., Khan, S., & Verma, N. (2014). Synthesis of PVA-CAP-based biomaterial in situ dispersed with cu nanoparticles and carbon micro-nanofibers for antibiotic drug delivery applications. *Biochemical Engineering Journal, 90*, 79–89.

Ashfaq, M., Verma, N., & Khan, S. (2017). Highly effective cu/Zn-carbon micro/nanofiber-polymer nanocomposite-based wound dressing biomaterial against the P. Aeruginosa multi- and extensively drug-resistant strains. *Materials Science and Engineering: C, 77*, 630–641.

Balaji, D., Basavaraja, S., Deshpande, R., Mahesh, D. B., Prabhakar, B., & Venkataraman, A. (2009). Extracellular biosynthesis of functionalized silver nanoparticles by strains of Cladosporium cladosporioides fungus. *Colloids and Surfaces B: Biointerfaces, 68*, 88–92.

Bansal, V., Poddar, P., Ahmad, A., & Sastry, M. (2006). Room-temperature biosynthesis of ferroelectric barium titanate nanoparticles. *Journal of the American Chemical Society, 128*, 11958–11963.

Bansal, V., Rautaray, D., Ahmad, A., & Sastry, M. (2004). Biosynthesis of zirconia nanoparticles using the fungus fusarium oxysporum. *Journal of Materials Chemistry, 14*, 3303–3305.

Bansal, V., Rautaray, D., Bharde, A., Ahire, K., Sanyal, A., Ahmad, A., & Sastry, M. (2005). Fungus-mediated biosynthesis of silica and titania particles. *Journal of Materials Chemistry, 15*, 2583–2589.

Basavaraja, S., Balaji, S., Lagashetty, A., Rajasab, A., & Venkataraman, A. (2008). Extracellular biosynthesis of silver nanoparticles using the fungus fusarium semitectum. *Materials Research Bulletin, 43*, 1164–1170.

Birla, S., Tiwari, V., Gade, A., Ingle, A., Yadav, A., & Rai, M. (2009). Fabrication of silver nanoparticles by Phoma glomerata and its combined effect against Escherichia coli, Pseudomonas aeruginosa and Staphylococcus aureus. *Letters in Applied Microbiology, 48*, 173–179.

Borkow, G., & Gabbay, J. (2005). Copper as a biocidal tool. *Current Medicinal Chemistry, 12*, 2163–2175.

Bouson, S., Krittayavathananon, A., Phattharasupakun, N., Siwayaprahm, P., & Sawangphruk, M. (2017). Antifungal activity of water-stable copper-containing metal-organic frameworks. *Royal Society Open Science, 4*, 170654.

Brown, G. D., Denning, D. W., Gow, N. A., Levitz, S. M., Netea, M. G., & White, T. C. (2012). Hidden killers: Human fungal infections. *Science Translational Medicine, 4*, 165rv13.

Brust, M., Walker, M., Bethell, D., Schiffrin, D. J., & Whyman, R. (1994). Synthesis of thiol-derivatised gold nanoparticles in a two-phase liquid–liquid system. *Journal of the Chemical Society, Chemical Communications, 7*, 801–802.

Chauhan, D., Talreja, N., & Ashfaq, M. (2024). Chapter 3 - Biogenic nanofungicides: Next-generation anti-microbial materials. In K. A. Abd-Elsalam (Ed.), *Nanofungicides*. Elsevier.

Chen, S. C.-A., Meyer, W., & Sorrell, T. C. (2014). Cryptococcus gattii infections. *Clinical Microbiology Reviews, 27*, 980–1024.

Cunha-Azevedo, E. P., Silva, J. R., Martins, O. P., Siqueira-Moura, M. P., Bocca, A. L., Felipe, M. S. S., Tedesco, A. C., & Azevedo, R. B. (2011). In vitro antifungal activity and toxicity of itraconazole in DMSA-PLGA nanoparticles. *Journal of Nanoscience and Nanotechnology, 11*, 2308–2314.

De Filpo, G., Palermo, A. M., Rachiele, F., & Nicoletta, F. P. (2013). Preventing fungal growth in wood by titanium dioxide nanoparticles. *International Biodeterioration & Biodegradation, 85*, 217–222.

Denning, D. W., & Hope, W. W. (2010). Therapy for fungal diseases: Opportunities and priorities. *Trends in Microbiology, 18*, 195–204.

Dhillon, G. S., Brar, S. K., Kaur, S., & Verma, M. (2012). Green approach for nanoparticle biosynthesis by fungi: Current trends and applications. *Critical Reviews in Biotechnology, 32*, 49–73.

Du, W., Gao, Y., Liu, L., Sai, S., & Ding, C. (2021). Striking back against fungal infections: The utilization of nanosystems for antifungal strategies. *International Journal of Molecular Sciences, 22*, 10104.

Durán, N., Marcato, P. D., Alves, O. L., De Souza, G. I., & Esposito, E. (2005). Mechanistic aspects of biosynthesis of silver nanoparticles by several fusarium oxysporum strains. *Journal of Nanobiotechnology, 3*, 1–7.

Durán, N., Marcato, P. D., De Souza, G. I., Alves, O. L., & Esposito, E. (2007). Antibacterial effect of silver nanoparticles produced by fungal process on textile fabrics and their effluent treatment. *Journal of Biomedical Nanotechnology, 3*, 203–208.

Edwards, J. A., Kemski, M. M., & Rappleye, C. A. (2013). Identification of an aminothiazole with antifungal activity against intracellular Histoplasma capsulatum. *Antimicrobial Agents and Chemotherapy, 57*, 4349–4359.

Fayaz, A. M., Balaji, K., Girilal, M., Yadav, R., Kalaichelvan, P. T., & Venketesan, R. (2010). Biogenic synthesis of silver nanoparticles and their synergistic effect with antibiotics: A study against gram-positive and gram-negative bacteria. *Nanomedicine: Nanotechnology Biology and Medicine, 6*, 103–109.

Fayaz, A. M., Balaji, K., Kalaichelvan, P., & Venkatesan, R. (2009). Fungal based synthesis of silver nanoparticles—An effect of temperature on the size of particles. *Colloids and Surfaces B: Biointerfaces, 74*, 123–126.

Fayaz, M., Tiwary, C., Kalaichelvan, P., & Venkatesan, R. (2010). Blue orange light emission from biogenic synthesized silver nanoparticles using Trichoderma viride. *Colloids and Surfaces B: Biointerfaces, 75*, 175–178.

Fisher, M. C., Hawkins, N. J., Sanglard, D., & Gurr, S. J. (2018). Worldwide emergence of resistance to antifungal drugs challenges human health and food security. *Science, 360*, 739–742.

Gade, A., Bonde, P., Ingle, A., Marcato, P., Duran, N., & Rai, M. (2008). Exploitation of aspergillus Niger for synthesis of silver nanoparticles. *Journal of Biobased Materials and Bioenergy, 2*, 243–247.

Garibotto, F. M., Garro, A. D., Masman, M. F., Rodríguez, A. M., Luiten, P. G., Raimondi, M., Zacchino, S. A., Somlai, C., Penke, B., & Enriz, R. D. (2010). New small-size peptides possessing antifungal activity. *Bioorganic & Medicinal Chemistry, 18*, 158–167.

Gericke, M., & Pinches, A. (2006). Biological synthesis of metal nanoparticles. *Hydrometallurgy, 83*, 132–140.

Ghasemian, E., Naghoni, A., Tabaraie, B., & Tabaraie, T. (2012). In vitro susceptibility of filamentous fungi to copper nanoparticles assessed by rapid XTT colorimetry and agar dilution method. *Journal de Mycologie Médicale, 22*, 322–328.

Ghosh, P., Han, G., De, M., Kim, C. K., & Rotello, V. M. (2008). Gold nanoparticles in delivery applications. *Advanced Drug Delivery Reviews, 60*, 1307–1315.

Giannousi, K., Avramidis, I., & Dendrinou-Samara, C. (2013). Synthesis, characterization and evaluation of copper based nanoparticles as agrochemicals against Phytophthora infestans. *RSC Advances, 3*, 21743–21752.

Gondal, A., Ijaz, M., Riaz, K., & Khan, A. (2012). Effect of different doses of fungicide (Mancozeb) against Alternaria leaf blight of tomato in tunnel. *International Journal of Plant Pathology and Microbiology, 3*, 3.

Gorczyca, A., Kasprowicz, M. J., & Lemek, T. (2009). Physiological effect of multi-walled carbon nanotubes (MWCNTs) on conidia of the entomopathogenic fungus, Paecilomyces fumosoroseus (Deuteromycotina: Hyphomycetes). *Journal of Environmental Science and Health Part A, 44*, 1592–1597.

Hao, Y., Cao, X., Ma, C., Zhang, Z., Zhao, N., Ali, A., Hou, T., Xiang, Z., Zhuang, J., & Wu, S. (2017). Potential applications and antifungal activities of engineered nanomaterials against gray mold disease agent Botrytis cinerea on rose petals. *Frontiers in Plant Science, 8*, 1332.

He, L., Liu, Y., Mustapha, A., & Lin, M. (2011). Antifungal activity of zinc oxide nanoparticles against Botrytis cinerea and Penicillium expansum. *Microbiological Research, 166*, 207–215.

Hernández-Lauzardo, A. N., Velázquez-Del Valle, M. G., & Guerra-Sanchez, M. G. (2011). Current status of action mode and effect of chitosan against phytopathogens fungi. *African Journal of Microbiology Research, 5*, 4243–4247.

Huang, R., Zhou, Z., Lan, X., Tang, F. K., Cheng, T., Sun, H., Leung, K. C.-F., Li, X., & Jin, L. (2023). Rapid synthesis of bismuth-organic frameworks as selective antimicrobial materials against microbial biofilms. *Materials Today Bio, 18*, 100507.

Ingle, A., Gade, A., Pierrat, S., Sonnichsen, C., & Rai, M. (2008). Mycosynthesis of silver nanoparticles using the fungus fusarium acuminatum and its activity against some human pathogenic bacteria. *Current Nanoscience, 4*, 141–144.

Ingle, A., Rai, M., Gade, A., & Bawaskar, M. (2009). Fusarium solani: A novel biological agent for the extracellular synthesis of silver nanoparticles. *Journal of Nanoparticle Research, 11*, 2079–2085.

Jayaseelan, C., Rahuman, A. A., Kirthi, A. V., Marimuthu, S., Santhoshkumar, T., Bagavan, A., Gaurav, K., Karthik, L., & Rao, K. B. (2012). Novel microbial route to synthesize ZnO nanoparticles using Aeromonas hydrophila and their activity against pathogenic bacteria and fungi. *Spectrochimica Acta Part A: Molecular and Biomolecular Spectroscopy, 90*, 78–84.

Jo, Y.-K., Kim, B. H., & Jung, G. (2009). Antifungal activity of silver ions and nanoparticles on phytopathogenic fungi. *Plant Disease, 93*, 1037–1043.

Kamat, P. V. (2002). Photophysical, photochemical and photocatalytic aspects of metal nanoparticles. *The Journal of Physical Chemistry B, 106*, 7729–7744.

Kanhed, P., Birla, S., Gaikwad, S., Gade, A., Seabra, A. B., Rubilar, O., Duran, N., & Rai, M. (2014). In vitro antifungal efficacy of copper nanoparticles against selected crop pathogenic fungi. *Materials Letters, 115*, 13–17.

Karimi, N., Minaei, S., Almassi, M., & Shahverdi, A. (2012). Application of silver nano-particles for protection of seeds in different soils. *African Journal of Agricultural Research, 7*, 1863–1869.

Kathiravan, M. K., Salake, A. B., Chothe, A. S., Dudhe, P. B., Watode, R. P., Mukta, M. S., & Gadhwe, S. (2012). The biology and chemistry of antifungal agents: A review. *Bioorganic & Medicinal Chemistry, 20*, 5678–5698.

Kathiresan, K., Manivannan, S., Nabeel, M., & Dhivya, B. (2009). Studies on silver nanoparticles synthesized by a marine fungus, Penicillium fellutanum isolated from coastal mangrove sediment. *Colloids and Surfaces B: Biointerfaces, 71*, 133–137.

Kauffman, C. A. (2007). Histoplasmosis: A clinical and laboratory update. *Clinical Microbiology Reviews, 20*, 115–132.

Kauffman, C. A., & Hedderwick, S. A. (2006). Fungal infections. In *Infection management for geriatrics in long-term care facilities* (pp. 479–498). CRC Press.

Kim, J. S., Kuk, E., Yu, K. N., Kim, J.-H., Park, S. J., Lee, H. J., Kim, S. H., Park, Y. K., Park, Y. H., & Hwang, C.-Y. (2007). Antimicrobial effects of silver nanoparticles. *Nanomedicine: Nanotechnology, Biology and Medicine, 3*, 95–101.

Kim, S. Y., Park, J., Koh, M., Park, S. B., & Hong, J.-I. (2009). Fluorescent probe for detection of fluoride in water and bioimaging in A549 human lung carcinoma cells. *Chemical Communications, 2009*, 4735–4737.

Kumar, C. V., & Mclendon, G. L. (1997). Nanoencapsulation of cytochrome c and horseradish peroxidase at the galleries of α-zirconium phosphate. *Chemistry of Materials, 9*, 863–870.

Kwon-Chung, K. J., Fraser, J. A., Doering, T. L., Wang, Z. A., Janbon, G., Idnurm, A., & Bahn, Y.-S. (2014). Cryptococcus neoformans and Cryptococcus gattii, the etiologic agents of cryptococcosis. *Cold Spring Harbor Perspectives in Medicine, 4*, a019760.

Lamsal, K., Kim, S. W., Jung, J. H., Kim, Y. S., Kim, K. S., & Lee, Y. S. (2011). Application of silver nanoparticles for the control of Colletotrichum species in vitro and pepper anthracnose disease in field. *Mycobiology, 39*, 194–199.

Leyva Salas, M., Mounier, J., Valence, F., Coton, M., Thierry, A., & Coton, E. (2017). Antifungal microbial agents for food biopreservation—A review. *Microorganisms, 5*, 37.

Lin, Z., Richardson, J. J., Zhou, J., & Caruso, F. (2023). Direct synthesis of amorphous coordination polymers and metal–organic frameworks. *Nature Reviews Chemistry, 7*, 273–286.

Liu, F., Chen, Y., Huang, Y., Jin, Q., & JI, J. (2024). Nanomaterial-based therapeutics for enhanced antifungal therapy. *Journal of Materials Chemistry B, 12*, 9173–9198.

Liu, Z., Zhang, M., Han, X., Xu, H., Zhang, B., Yu, Q., & Li, M. (2016). TiO2 nanoparticles cause cell damage independent of apoptosis and autophagy by impairing the ROS-scavenging system in Pichia pastoris. *Chemico-Biological Interactions, 252*, 9–18.

Macdonald, I., & Smith, W. (1996). Orientation of cytochrome c adsorbed on a citrate-reduced silver colloid surface. *Langmuir, 12*, 706–713.

Maliszewska, I., Lisiak, B., Popko, K., & Matczyszyn, K. (2017). Enhancement of the efficacy of photodynamic inactivation of Candida albicans with the use of biogenic gold nanoparticles. *Photochemistry and Photobiology, 93*, 1081–1090.

Mansoori, G. A. (2013). *Synthesis of nanoparticles by fungi*. Google.

Márquez, I. G., Akuaku, J., Cruz, I., Cheetham, J., Golshani, A., & Smith, M. L. (2013). Disruption of protein synthesis as antifungal mode of action by chitosan. *International Journal of Food Microbiology, 164*, 108–112.

Marr, K. A., Datta, K., Pirofski, L.-A., & BARNES, R. (2012). Cryptococcus gattii infection in healthy hosts: A sentinel for subclinical immunodeficiency? *Clinical Infectious Diseases, 54*, 153–154.

Midander, K., Cronholm, P., Karlsson, H. L., Elihn, K., Möller, L., Leygraf, C., & Wallinder, I. O. (2009). Surface characteristics, copper release, and toxicity of nano-and micrometer-sized copper and copper (II) oxide particles: A cross-disciplinary study. *Small, 5*, 389–399.

Min, J. S., Kim, K. S., Kim, S. W., Jung, J. H., Lamsal, K., Bin Kim, S., Jung, M., & Lee, Y. S. (2009). Effects of colloidal silver nanoparticles on sclerotium-forming phytopathogenic fungi. *The Plant Pathology Journal, 25*, 376–380.

Mishra, V., Singh, M., Mishra, Y., Charbe, N., Nayak, P., Sudhakar, K., Aljabali, A. A., Shahcheraghi, S. H., Bakshi, H., & Serrano-Aroca, Á. (2021). Nanoarchitectures in management of fungal diseases: An overview. *Applied Sciences, 11*, 7119.

Mukherjee, P., Ahmad, A., Mandal, D., Senapati, S., Sainkar, S. R., Khan, M. I., Parishcha, R., Ajaykumar, P., Alam, M., & Kumar, R. (2001). Fungus-mediated synthesis of silver nanoparticles and their immobilization in the mycelial matrix: A novel biological approach to nanoparticle synthesis. *Nano Letters, 1*, 515–519.

Mukherjee, P., Senapati, S., Mandal, D., Ahmad, A., Khan, M. I., Kumar, R., & Sastry, M. (2002). Extracellular synthesis of gold nanoparticles by the fungus fusarium oxysporum. *Chembiochem, 3*, 461–463.

Nami, S., Aghebati-Maleki, A., & Aghebati-Maleki, L. (2021). Current applications and prospects of nanoparticles for antifungal drug delivery. *EXCLI Journal, 20*, 562.

Oh, S.-D., Lee, S., Choi, S.-H., Lee, I.-S., Lee, Y.-M., Chun, J.-H., & Park, H.-J. (2006). Synthesis of Ag and Ag–SiO2 nanoparticles by γ-irradiation and their antibacterial and antifungal efficiency against salmonella enterica serovar typhimurium and Botrytis cinerea. *Colloids and Surfaces A: Physicochemical and Engineering Aspects, 275*, 228–233.

Omar, R. A., Afreen, S., Talreja, N., Chauhan, D., & Ashfaq, M. (2019). Impact of nanomaterials in plant systems. In R. Prasad (Ed.), *Plant nanobionics: Volume 1, advances in the understanding of nanomaterials research and applications*. Springer International Publishing.

Pal, S., Tak, Y. K., & Song, J. M. (2007). Does the antibacterial activity of silver nanoparticles depend on the shape of the nanoparticle? A study of the gram-negative bacterium Escherichia coli. *Applied and Environmental Microbiology, 73*, 1712–1720.

Panáček, A., Kolář, M., Večeřová, R., Prucek, R., Soukupová, J., Kryštof, V., Hamal, P., Zbořil, R., & Kvítek, L. (2009). Antifungal activity of silver nanoparticles against Candida spp. *Biomaterials, 30*, 6333–6340.

Pandey, S., Giri, K., Kumar, R., Mishra, G., & Raja Rishi, R. (2018). Nanopesticides: Opportunities in crop protection and associated environmental risks. *Proceedings of the National Academy of Sciences, India Section B: Biological Sciences, 88*, 1287–1308.

Park, H. J., Kim, S. H., Kim, H. J., & Choi, S. H. (2006). A new composition of nanosized silica-silver for control of various plant diseases. *Plant Pathology Journal, 22*, 295–302.

Patra, P., Mitra, S., Debnath, N., & Goswami, A. (2012). Biochemical-, biophysical-, and microarray-based antifungal evaluation of the buffer-mediated synthesized nano zinc oxide: An in vivo and in vitro toxicity study. *Langmuir, 28*, 16966–16978.

Patterson, T. (2007). *Focus on candidemia*. Applied Clinical Education.

Paul, S., Mohanram, K., & Kannan, I. (2018). Antifungal activity of curcumin-silver nanoparticles against fluconazole-resistant clinical isolates of Candida species. *AYU (An International Quarterly Journal of Research in Ayurveda), 39*, 182–186.

Perfect, J. R. (2012). The impact of the host on fungal infections. *The American Journal of Medicine, 125*, S39–S51.

Perfect, J. R., Dismukes, W. E., Dromer, F., Goldman, D. L., Graybill, J. R., Hamill, R. J., Harrison, T. S., Larsen, R. A., Lortholary, O., & Nguyen, M.-H. (2010). Clinical practice guidelines for the management of cryptococcal disease: 2010 update by the Infectious Diseases Society of America. *Clinical Infectious Diseases, 50*, 291–322.

Petica, A., Gavriliu, S., Lungu, M., Buruntea, N., & Panzaru, C. (2008). Colloidal silver solutions with antimicrobial properties. *Materials Science and Engineering B, 152*, 22–27.

Pfaller, M. A., & Diekema, D. J. (2010). Epidemiology of invasive mycoses in North America. *Critical Reviews in Microbiology, 36*, 1–53.

Priyanka, K. P., Harikumar, V. S., Balakrishna, K. M., & Varghese, T. (2017). Inhibitory effect of TiO2 NPs on symbiotic arbuscular mycorrhizal fungi in plant roots. *IET Nanobiotechnology, 11*, 66–70.

Rubina, M. S., Vasil'kov, A. Y., Naumkin, A. V., Shtykova, E. V., Abramchuk, S. S., Alghuthaymi, M. A., & Abd-Elsalam, K. A. (2017). Synthesis and characterization of chitosan–copper nanocomposites and their fungicidal activity against two sclerotia-forming plant pathogenic fungi. *Journal of Nanostructure in Chemistry, 7*, 249–258.

Ruenraroengsak, P., Cook, J. M., & Florence, A. T. (2010). Nanosystem drug targeting: Facing up to complex realities. *Journal of Controlled Release, 141*, 265–276.

Sachin, K., & Karn, S. K. (2021). Microbial fabricated nanosystems: Applications in drug delivery and targeting. *Frontiers in Chemistry, 9*, 617353.

Sadhasivam, S., Shanmugam, P., & Yun, K. (2010). Biosynthesis of silver nanoparticles by Streptomyces hygroscopicus and antimicrobial activity against medically important pathogenic microorganisms. *Colloids and Surfaces B: Biointerfaces, 81*, 358–362.

Sadowski, Z., Maliszewska, I., Grochowalska, B., Polowczyk, I., & Kozlecki, T. (2008). Synthesis of silver nanoparticles using microorganisms. *Materials Science-Poland, 26*, 419–424.

Saharan, V., Mehrotra, A., Khatik, R., Rawal, P., Sharma, S., & Pal, A. (2013). Synthesis of chitosan based nanoparticles and their in vitro evaluation against phytopathogenic fungi. *International Journal of Biological Macromolecules, 62*, 677–683.

Salavati-Niasari, M., Davar, F., & Mir, N. (2008). Synthesis and characterization of metallic copper nanoparticles via thermal decomposition. *Polyhedron, 27*, 3514–3518.

Sanghi, R., & Verma, P. (2009). Biomimetic synthesis and characterisation of protein capped silver nanoparticles. *Bioresource Technology, 100*, 501–504.

Saraf, M., Pandya, U., & Thakkar, A. (2014). Role of allelochemicals in plant growth promoting rhizobacteria for biocontrol of phytopathogens. *Microbiological Research, 169*, 18–29.

Saravanan, M., & Nanda, A. (2010). Extracellular synthesis of silver bionanoparticles from aspergillus clavatus and its antimicrobial activity against MRSA and MRSE. *Colloids and Surfaces B: Biointerfaces, 77*, 214–218.

Sastry, M., Ahmad, A., Khan, M. I., & Kumar, R. (2003). Biosynthesis of metal nanoparticles using fungi and actinomycete. *Current Science, 82*, 162–170.

Sawangphruk, M., Srimuk, P., Chiochan, P., Sangsri, T., & Siwayaprahm, P. (2012). Synthesis and antifungal activity of reduced graphene oxide nanosheets. *Carbon, 50*, 5156–5161.

Sawant, B., & Khan, T. (2017). Recent advances in delivery of antifungal agents for therapeutic management of candidiasis. *Biomedicine & Pharmacotherapy, 96*, 1478–1490.

Seaberg, J., Montazerian, H., Hossen, M. N., Bhattacharya, R., Khademhosseini, A., & Mukherjee, P. (2021). Hybrid nanosystems for biomedical applications. *ACS Nano, 15*, 2099–2142.

Senapati, S., Mandal, D., Ahmad, A., Khan, M. I., Sastry, M., & Kumar, R. (2004). Fungus mediated synthesis of silver nanoparticles: A novel biological approach. *Indian Journal of Physics, 78*, 101–105.

Seo, Y.-C., Cho, J.-S., Jeong, H.-Y., Yim, T.-B., Cho, K.-S., Lee, T.-W., Jeong, M.-H., Lee, G.-H., Kim, S.-I., & Yoon, W.-B. (2011). Enhancement of antifungal activity of anthracnose in pepper by nanopaticles of thiamine di-lauryl sulfate. *Korean Journal of Medicinal Crop Science, 19*, 198–204.

Shaligram, N. S., Bule, M., Bhambure, R., Singhal, R. S., Singh, S. K., Szakacs, G., & Pandey, A. (2009). Biosynthesis of silver nanoparticles using aqueous extract from the compact in producing fungal strain. *Process Biochemistry, 44*, 939–943.

Sharon, M., Choudhary, A. K., & Kumar, R. (2010). Nanotechnology in agricultural diseases and food safety. *Journal of Phytology, 2*, 83.

Shikanai-Yasuda, M. A., Mendes, R. P., Colombo, A. L., Queiroz-Telles, F. D., Kono, A. S. G., Paniago, A. M. M., Nathan, A., Valle, A. C. F. D., Bagagli, E., & Benard, G. (2017). Brazilian guidelines for the clinical management of paracoccidioidomycosis. *Revista da Sociedade Brasileira de Medicina Tropical, 50*, 715–740.

Shrivastava, S., Prasad, R., & Varma, A. (2014). Anatomy of root from eyes of a microbiologist. In *Root engineering: Basic and applied concepts* (pp. 3–22). Springer.

Sirelkhatim, A., Mahmud, S., Seeni, A., Kaus, N. H. M., Ann, L. C., Bakhori, S. K. M., Hasan, H., & Mohamad, D. (2015). Review on zinc oxide nanoparticles: Antibacterial activity and toxicity mechanism. *Nano-Micro Letters, 7*, 219–242.

Souza, A. C., & Amaral, A. C. (2017). Antifungal therapy for systemic mycosis and the nanobiotechnology era: Improving efficacy, biodistribution and toxicity. *Frontiers in Microbiology, 8*, 336.

Spirescu, V. A., Chircov, C., Grumezescu, A. M., & Andronescu, E. (2021). Polymeric nanoparticles for antimicrobial therapies: An up-to-date overview. *Polymers, 13*, 724.

Tampieri, M. (2004). Update on the diagnosis of dermatomycosis. *Parassitologia, 46*, 183–186.

Thakkar, K. N., Mhatre, S. S., & Parikh, R. Y. (2010). Biological synthesis of metallic nanoparticles. *Nanomedicine: Nanotechnology, Biology and Medicine, 6*, 257–262.

Tiwari, D. K., Behari, J., & Prasenjit Sen, P. S. (2008). Application of nanoparticles in waste water treatment. *World Applied Sciences Journal, 3*, 3.

Tsivileva, O., Pozdnyakov, A., & Ivanova, A. (2021). Polymer nanocomposites of selenium biofabricated using fungi. *Molecules, 26*, 3657.

Tufa, T. B., & Denning, D. W. (2019). The burden of fungal infections in Ethiopia. *Journal of Fungi, 5*, 109.

Tumbarello, M., Posteraro, B., Trecarichi, E. M., Fiori, B., Rossi, M., Porta, R., De Gaetano Donati, K., La Sorda, M., Spanu, T., & Fadda, G. (2007). Biofilm production by Candida species and inadequate antifungal therapy as predictors of mortality for patients with candidemia. *Journal of Clinical Microbiology, 45*, 1843–1850.

Ul Haq, I., & Ijaz, S. (2019). Use of metallic nanoparticles and nanoformulations as nanofungicides for sustainable disease management in plants. In *Nanobiotechnology in bioformulations* (pp. 289–316). Springer.

Velmurugan, N., Kumar, G. G., Han, S. S., Nahm, K. S., & Lee, Y. S. (2009). Synthesis and characterization of potential fungicidal silver nano-sized particles and chitosan membrane containing silver particles. *Iranian Polymer Journal, 18*, 383–392.

Vigneshwaran, N., Ashtaputre, N., Varadarajan, P., Nachane, R., Paralikar, K., & Balasubramanya, R. (2007). Biological synthesis of silver nanoparticles using the fungus aspergillus flavus. *Materials Letters, 61*, 1413–1418.

Vigneshwaran, N., Kathe, A. A., Varadarajan, P., Nachane, R. P., & Balasubramanya, R. (2006). Biomimetics of silver nanoparticles by white rot fungus, Phaenerochaete chrysosporium. *Colloids and Surfaces B: Biointerfaces, 53*, 55–59.

Wang, X., Liu, X., Chen, J., Han, H., & Yuan, Z. (2014). Evaluation and mechanism of antifungal effects of carbon nanomaterials in controlling plant fungal pathogen. *Carbon, 68*, 798–806.

Wang, X., Zhou, Z., & Chen, F. (2017). Surface modification of carbon nanotubes with an enhanced antifungal activity for the control of plant fungal pathogen. *Materials, 10*, 1375.

Wright, G. D. (2005). Bacterial resistance to antibiotics: Enzymatic degradation and modification. *Advanced Drug Delivery Reviews, 57*, 1451–1470.

Wright, G. D., & Sutherland, A. D. (2007). New strategies for combating multidrug-resistant bacteria. *Trends in Molecular Medicine, 13*, 260–267.

Xie, Y., He, Y., Irwin, P. L., Jin, T., & Shi, X. (2011). Antibacterial activity and mechanism of action of zinc oxide nanoparticles against campylobacter jejuni. *Applied and Environmental Microbiology, 77*, 2325–2331.

Xin, Y., Quan, L., Zhang, H., & Ao, Q. (2023). Emerging polymer-based nanosystem strategies in the delivery of antifungal drugs. *Pharmaceutics, 15*, 1866.

Yuan, K., Huang, K., Yang, Y., Lin, Y., Liu, Y., Li, F., Liang, Y., Chang, H., Chen, Y., & Tang, T. (2022). Multi-roles of nanoscale bismuth metal-organic frameworks: Infectious photoacoustic probe and inhibitor of antibiotics tolerant bacteria via targeting endogenous H2S. *Nano Today, 47*, 101683.

Zarzuela, R., Carbú, M., Gil, M. A., Cantoral, J. M., & Mosquera, M. J. (2017). Cuo/SiO2 nanocomposites: A multifunctional coating for application on building stone. *Materials & Design, 114*, 364–372.

Zhang, L., Ouyang, M., Zhang, Y., Zhang, L., Huang, Z., He, L., Lei, Y., Zou, Z., Feng, F., & Yang, R. (2021). The fluorescence imaging and precise suppression of bacterial infections in chronic wounds by porphyrin-based metal–organic framework nanorods. *Journal of Materials Chemistry B, 9*, 8048–8055.

Chapter 13
Challenges and Embracing Opportunities, Sustainability and Emerging Trends Fungi in Wastewater Treatment

Peeyush Phogat, Anshi Pandey, Shreya, Ranjana Jha, and Sukhvir Singh

Abstract The application of fungi in wastewater treatment presents a novel and eco-friendly approach to addressing global water pollution challenges. Fungi, with their unique enzymatic capabilities, have demonstrated potential in degrading complex pollutants, including heavy metals, pharmaceuticals, and persistent organic compounds. This chapter delves into the key challenges and opportunities associated with employing fungi in wastewater treatment. It explores the technical, economic, and environmental barriers to large-scale adoption, while also highlighting the opportunities that fungal treatment offers for sustainability. Additionally, the chapter examines emerging trends in the field, such as genetic engineering, the use of fungal consortia, and the integration of fungal processes with other treatment technologies. By embracing these opportunities and overcoming existing challenges, fungi could play a pivotal role in the future of sustainable wastewater management.

Keywords Fungi · Wastewater treatment · Sustainability

P. Phogat (✉) · Shreya
Research Lab for Energy Systems, Department of Physics, Netaji Subhas University of Technology, New Delhi, India

Council of Scientific and Industrial Research-National Institute of Science Communication and Policy Research, New Delhi, India

A. Pandey · R. Jha
Research Lab for Energy Systems, Department of Physics, Netaji Subhas University of Technology, New Delhi, India

S. Singh
Department of Nanotechnology, Delhi Skill Entrepreneurship University, New Delhi, India

C. M. Hussain et al. (eds.), *Fungi in Waste Water Treatment*, Springer Water,
https://doi.org/10.1007/978-3-032-12693-1_13

13.1 Introduction

The global issue of industrial wastewater-induced environmental contamination is becoming increasingly severe. Fungi and yeasts have been utilized in wastewater treatment since the 1960 (Ojha et al., 2021). These organisms, which have existed for millions of years, possess a remarkable capacity to extract harmful components from their surroundings. The widespread distribution of fungi across the globe, their ease of cultivation, and their resilience to extreme conditions make them a compelling option for waste management (Bhatnagar & Kumari, 2013). The question arises: what pollutants can fungi eliminate from contaminated or wastewater?

Fungi are the most important waste decomposers in both soil and aquatic ecosystems (Hyde et al., 1998). It has recently come to our attention that fungi may also play key roles in the purification of wastewater (Leng et al., 2021). In this chapter, I have marshaled information from an array of scientific disciplines to better understand fungi associated with wastewater treatment systems, particulate matter with them, and educate the reader about what may be some currently overlooked activities of fungi in wastewater treatment. There are quite a lot of opportunities and significant challenges associated with the unique and promising use of fungus in wastewater treatment. Fungi offers unique advantages in biodegrading complex organic pollutants and remediating heavy metals, potentially enhancing the efficiency of wastewater treatment processes (Chen et al., 2022). Their ability to decompose a range of contaminants, from dyes to pharmaceuticals, and their capacity to remove excess nutrients like nitrogen and phosphorus make them an asset in achieving more sustainable and cost-effective treatment solutions (Gomaa et al., 2023). However, the practical application of fungi in wastewater treatment is not without obstacles. Maintaining optimal growth conditions for fungi, managing their relatively slow growth rates, and addressing potential pathogenic risks pose significant challenges. Additionally, scaling up fungal-based systems from laboratory settings to industrial applications requires careful consideration and adaptation. Emerging trends in genetic engineering, hybrid treatment systems, and novel applications such as microbial fuel cells highlight the evolving landscape of fungal wastewater treatment, underscoring the need for continued research and innovation to fully leverage their potential in addressing complex wastewater challenges.

Wastewater management has become a growing global concern as population growth, urbanization, and industrialization place unprecedented pressure on water resources (Larsen et al., 2016). Water is an essential resource, and its quality and availability are critical to human health, economic development, and ecosystem sustainability. However, the effective treatment and management of wastewater comprising both domestic and industrial effluents pose significant challenges worldwide. The global community faces a daunting task as it works to address these issues and ensure the sustainable management of this vital resource. The world generates a massive volume of wastewater daily, with an estimated 80% of all wastewaters generated globally being released into the environment without adequate treatment (Sato et al., 2013). This untreated wastewater contaminates freshwater resources, affecting the health and well-being of over 2 billion people who rely on these water sources. In many developing countries, particularly in regions like Sub-Saharan Africa, South Asia, and parts of Latin America, the lack of wastewater treatment facilities is stark (Kookana et al., 2020). The direct discharge of sewage and industrial effluents into rivers, lakes, and oceans is common, leading to increasingly polluted water bodies and severe health risks from waterborne diseases like cholera, diarrhoea, and dysentery (Tariq & Mushtaq, 2023). A major obstacle in effective wastewater management is the lack of adequate infrastructure. Numerous regions worldwide, particularly in developing nations, suffer from either absent or obsolete sewage systems. Rapid urbanization exacerbates the situation, with cities growing faster than the development of essential infrastructure. Even in developed countries, where wastewater treatment infrastructure is more advanced, challenges remain. Aging treatment plants require significant upgrades to handle increased wastewater volumes and to meet stricter environmental regulations, but funding for maintenance and operations is often a challenge.

Traditional wastewater treatment methods are often not equipped to deal with emerging contaminants, such as pharmaceuticals, personal care products, microplastics, and industrial chemicals (Morin-Crini et al., 2022). These substances are increasingly detected in water bodies and pose risks to human health and the environment. For instance, pharmaceutical residues can disrupt ecosystems and contribute to the development of antibiotic-resistant bacteria, while microplastics can be ingested by aquatic organisms and eventually enter the human food chain (Sharma et al., 2022). The presence of these contaminants underscores the need for advancements in wastewater treatment technologies to effectively address a broader range of pollutants. Climate change adds another layer of complexity to the global wastewater challenge. Changing precipitation patterns, rising temperatures, and the increasing frequency of extreme weather events like floods and droughts all impact wastewater management (Singh & Tiwari, 2019). Flooding can overwhelm sewage systems, leading to the release of untreated or partially treated wastewater into the environment, while droughts can concentrate pollutants in water bodies, exacerbating water quality issues. The economic costs of inadequate wastewater management are considerable, affecting public health, industries, and overall economic productivity. Conversely, investments in wastewater treatment and management can yield significant economic benefits, contributing to water security and economic resilience.

Tackling the worldwide issue of wastewater management necessitates creative approaches and the implementation of cutting-edge technologies. Recent developments in wastewater treatment methods, including membrane bioreactors, advanced oxidation processes, and engineered wetlands, show promise for more effective and efficient wastewater processing. Additionally, there is increasing attention being paid to the circular economy model, which considers wastewater as a valuable asset rather than a disposable byproduct (Gherghel et al., 2019). This approach involves recovering and reusing water, energy, and nutrients from wastewater, reducing the demand for freshwater and minimizing environmental impacts. While the challenges are complex, the benefits of improved wastewater management are clear:

better public health, enhanced environmental sustainability, and greater economic resilience.

Fungi are gaining recognition as powerful agents in wastewater treatment due to their unique biochemical capabilities, which complement traditional bacterial processes (Harms et al., 2011). Unlike bacteria, which primarily break down simple organic molecules, fungi excel at decomposing complex and recalcitrant compounds. This ability stems from their production of extracellular enzymes such as laccases, peroxidases, and cellulases, which can degrade substances like lignin, dyes, pharmaceuticals, and persistent organic pollutants (POPs) that are often resistant to bacterial degradation (Chandra & Enespa., 2019). In wastewater treatment, fungi can be utilized in various processes. For instance, white-rot fungi are particularly effective at breaking down lignin and other complex aromatic compounds found in industrial effluents, such as those from paper mills, textile factories, and pharmaceutical industries (Periasamy et al., 2019). Fungi are also adept at bioaccumulating heavy metals from contaminated water, which is crucial in the treatment of industrial wastewater containing toxic metals like lead, mercury, and cadmium. It's ability to operate under extreme environmental conditions such as low pH, high salinity, and low nutrient levels makes them suitable for treating a wide range of wastewater types, including those where bacteria might struggle to survive (Chatterjee & Venkata Mohan, 2022). Its robustness and metabolic diversity allow it to drastically lower the number of pollutants in wastewater, producing cleaner effluents that may be reused or safely released into the environment.

Moreover, the use of fungi in wastewater treatment aligns with sustainable and eco-friendly practices. Fungi can be cultivated on low-cost, organic waste materials, reducing the need for expensive chemical inputs. Their application in bioremediation also supports the circular economy by transforming pollutants into less harmful substances or even valuable byproducts.

Sustainable and eco-friendly treatment methods are essential in managing environmental challenges, particularly in waste and wastewater management (Horst et al., 2020). These approaches prioritize reducing environmental harm, conserving resources, and supporting long-term ecological balance. Traditional treatment methods often rely on energy-intensive processes and chemicals that can generate harmful byproducts. In contrast, sustainable methods utilize natural processes such as biological treatments, phytoremediation, and constructed wetlands that are less resource-intensive and minimize ecological disruption (Kesari et al., 2021). These methods help reduce pollution, conserve energy, and protect ecosystems. Moreover, eco-friendly treatment methods often incorporate the principles of the circular economy by recovering valuable resources from waste. For instance, organic waste can be turned into compost or biogas, and treated wastewater can be used again for industrial or agricultural applications. This not only reduces the strain on natural resources but also contributes to economic sustainability (Sankaran et al., 2010). As global challenges like climate change and water scarcity intensify, the adoption of sustainable and eco-friendly treatment methods becomes increasingly critical. These approaches ensure that waste management contributes to environmental protection, resource conservation, and the overall well-being of communities, making them vital for a sustainable future.

13.2 Fungal Mechanisms in Wastewater Treatment

Since fungi have special metabolic and enzymatic properties, they are important in the treatment of wastewater (Dave & Das, 2021). Unlike bacteria, fungi produce a variety of extracellular enzymes, such as laccases, peroxidases, and cellulases, which enable them to break down complex organic compounds that are often resistant to bacterial degradation. These include lignin, dyes, pharmaceuticals, and other persistent organic pollutants as shown in Fig. 13.1b. One key fungal mechanism is biodegradation, where fungi decompose complex pollutants into simpler, less harmful substances. White-rot fungi, for instance, are particularly effective at degrading lignin and other aromatic compounds found in industrial effluents (Wesenberg et al., 2003). This makes fungi highly valuable in treating wastewater from industries such as textiles, paper mills, and pharmaceuticals. Another important mechanism is biosorption, where fungi use their cell walls to absorb and accumulate heavy metals and other toxic substances from wastewater (Gouda & Taha, 2023). This process is crucial for removing pollutants like lead, mercury, and cadmium, which are difficult to treat using conventional methods. Fungi also contribute to bioremediation by enhancing the breakdown of pollutants in contaminated environments (Vaksmaa et al., 2023). They can thrive in harsh conditions, such as low pH and high salinity, where other microorganisms might not survive, making them versatile

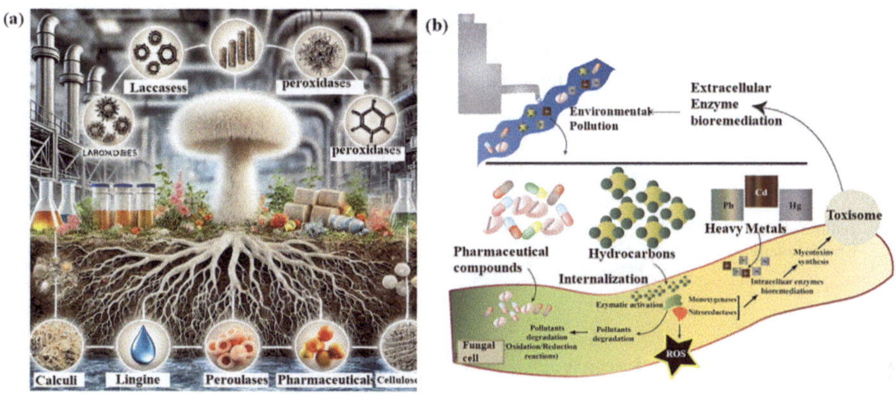

Fig. 13.1 (**a**) Fungal mechanisms in wastewater treatment. This image is generated by Lexica.art, (**b**) Filamentous fungi for sustainable remediation utilizing extracellular and intracellular enzymes for the bioremediation of environmental pollutants, including hydrocarbons, heavy metals and pharmaceutical compounds while tackling toxisome accumulation and promoting ecosystem health restoration reproduced from Ghosh et al. (2023). Copyright © 2023 Ghosh, Rusyn, Dmytruk, Dmytruk, Onyeaka, Gryzenhout and Gafforov

agents in wastewater treatment. Overall, the fungal mechanisms in wastewater treatment offer effective, sustainable solutions for managing complex pollutants, contributing to cleaner water and reduced environmental impact.

13.2.1 *Enzymatic Degradation of Pollutants*

Enzymatic degradation of pollutants is emerging as a revolutionary approach to wastewater treatment (Dave & Das, 2021), drawing on the natural abilities of microorganisms to break down harmful substances. Among these microorganisms, fungi, particularly white-rot fungus, have gained substantial attention due to their extraordinary enzymatic capabilities. Numerous extracellular enzymes, including cellulases, peroxidases, and laccases, are secreted by these fungi and are essential for the breakdown of complex organic contaminants. As depicted in research studies refer to Fig. 13.2, these enzymes catalyze reactions that break down intricate molecules into simpler, less toxic substances, which are far easier to manage and dispose of in an environmentally responsible manner. White-rot fungi, which belong to the Basidiomycota phylum, distinguished by their capacity to break down the complex polymer called lignin (Levasseur et al., 2014). They can break down a wide range of environmental pollutants, such as pesticides, industrial dyes, and pharmaceutical residues, thanks to their special enzymatic system. These pollutants, which often persist in the environment due to their resistance to conventional degradation methods, pose a severe threat to ecosystems and human health. Laccases, peroxidases, and cellulases, the key enzymes produced by white-rot fungi, exhibit a wide range

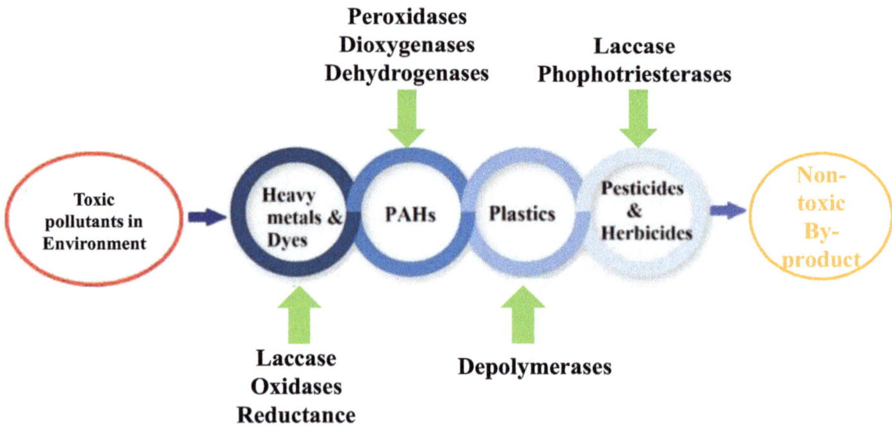

Fig. 13.2 Mechanism of catalytic enzymatic degradation of toxic environmental pollutants. Fungi utilizes reductases, peroxidases and laccases for the enzymatic degradation of harmful substances like heavy metals and dyes, polycyclic aromatic hydrocarbons (PAHs), plastics and pesticides. These substances, after being broken down by enzymes, are converted into non-toxic by-products modified from Saravanan et al. (2021)

of substrate specificities (Wesenberg et al., 2003). Laccases, for example, catalyze the oxidation of phenolic compounds, which are commonly found in industrial waste streams (Chen, Da Oh, & Yap, 2022). They achieve this by removing electrons from the pollutants, resulting in the breakdown of complex molecules. Peroxidases, another class of enzymes secreted by fungi, utilize hydrogen peroxide to oxidize a wide variety of organic and inorganic substrates, such as dyes and pharmaceuticals (Sellami et al., 2022). Cellulases, on the other hand, break down cellulose and other polysaccharides, which are prevalent in agricultural and industrial waste (Imran et al., 2016). The combined action of these enzymes results in the comprehensive degradation of various types of pollutants, contributing to the restoration of water quality. By leveraging the enzymatic machinery of fungi, we can treat industrial and municipal wastewater in a much more effective and sustainable way (Demarche et al., 2012). This strategy has a lot of potential to address one of the biggest issues facing the world today: the pollution of water bodies caused by industrial pollutants and dangerous chemicals. The capacity of fungal enzymes to break down recalcitrant compound pollutants that are difficult for traditional chemical or biological processes to break down—is a key benefit of using them in wastewater treatment. Conventional wastewater treatment methods, such as chemical coagulation or activated carbon filtration, are often limited in their effectiveness, especially when dealing with persistent organic pollutants (POPs). These methods can also require high energy input and may produce secondary pollutants that necessitate additional treatment. In contrast, enzymatic degradation operates under milder conditions, such as neutral pH and moderate temperatures, significantly reducing energy consumption and the generation of harmful by-products. Furthermore, enzymatic processes can target specific pollutants, which enhances the precision and efficacy of treatment while minimizing the overall environmental footprint.

In addition to the ecological benefits, the use of fungal enzymatic systems also holds economic advantages. Traditional wastewater treatment methods can be costly due to the need for large-scale infrastructure and the high energy demands of mechanical and chemical processes. In contrast, enzymatic degradation can be implemented at a lower cost, as it requires fewer resources and less energy input. Moreover, fungal enzymes are biodegradable, further contributing to the sustainability of the process. This makes enzymatic treatment an attractive option for industries looking to reduce operational costs while adhering to stricter environmental regulations. However, despite the many advantages of enzymatic degradation, several challenges need to be overcome prior to its widespread implementation in the industrial setting. The stability of the enzymes is a major problem. Enzymes are very sensitive to external factors, such as temperature, pH, and the presence of inhibitors (Rao et al., 2014), which can impact their activity and longevity. In suboptimal conditions, enzymes can denature or lose their catalytic efficiency, rendering them ineffective for pollutant degradation. The extensive application of fungal enzymatic systems is severely hampered by this, especially in industrial wastewater treatment facilities where environmental conditions might differ greatly. To overcome these limitations, researchers are exploring various strategies to enhance enzyme stability and performance. One promising approach is enzyme immobilization, which involves attaching enzymes to solid supports such as silica gels, polymeric matrices, or nanoparticles (Sirisha et al., 2016). Immobilization can increase enzyme stability, allowing them to retain their activity over longer periods and under a wider range of environmental conditions. Furthermore, immobilized enzymes can be easily recovered and reused, reducing the overall cost of the treatment process. Genetic engineering is another avenue being investigated to improve the robustness of fungal strains. By modifying the genetic makeup of fungi, scientists can improve their ability to produce more resilient and efficient enzymes, further optimizing the degradation process.

13.2.2 Fungal Bioaccumulation and Biosorption of Heavy Metals

Fungal bioaccumulation and biosorption of heavy metals are emerging as highly promising and sustainable solutions to the problem of heavy metal contamination in wastewater (Dhankhar & Hooda, 2011). Heavy metals, including lead (Pb), mercury (Hg), cadmium (Cd), and chromium (Cr), are among the most persistent environmental pollutants due to their non-biodegradable nature and toxicity to both ecosystems and human health. These pollutants can accumulate in water bodies, soils, and living organisms, leading to long-term ecological damage and posing significant risks to public health. Especially when working with low metal concentrations, traditional heavy metal removal techniques including chemical precipitation, ion exchange, and membrane filtration are frequently expensive and ineffective (Fu & Wang, 2011). In contrast, fungi offer an eco-friendly and cost-effective alternative through two key processes: bioaccumulation and biosorption.

Bioaccumulation is the process by which live fungi actively absorb heavy metals. Through this process, fungi metabolize and incorporate heavy metals into their cellular structure, often sequestering them in vacuoles or binding them to proteins within the cell. This allows the fungus to tolerate and neutralize the toxic effects of heavy metals. Several fungal species, including Aspergillus, Penicillium, and Trichoderma, have demonstrated the ability to bioaccumulate significant concentrations of heavy metals (Chen et al., 2020), as illustrated in experimental studies (refer to Fig. 13.3). These fungi have evolved sophisticated mechanisms to deal with toxic metal ions, including the expression of metal transporters and chelating agents that help to immobilize and store metals safely within the fungal cells. This bioaccumulation process is highly efficient, with some fungi capable of removing up to 90% of certain heavy metals from their environment (Maqsood et al., 2022). One of the most intriguing aspects of bioaccumulation is the adaptability of fungi to different environmental conditions. Many fungi can thrive in polluted environments where metal concentrations are high, and they continue to metabolize and sequester these metals over time. However, bioaccumulation requires that the fungal biomass remain viable throughout the process, which can pose challenges for large-scale industrial applications. Maintaining a living fungal population in a controlled environment can be resource-intensive and may require careful regulation of

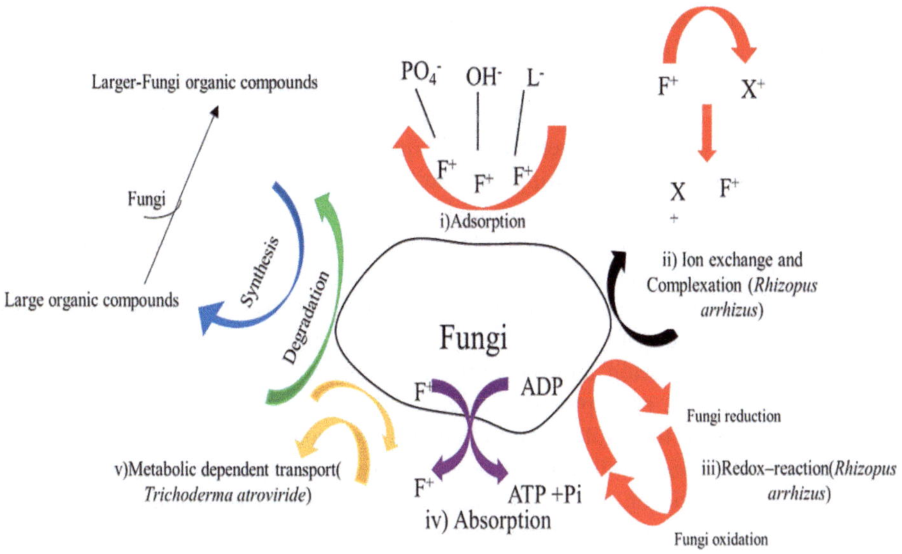

Fig. 13.3 Biosorption of heavy metals like copper, cobalt and other metals by filamentous fungi. Copyright © 2020 Dusengemungu, Kasali, Gwanama and Ouma

parameters such as pH, nutrient availability, and oxygen levels to ensure optimal performance.

Biosorption offers a more practical approach for industrial applications, as it does not require living cells. Biosorption is a passive process where the cell walls of fungi, which are composed of chitin, proteins, and polysaccharides, bind and immobilize heavy metals from the surrounding environment (Zabochnicka-Świątek & Krzywonos, 2014). This process allows for the use of dead or immobilized fungal biomass, making it a versatile and cost-effective solution for wastewater treatment. The fungal cell walls contain functional groups such as hydroxyl, carboxyl, and amine groups that interact with metal ions, enabling the binding and removal of metals from water (Naveena & Latha, 2012). This binding occurs through a variety of mechanisms, including ion exchange, complexation, and surface adsorption, which allow fungi to capture metals from solution with high efficiency. Fungi like Saccharomyces cerevisiae (commonly known as baker's yeast), Rhizopus arrhizus, and Mucor species have been extensively studied for their ability to biosorb heavy metals (Dhankhar & Hooda, 2011). Due to their high surface area and the composition of their cell walls, these fungi can remove substantial amounts of metals from wastewater, even at low concentrations. The advantage of using dead or immobilized fungal biomass is that it eliminates the need for maintaining viable cells, simplifying the treatment process. Additionally, once the metal is adsorbed onto the fungal biomass, the metal-laden biomass can be collected and treated, allowing for the recovery of valuable metals or safe disposal of hazardous materials.

Despite the effectiveness of fungal bioaccumulation and biosorption, several challenges remain in optimizing these processes for large-scale wastewater treatment. One of the key factors influencing the efficiency of these methods is the environmental conditions under which they are applied. Factors such as pH, metal concentration, temperature, and the presence of competing ions can significantly affect the rate and extent of metal uptake by fungal biomass. For example, low pH can increase the solubility of certain metals, enhancing their availability for adsorption, while high temperatures may denature fungal enzymes involved in metal transport (Priya et al., 2022). Another major challenge is the regeneration and reuse of fungal biomass. In industrial settings, the ability to reuse fungal biomass after it has adsorbed heavy metals is crucial for making the process economically viable (Sağ, 2001). Research is ongoing into methods for regenerating fungal biomass, including desorption techniques that release the bound metals without damaging the fungal cell structure. However, this remains an area that requires further exploration to ensure the sustainability of biosorption systems. To improve the overall efficiency of fungal bioaccumulation and biosorption, advances in genetic engineering and strain selection are being explored. By genetically modifying fungal strains, scientists aim to enhance their tolerance to heavy metals and increase their biosorptive capacities. For instance, introducing genes that code for metal-binding proteins or enhancing the expression of metal transporters could improve a fungus's ability to accumulate or adsorb metals more effectively. Additionally, selecting and cultivating fungal strains that naturally exhibit high metal tolerance and biosorptive capabilities could further optimize the treatment process.Fungal bioaccumulation and biosorption align closely with green technology and circular economy, offering a sustainable approach to heavy metal remediation. These processes reduce the need for energy-intensive chemical treatments and minimize the generation of toxic waste, making them environmentally friendly alternatives. Moreover, by using fungal biomass to capture and remove metals, there is potential for the recovery of valuable metals, which can be recycled and reused in industrial applications. This contributes to a circular economy by turning waste products into valuable resources, reducing environmental pollution, and promoting resource efficiency.

13.2.3 Case Studies of Fungal Species Used in Treatment Processes

Several fungal species have been successfully employed in wastewater treatment due to their unique abilities to degrade pollutants and remove heavy metals. For instance,

(a) Trametes versicolor, a white-rot fungus as shown in Fig. 13.4a, also known as the turkey tail fungus, is widely studied for its potential in wastewater treatment due to its robust enzymatic system. This white-rot fungus produces lignin-degrading enzymes like laccase and manganese peroxidase, which can break down complex organic pollutants such as dyes, pesticides, and pharmaceuticals (Periasamy et al., 2019). Its ability to degrade lignin also makes it useful in bioremediation, as lignin shares structural similarities with many environmental pollutants. Trametes versicolor has shown significant success in removing

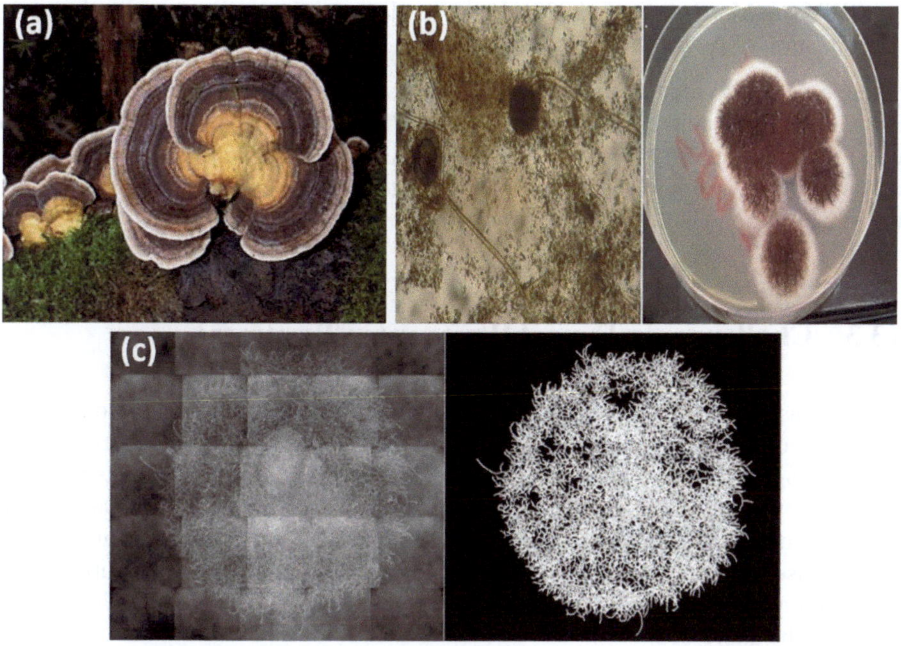

Fig. 13.4 (**a**) Trametes versicolor, (**b**) Aspergillus niger (Ibrahim et al., 2017) Copyright © 2017 Net Journals, (**c**) Phanerochaete chrysosporium (Khalil et al., 2021) © 2021, The Author(s)

heavy metals through biosorption, making it a versatile and eco-friendly option for treating industrial and municipal wastewater.

(b) Aspergillus niger, which is a filamentous fungus as shown in Fig. 13.4b extensively used in wastewater treatment due to its powerful enzyme production and high tolerance to pollutants. Known for producing enzymes like amylases, cellulases, and proteases, it plays a significant role in breaking down organic pollutants. Additionally, A. niger demonstrates effective biosorption capabilities, particularly in removing heavy metals such as lead, cadmium, and chromium from contaminated water. Its cell wall components, such as chitin and glucans, facilitate heavy metal binding, making it an efficient agent for bioremediation (Shakya et al., 2015). Its widespread availability and adaptability make A. niger a valuable resource in wastewater treatment processes.

(c) Phanerochaete chrysosporium, another white-rot fungus as shown in Fig. 13.4c, being the most studied species for wastewater treatment. Known for its lignin-degrading ability, this fungus produces powerful oxidative enzymes like lignin peroxidase and manganese peroxidase, which can break down a wide range of organic pollutants, including dyes, pesticides, and phenolic compounds. Additionally, Phanerochaete species are capable of bioaccumulating and biosorbing heavy metals, making them valuable for bioremediation. Their metabolic flexibility and capacity to function under diverse environmental conditions make them ideal candidates for treating industrial and municipal wastewater

(Bibbins-Martínez et al., 2023). However, large-scale applications require optimization of fungal growth conditions.

These case studies demonstrate the effectiveness of specific fungal species in treating various pollutants, highlighting their potential in developing sustainable and eco-friendly wastewater treatment processes. As research advances, the application of fungi in environmental cleanup continues to expand, offering promising solutions to pollution challenges.

13.3 Challenges in Fungal Wastewater Treatment

Fungal wastewater treatment holds significant potential due to its ability to degrade complex pollutants, including heavy metals and pharmaceuticals. However, despite its advantages, several technical, economic, and environmental challenges must be addressed to make fungal systems a viable alternative to traditional wastewater treatment methods. These challenges are best presented in a tabular format for clarity and ease of understanding shown in Table 13.1. The following table outlines the key challenges within each category, providing a concise overview of the obstacles that must be addressed.

This table provides a detailed overview of the various challenges involved in advancing fungal wastewater treatment, highlighting the technical, economic, environmental, and regulatory complexities that must be addressed to make fungal bioremediation a viable and widespread solution in wastewater management.

13.4 Embracing Opportunities for Sustainability

The integration of fungal wastewater treatment systems presents significant opportunities for promoting sustainability in environmental management. Fungal systems, with their unique enzymatic abilities, offer a sustainable and efficient means of degrading complex pollutants that are often resistant to traditional treatment methods (Kües, 2015). By turning garbage into useful resources, this creative strategy not only promotes the circular economy but also reduce environmental contamination. Heavy metals, pharmaceuticals, and other persistent pollutants can be effectively treated using fungal wastewater systems, providing a significant advantage over conventional chemical and physical treatments. One of the most critical steps in maximizing the potential of fungal wastewater treatment is investing in research and development. Optimizing fungal strains for specific pollutants can significantly enhance degradation efficiency, ensuring that even the most complex and recalcitrant compounds are broken down. Hybrid systems that combine fungal treatments with other biological or physical processes can further improve the effectiveness of wastewater treatment. These systems can capture valuable by-products, such

Table 13.1 Challenges in fungal wastewater treatment

Category	Challenges	Description
Technical	Slow degradation rates	Fungi, though efficient at breaking down complex pollutants, often work more slowly compared to bacteria, extending treatment times and potentially reducing throughput (Espinosa-Ortiz et al., 2022)
	Optimal growth conditions	Fungi require specific environmental conditions (e.g., pH, temperature, oxygen levels) for optimal growth, which may be challenging to maintain in large-scale operations
	Reactor design limitations	Existing bioreactor designs are often optimized for bacterial growth, not fungal activity, which can limit the efficiency of fungal-based wastewater treatments (Sankaran et al., 2010)
	Fungal biomass disposal	After treatment, fungal biomass accumulates and needs to be safely disposed of or treated, adding complexity to waste management
	Synergy with other organisms	Creating effective hybrid systems with bacteria or algae can be complex, requiring careful balance of conditions to avoid competition between organisms (Hu et al., 2019)
Economic	High initial costs	The costs of developing, scaling, and maintaining fungal-based treatment systems, particularly bioreactors, are often higher compared to traditional bacterial systems
	Limited commercial availability	Fungal bioremediation technologies are still emerging, and there are few commercially available systems, making adoption costly and less widespread
	Energy consumption	Maintaining the specific environmental conditions needed for fungal activity (e.g., temperature, aeration) can lead to higher energy costs in large-scale wastewater plants
	Investment in R&D	Significant financial investment is required for research and development to better understand fungi's capabilities and optimize their use in bioremediation
Environmental	Potential pathogen spread	Some fungi may pose risks of pathogenicity, requiring careful strain selection and management to avoid contamination risks to the environment or human health (Nji et al., 2023)
	Secondary pollution	Fungal degradation can produce secondary byproducts that may require additional treatment, complicating the overall wastewater treatment process
	Resource intensity	Fungi require a continuous supply of nutrients, oxygen, and controlled environments, which can consume significant natural resources during large-scale operations
	Ecosystem impact	The introduction of non-native or genetically modified fungi into wastewater systems could potentially disrupt local ecosystems if not carefully controlled
Regulatory	Lack of regulatory standards	Fungal bioremediation is not yet widely regulated, and there are limited standards for its implementation, which could slow adoption and increase regulatory hurdles

as biogas from organic waste, that can be repurposed as energy. Investing in research to develop robust fungal strains capable of thriving in diverse wastewater conditions also reduces the overall environmental impact of treatment plants. Additionally, by increasing the efficiency of pollutant removal, operational costs can be lowered, making fungal systems a more viable and attractive option for both industrial and municipal wastewater management. Developing advanced bioreactors specifically designed for fungal applications is another key aspect of promoting sustainability through fungal wastewater treatment. Bioreactors provide the controlled environments necessary for fungi to thrive, allowing for the continuous treatment of wastewater while maintaining optimal conditions for fungal growth and activity (Visvanathan et al., 2000). These bioreactors can be equipped with real-time monitoring systems, which ensure that any fluctuations in environmental conditions are immediately addressed, thereby improving overall system efficiency. Scaling these bioreactors for industrial use presents technical challenges, but with continuous innovation, such systems can become more adaptable and capable of handling larger volumes of wastewater.

Furthermore, bioreactors designed with sustainability in mind can operate with lower energy consumption, further reducing the carbon footprint of wastewater treatment operations. By integrating fungal wastewater treatment systems into existing infrastructure, industries and municipalities can create a more resilient and environmentally friendly approach to managing wastewater. Fostering collaboration between academia, industry, and policymakers is essential for ensuring the widespread adoption of fungal wastewater treatment technologies. Researchers can provide the scientific insights needed to optimize fungal applications, while industries can help develop and implement the necessary infrastructure for large-scale operations. Policymakers play a crucial role in creating regulatory frameworks that encourage sustainable practices and provide incentives for adopting innovative technologies. This collaborative approach not only helps to overcome the technical and economic challenges associated with scaling fungal wastewater treatment but also ensures that environmental sustainability remains a top priority. Embracing the potential of fungal wastewater treatment offers an opportunity to address significant environmental challenges while contributing to a more sustainable future. By investing in research and development, advancing bioreactor technology, and fostering collaboration across sectors, fungal systems can become a mainstream solution for water management. This approach aligns with global sustainability goals, creating healthier ecosystems, improving public health, and building more resilient communities.

13.4.1 Potential for Circular Economy in Wastewater Treatment

The circular economy paradigm presents a transformative methodology to wastewater treatment, prioritizing resource recovery and waste minimization as shown in Fig. 13.5. Unlike the conventional linear model, this treats wastewater as a waste

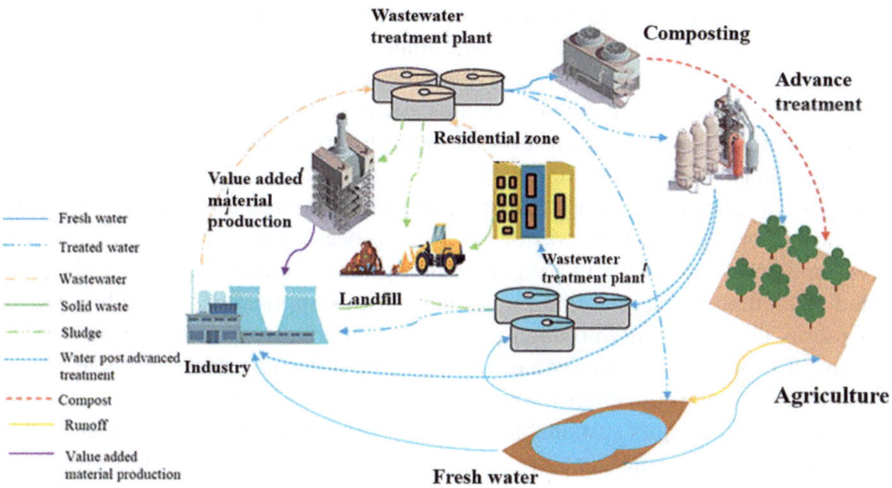

Fig. 13.5 Illustration of implementation of circular economy in wastewater treatment sector reproduced from Philip et al. (n.d.) Copyright © 2023 The Authors

product that requires disposal, the circular economy views wastewater as a valuable resource from which essential materials can be recovered (Smol et al., 2020). By turning waste into resources, this change not only reduces its negative effects on the environment but also creates value economically. The principles of a circular economy align wastewater treatment with sustainability, resource conservation, and waste reduction, making it a key strategy in addressing environmental challenges and supporting economic growth. Nutrient recovery is one of the main opportunities for wastewater treatment under the circular economy paradigm. Wastewater often contains high levels of nutrients such as nitrogen and phosphorus. In traditional wastewater treatment systems, these nutrients are treated as contaminants and are removed through processes that often generate waste and consume energy. However, in the circular economy model, nitrogen and phosphorus are seen as valuable resources that can be recovered and reused, particularly in agriculture (Robles et al., 2020). These nutrients are critical for plant growth and recovering them from wastewater to produce fertilizers can help reduce the need for synthetic fertilizers. Synthetic fertilizers are not only energy-intensive to produce but also contribute to environmental issues like soil degradation and water pollution. Nutrient recovery from wastewater can have significant benefits for agriculture and the environment. By repurposing nitrogen and phosphorus as fertilizers, wastewater treatment plants can provide a sustainable source of nutrients for crops, reducing the dependence on industrially produced fertilizers. This is particularly important in regions facing fertilizer shortages or where the costs of agricultural inputs are high. Moreover, recovering nutrients from wastewater helps address the problem of eutrophication—a process where excess nutrients, particularly phosphorus and nitrogen, enter water bodies and lead to the growth of harmful algal blooms. These blooms can result in oxygen depletion in water systems, which harms aquatic life and leads to ecosystem degradation. By capturing these nutrients before they reach natural water bodies, wastewater treatment within a circular economy framework helps protect ecosystems and promote water quality. Organic matter present in wastewater is another valuable resource that can be harnessed through circular economy principles. Traditional wastewater treatment methods focus on removing organic matter to reduce pollution, but they often miss the opportunity to convert this material into useful products (Puyol et al., 2017).

Biogas, a renewable energy resource, can be generated through the anaerobic digestion of organic waste in wastewater. This process involves the decomposition of organic substances without oxygen, yielding biogas that is predominantly composed of methane. The resulting biogas has various applications, including power generation, thermal energy production, and as an alternative fuel for transportations (Hakawati et al., 2017). Wastewater treatment facilities can decrease their dependence on non-renewable energy sources, cut down on greenhouse gas output, and support the shift towards sustainable energy by generating biogas from wastewater. Furthermore, the organic content in wastewater can be converted into biochar, a substance rich in carbon that offers various environmental advantages. In agriculture, biochar serves as a valuable resource for enhancing soil quality and moisture retention, promoting sustainable farming practices (Ayaz et al., 2021). Additionally, it can be utilized as a filtering agent in water purification processes, improving the removal of contaminants. The production of biochar from organic waste aligns with circular economy principles by creating value from what would otherwise be discarded and promoting a closed-loop system where waste is continuously recycled into useful products. Implementing the circular economy in wastewater treatment requires rethinking traditional processes and adopting new technologies that prioritize resource recovery. Bioreactors, for example, can be designed not only to treat wastewater but also to recover valuable by-products such as nutrients, biogas, and biochar. These bioreactors can integrate multiple processes, such as nutrient capture and biogas production, within a single system, making wastewater treatment more efficient and sustainable.

By adopting such innovations, wastewater treatment plants can significantly reduce waste generation, lower energy consumption, and create economic value by producing marketable products from wastewater. Adopting circular economy principles in wastewater treatment also supports broader environmental and sustainability goals. The reduction of waste and pollution, coupled with the generation of renewable energy and valuable materials, aligns with global efforts to fight climate change, encourage resource efficiency, and transition toward sustainable development. By focusing on resource recovery and waste minimization, the circular economy model promotes a more resilient and sustainable wastewater management system that can adapt to future environmental challenges, such as water scarcity and energy shortages. Furthermore, there are significant financial advantages to treating wastewater using a circular economy strategy. Treatment facilities can lower operating costs and create new revenue streams by recovering materials from wastewater. For example, selling recovered nutrients as fertilizers or producing biogas for energy generation can offset the costs of wastewater treatment, making the process more economically viable. This creates a self-sustaining cycle where the value generated from recovered resources supports the continued operation and expansion of wastewater treatment infrastructure.

13.4.2 Integration with Other Bioremediation Technologies

The increasing complexity of pollutants in wastewater has driven researchers to explore innovative bioremediation methods. Traditional methods alone often fall short in addressing the diverse range of contaminants, especially those that are persistent and difficult to degrade. Integrating fungal wastewater treatment with other bioremediation technologies offers a promising strategy to improve both the efficacy and scope of wastewater treatment (Jain et al., 2022). This combination not only enhances the breakdown of complex pollutants but also targets other contaminants like nutrients and pathogens, resulting in a more efficient and comprehensive solution for wastewater management. Fungi, particularly certain species such as white-rot fungi, have demonstrated exceptional potential in breaking down a wide array of organic pollutants, including complex and stubborn compounds that conventional treatments struggle with. Fungi have demonstrated remarkable capabilities in breaking down complex organic pollutants, including recalcitrant compounds such as lignin, pesticides, and pharmaceuticals. The enzymatic arsenal of fungi, such as laccases and peroxidases, enables them to degrade these pollutants effectively, making them an invaluable tool in advanced wastewater treatment. However, despite their remarkable abilities, fungal treatments cannot address every type of pollutant found in wastewater. Fungal treatment alone may not address all contaminants present in wastewater, such as nutrients (nitrogen and phosphorus) or pathogens, which require complementary treatment methods. Nutrients, particularly nitrogen and phosphorus, can contribute to eutrophication if released untreated into aquatic ecosystems. Pathogens, on the other hand, pose significant public health risks, making it essential to remove them before wastewater is discharged into the environment. To create a more holistic approach to wastewater treatment, it becomes necessary to combine fungal bioremediation with other established technologies. When combined with conventional bioremediation processes like activated sludge systems, fungal treatment can fill critical gaps. One of the most popular techniques in wastewater treatment facilities is activated sludge systems, which are effective in eliminating nutrients and organic materials (Gernaey et al., 2004), but they may struggle with breaking down persistent pollutants as shown in Fig. 13.6. These processes work by using aerobic bacteria to metabolize organic compounds, but their limitations become evident when dealing with recalcitrant pollutants like pesticides, pharmaceuticals, and other industrial chemicals. Fungal species, particularly white-rot fungi, excel in degrading these compounds through enzymatic actions like laccases and peroxidases. Fungi employ these enzymes to decompose intricate organic contaminants into simpler, less toxic substances. This makes them particularly effective in tackling pollutants that persist in the environment despite traditional treatment methods. Therefore, incorporating fungi as part of a wastewater treatment system can address some of the shortcomings of activated sludge processes. A

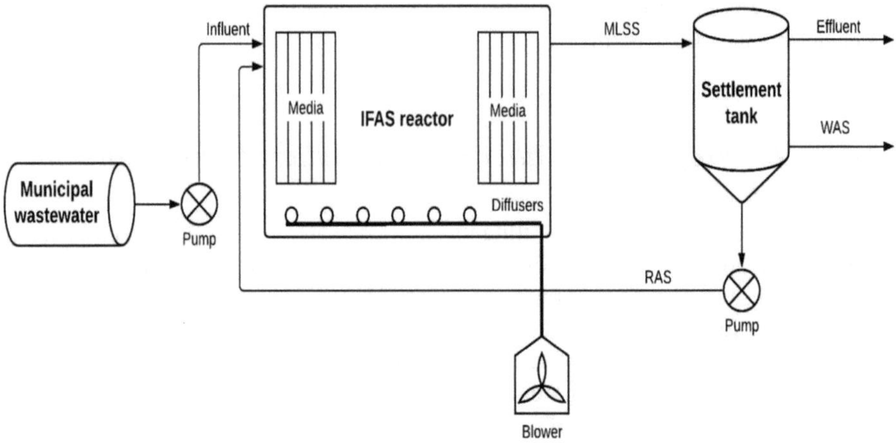

Fig. 13.6 An integrated fixed-film activated sludge (IFAS) system reproduced from Pryce et al. (2022) Copyright © 2022, The Author(s)

possible approach to integration could involve using fungal treatment as a pre- or post-step to the activated sludge system. By integrating fungal treatment as a pre- or post-step in activated sludge systems, a more complete breakdown of contaminants can be achieved. For example, in a pre-treatment stage, fungi could degrade recalcitrant compounds, making the wastewater more amenable to further biological treatment in the activated sludge system (Ghosh Ray & Ghangrekar, 2018). In a post-treatment scenario, fungi could tackle any persistent pollutants that remain after the primary treatment process. This combination ensures that even the most stubborn pollutants are broken down, resulting in cleaner and safer water discharge. In addition to combining fungal treatment with activated sludge, other technologies like constructed wetlands can be used to further enhance wastewater treatment. Similarly, constructed wetlands can be coupled with fungal bioremediation to harness the natural filtration and nutrient-removal capabilities of wetland plants, while fungi degrade more complex or hazardous substances. Constructed wetlands use natural processes including plants and microorganism to effectively remove suspended particles and nutrients (Stottmeister et al., 2003). When fungi are added to this system, they complement the nutrient-removal capability by degrading organic pollutants that plants or bacteria might not handle effectively. This partnership allows for a more complete treatment process that addresses both organic pollutants and nutrient removal in an environmentally friendly manner.

Another promising integration involves Advanced Oxidation Processes (AOPs), which uses chemical oxidants to degrade organic contaminants, can also be integrated with fungal treatment to target both organic pollutants and recalcitrant chemicals effectively. AOPs utilize strong oxidizing agents like ozone or hydrogen peroxide to break down pollutants that resist biological treatment (Al Mayyahi & Ali Abed Al-Asadi, n.d.). When combined with fungal bioremediation, AOPs can act as an additional line of defense, ensuring that even the most resistant contaminants are degraded. This synergy between fungal bioremediation and AOPs offers a powerful solution for wastewater treatment plants facing particularly challenging contaminants. The result of integrating fungal treatment with other bioremediation technologies is a robust, multi-faceted system capable of tackling a wide array of pollutants. This integrated approach results in a synergistic system where fungal bioremediation handles complex pollutants, and complementary technologies manage other contaminants like nutrients and pathogens. By targeting different types of contaminants with specialized treatments, the overall efficiency of the wastewater treatment process is greatly improved. The benefits of this integrated approach go beyond just enhanced pollutant removal. The outcome is a more robust, efficient, and sustainable wastewater treatment process, optimizing pollutant removal, minimizing environmental impacts, and enhancing resource recovery efforts, such as biogas production or nutrient recycling. This not only helps in maintaining clean water but also supports environmental sustainability by recovering valuable resources from wastewater. Whether it is through the production of biogas from organic matter or the recycling of nutrients for agricultural use, these additional benefits further enhance the value of integrated bioremediation systems.

13.4.3 Advances in Fungal Biotechnology and Genetic Engineering

Recent developments in genetic engineering and fungal biotechnology have opened up new possibilities for optimizing wastewater treatment. Traditional fungal treatment methods are already effective at breaking down complex organic pollutants, but the integration of modern biotechnological techniques promises to significantly enhance the capabilities of fungi in wastewater bioremediation. One of the key breakthroughs in this field is the use of genetic engineering to develop fungi with enhanced capabilities for degrading specific contaminants. By introducing targeted modifications into the fungal genome, researchers can create genetically modified fungi that possess superior enzymatic activity or metabolic pathways tailored to break down complex and persistent pollutants (Harms et al., 2011). For example, genes coding for enzymes like laccases, lignin peroxidases, and manganese peroxidases can be overexpressed in certain fungi, improving their ability to degrade difficult-to-treat pollutants such as pharmaceuticals, industrial chemicals, and microplastics. These enzymes are vital in oxidative degradation, making fungi valuable assets in wastewater treatment processes. One of the most significant tools in genetic engineering today is CRISPR-Cas9 technology, which allows for precise, targeted modifications of the fungal genome. This system can be used to knock out unwanted genes or introduce new genetic material, creating strains with optimized metabolic functions for pollutant degradation. For example, researchers have used CRISPR to modify white-rot fungi strains, enabling them to break down not only lignin but also synthetic chemicals that are commonly found in industrial wastewater. By focusing on enhancing the pathways that break down specific pollutants, scientists can reduce treatment times, increase efficiency, and broaden the range of contaminants that can be effectively managed using fungal bioremediation.

Another exciting development is the application of synthetic biology to engineer novel fungal strains with tailored abilities. This field involves designing and building new biological components, mechanisms, and systems, or modifying existing biological structures for practical applications (Rollié et al., 2012). Through this approach, researchers are developing synthetic fungal strains that can tackle pollutants that were previously untreatable through natural means. These synthetic fungi may have increased resistance to harsh environmental conditions or enhanced degradation pathways that allow them to thrive in complex wastewater matrices. Beyond the development of genetically engineered fungi, these innovations enable metabolic pathway engineering, where the entire metabolism of a fungal cell can be redesigned for optimal performance in wastewater treatment. This could involve rerouting metabolic fluxes to ensure that a higher proportion of energy and resources are dedicated to breaking down pollutants or engineering new pathways that can target contaminants for which fungi would otherwise have no natural affinity. These advances in fungal biotechnology are also allowing for the creation of fungi that can target a wider range of contaminants, including emerging pollutants such as endocrine disruptors, personal care products, and antibiotic residues. As new environmental challenges arise, the adaptability of genetically modified fungi ensures that they can be tailored to address specific problems, making them a versatile tool in modern wastewater treatment strategies.

The implications of these biotechnological advancements are far-reaching. As research continues, the integration of genetically modified fungi into existing wastewater treatment infrastructure will make processes faster, more efficient, and more sustainable. Moreover, these innovations will expand the potential applications of fungal bioremediation to industries beyond municipal wastewater, including pharmaceuticals, textiles, and agriculture, where complex wastewater streams are common. In summary, advances in genetic engineering and fungal biotechnology have the potential to revolutionize the field of wastewater treatment. By harnessing tools like CRISPR and synthetic biology, researchers are developing fungi that are not only more effective at breaking down pollutants but are also highly adaptable to emerging environmental challenges. These developments promise to make wastewater treatment more efficient, scalable, and environmentally sustainable.

13.5 Emerging Trends and Innovations

13.5.1 Fungal Consortia and Synergistic Interactions

Fungal consortia, the combination of multiple fungal species working in tandem, are emerging as a powerful tool in wastewater treatment, particularly in enhancing the biodegradation of complex pollutants. The idea behind using fungal consortia is to harness the unique capabilities of different fungal species, which, when combined, create a synergistic effect that is more effective than using individual species alone (Thirumalaivasan et al., 2024). This collaborative approach enables the breakdown of pollutants that are resistant to traditional treatment methods and optimizes the overall performance of bioremediation systems. In a fungal consortium, each species often brings its own set of enzymes and metabolic pathways to the process. For instance, ligninolytic enzymes like laccases and peroxidases, which are produced by white-rot fungus, are well known for their ability at breaking down recalcitrant organic contaminants including lignin, dyes, and some pharmaceuticals as shown in Fig. 13.7a. Other fungi may excel at degrading simpler organic molecules or play supportive roles by producing enzymes that help in breaking down the

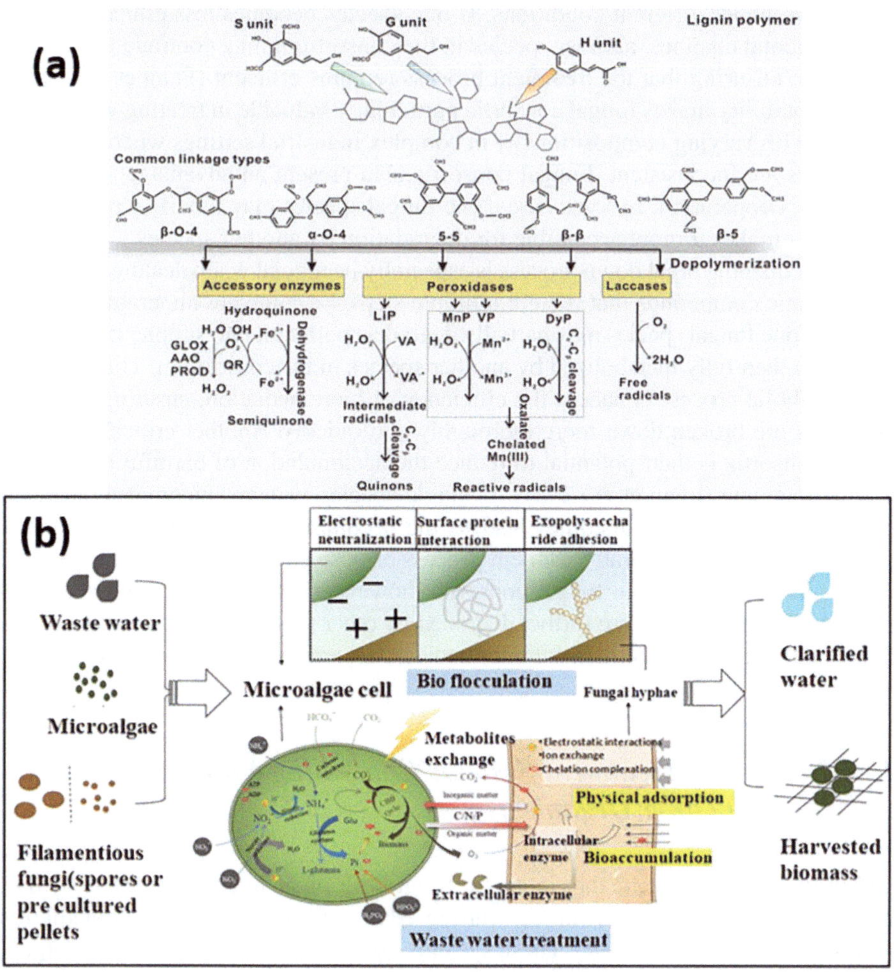

Fig. 13.7 (**a**) Study of a bottom-up approach in synthetic ecology focusing on microbial consortia to improve lignocellulose bioconversion efficiency reproduced from Hayat et al. (2012) Copyright © 2022, The Author(s). (**b**) Schematic diagram of process of the fungi-microalgae consortium for wastewater treatment reproduced from Leng et al. (2021) Copyright © 2021 Elsevier Ltd

byproducts of these reactions. This division of labor allows for a more comprehensive degradation process, targeting a broader range of pollutants simultaneously.

The interaction between different fungi can also lead to the production of complementary enzymes, enhancing the breakdown of pollutants that would be difficult for a single fungal species to tackle alone. Beyond improving pollutant degradation, fungal consortia increase the resilience and adaptability of wastewater treatment systems. Environmental conditions in wastewater treatment plants can vary remarkably, including changes in temperature, pH, and the composition of contaminants. Single-species fungal systems may struggle to adapt to these fluctuations, leading to reduced efficiency. In contrast, fungal consortia offer greater flexibility, as different species may thrive under different conditions. If one species becomes less effective due to environmental changes, another species in the consortium may continue to function optimally, ensuring that the treatment process remains efficient (Peng et al., 2021). This adaptability makes fungal consortia particularly valuable in treating wastewater streams with varying compositions or in complex industrial settings where contaminant loads are inconsistent. Fungal consortia also present an advantage in terms of metabolic cooperation. In some cases, one fungal species may modify a pollutant in a way that makes it more accessible for degradation by another species.

This sequential breakdown process is especially beneficial when dealing with complex organic compounds that require multiple steps for complete mineralization. For instance, one fungal species may partially degrade a pollutant into simpler byproducts, which are then fully metabolized by another species in the consortium. This cooperative metabolic process enhances the efficiency of bioremediation, ensuring that contaminants are broken down more thoroughly and quickly. Another critical aspect of fungal consortia is their potential to reduce the accumulation of harmful byproducts during treatment (Leng et al., 2021). In single-species systems, incomplete degradation of pollutants can sometimes lead to the accumulation of toxic intermediates, which can hinder the overall treatment process or pose environmental risks. The synergistic interactions within fungal consortia, however, help to mitigate this issue by ensuring that byproducts are further degraded by other species within the consortium, reducing the likelihood of harmful intermediates accumulating in the system.

13.5.2 Hybrid Treatment Systems Combining Fungi with Other Organisms

Hybrid treatment systems that integrate fungi with other microorganisms, such as bacteria or algae, represent a cutting-edge innovation in bioremediation (Thirumalaivasan et al., 2024). These systems use the special abilities of each organism to produce a more thorough and effective method of removing pollutants from the environment. Fungi are especially good at breaking down recalcitrant compounds, like lignin, pharmaceuticals, and synthetic chemicals, due to their powerful enzymatic activity. However, they often work slowly compared to other microorganisms. This is where bacteria, with their fast growth rates and ability to rapidly degrade simpler organic compounds, complement fungal activity, speeding up the overall process. In these hybrid systems, fungi and bacteria form a synergistic partnership that maximizes pollutant degradation. Fungi specialize in breaking down large, complex pollutants into smaller molecules. These smaller molecules are then quickly metabolized by bacteria, which can process them more rapidly than fungi. In this way, both organisms play distinct but complementary roles, ensuring that a wider variety of contaminants are broken down more efficiently. This division of labour between fungi and bacteria not only accelerates the overall treatment process but also improves its efficacy by ensuring a more complete degradation of pollutants.

In a hybrid system, fungi and bacteria work together in a synergistic relationship as shown in Fig. 13.8. Complex pollutants can be decomposed by fungi into smaller molecular components, which are then easily metabolized by bacteria. This division of labour ensures a more complete degradation of contaminants, improving the efficiency of treatment. Algae can also be introduced into these systems to address other environmental challenges. Algae are particularly useful in wastewater treatment, where they play a critical role in nutrient removal. Excessive nutrients like nitrogen and phosphorus can cause eutrophication, which damages aquatic ecosystems by reducing oxygen levels in bodies of water. Algae, through photosynthesis, remove these nutrients from wastewater, preventing eutrophication and contributing to the overall sustainability of the treatment process (Cuellar-Bermudez et al., 2017). Algae can also produce oxygen during photosynthesis, which can further support the bacterial degradation processes by providing oxygen to aerobic bacteria. These hybrid systems are versatile and adaptable, making them applicable to various forms of environmental remediation, including wastewater treatment, soil

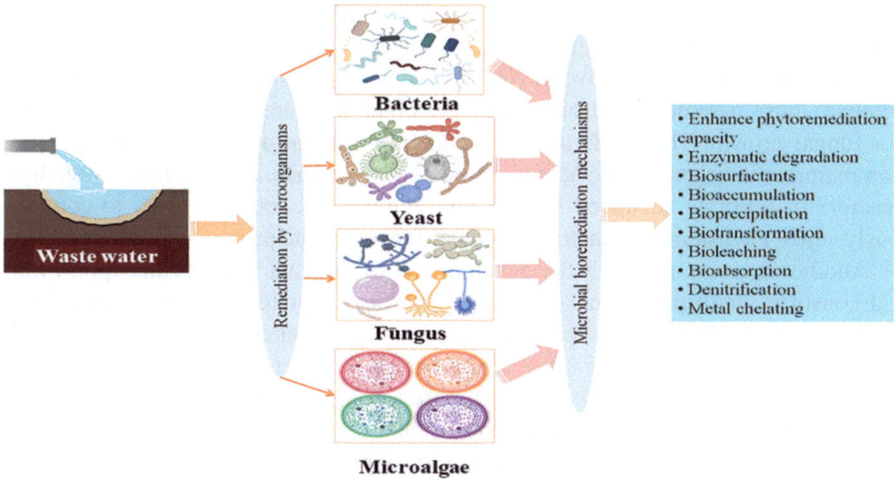

Fig. 13.8 Microbial engineering approaches for wastewater treatment reproduced from Sharma et al. (2023) Copyright © 2023 The Author(s)

remediation, and even air purification. For instance, in soil remediation, fungi can degrade persistent organic pollutants, while bacteria work on more easily degradable compounds. Algae, which can grow on contaminated soils, can help capture nutrients or heavy metals, aiding in soil recovery. Similarly, in air purification systems, microorganisms like fungi, bacteria, and algae can be used together to capture and break down airborne pollutants, offering a more comprehensive solution. By combining fungi's enzymatic capabilities with the complementary strengths of bacteria and algae, hybrid treatment systems provide a powerful, holistic tool for sustainable environmental management. These systems are a potential strategy for the future of bioremediation since they provide a more comprehensive and flexible solution for the remediation of different environmental contaminants, from polluted soils to wastewater.

13.5.3 Development of Novel Bioreactors and Treatment Systems

The development of novel bioreactors specifically designed for fungal-based treatments is a key advancement in environmental biotechnology. These bioreactors are engineered to create optimal conditions for fungi to grow, thrive, and efficiently break down pollutants. Given fungi's unique ability to degrade complex and recalcitrant contaminants, such as pharmaceuticals, pesticides, and industrial chemicals, designing bioreactors that harness and enhance these capabilities is essential for improving the overall efficiency of wastewater treatment and bioremediation processes (Kumar Singh et al., 2020) as shown in Fig. 13.9. One of the key innovations in bioreactor design is the focus on maintaining optimal environmental conditions for fungal activity. Fungi often require specific temperature, pH, oxygen, and nutrient levels to function effectively. Novel bioreactors are designed to monitor and regulate these parameters, ensuring that the conditions are consistently favourable for fungal growth. This leads to more efficient pollutant degradation, as the fungi can maintain their enzymatic activity at peak performance. For example, some bioreactors incorporate advanced sensors and automated control systems to continuously adjust environmental factors, allowing for stable and prolonged fungal activity.

Another important trend is the development of bioreactors that can support fungal consortia, multiple fungal species working together to break down pollutants (Vaksmaa et al., 2023). These consortia are more resilient and versatile than single-species systems, allowing for the treatment of a wider range of contaminants. Bioreactors designed for fungal consortia often have multiple chambers or sections where different fungal species can thrive in their ideal environments while contributing to the overall degradation process. This approach maximizes the diversity of enzymes and metabolic pathways involved, making the treatment system more effective at handling complex and diverse pollutants. Continuous operation is another critical innovation in bioreactor design. Traditional batch reactors, which operate in cycles, require downtime for cleaning, reloading, or other maintenance

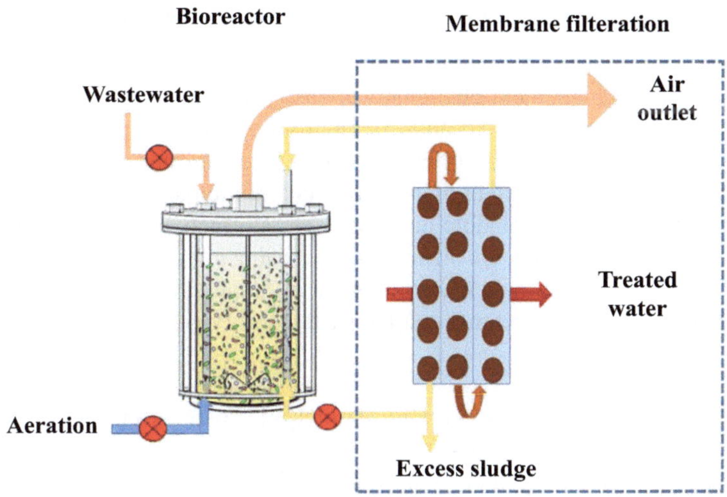

Fig. 13.9 Illustration for the water treatment using bioreactor systems reproduced from Chandran et al. (2023) Copyright © 2023, The Author(s)

tasks. In contrast, modern fungal-based bioreactors are being developed to allow for continuous operation, where wastewater or contaminated material flows steadily through the system without interruption. This not only increases efficiency by reducing downtime but also enables a more scalable and sustainable solution for large-scale wastewater treatment. Continuous flow systems often incorporate features like self-cleaning mechanisms, reducing the need for manual maintenance and minimizing system disruptions. Moreover, bioreactors are becoming increasingly modular and scalable, making them more adaptable to various treatment needs (de Mello et al., 2023). Whether for small-scale, decentralized wastewater treatment or large industrial applications, these reactors can be scaled up or down depending on the volume of wastewater and the nature of the pollutants being treated.

The development of novel bioreactors for fungal-based treatments is revolutionizing the field of bioremediation. These innovations optimize fungal growth and pollutant degradation, promote the use of fungal consortia, enable continuous operation, and offer scalable solutions. These advances are paving the way for more effective, sustainable, and adaptable bioremediation technologies that can meet the growing challenges of environmental pollution.

13.6 Future Perspective and Applications

Fungal wastewater treatment holds immense potential for the future, emerging as a key technology in addressing global water management and environmental sustainability challenges.

13.6.1 Growing Importance in Global Water Management

As global water management challenges intensify, fungal wastewater treatment is becoming a vital technology to address environmental sustainability. Traditional wastewater treatment methods often fall short in degrading complex pollutants, which are increasingly prevalent in industrial and municipal waste streams. Fungal-based systems offer a sustainable and adaptable alternative, as they can handle a wide range of pollutants that are difficult to treat using conventional methods. These systems are likely to become a key component in future strategies to improve water quality, reduce pollution, and meet global sustainability goals.

13.6.1.1 Adoption in Industrial and Municipal Wastewater Treatment

The versatility of fungal systems makes them suitable for a wide range of applications, particularly in industries that generate wastewater with complex or hazardous contaminants.

Industrial applications: Industries such as textiles, pharmaceuticals, agriculture, and petrochemicals produce wastewater streams that are often laden with synthetic organic compounds, heavy metals, and other recalcitrant pollutants (Pereira et al., 2021). These pollutants are resistant to conventional biological treatments but can be broken down by fungi through enzymatic processes. The adoption of fungal treatment in these industries offers a solution to meet stringent environmental regulations, reduce ecological impact, and improve the overall sustainability of industrial wastewater management as shown in Fig. 13.10b.

Municipal applications: In municipal wastewater treatment, fungi can complement or even replace conventional bacterial systems in dealing with emerging contaminants such as personal care products, pharmaceuticals, and microplastics (Morin-Crini et al., 2022). These pollutants are increasingly found in urban wastewater, posing challenges for conventional systems as shown in Fig. 13.10a. Fungi's ability to degrade complex pollutants makes them an attractive option for enhancing municipal wastewater treatment capabilities.

13.6.1.2 Potential in Degrading Recalcitrant Pollutants

The potential of fungal wastewater treatment to break down recalcitrant pollutants that are resistant to traditional treatment methods. Many different enzymes, including laccases and peroxidases, are produced by fungi, especially white-rot fungi, which are capable of dissolving complex organic compounds into simpler, less toxic byproducts.

Pharmaceuticals and Synthetic Compounds: Pharmaceuticals and synthetic chemicals, which are commonly found in wastewater, are particularly challenging for conventional treatments. Fungi's enzymatic diversity allows for the breakdown

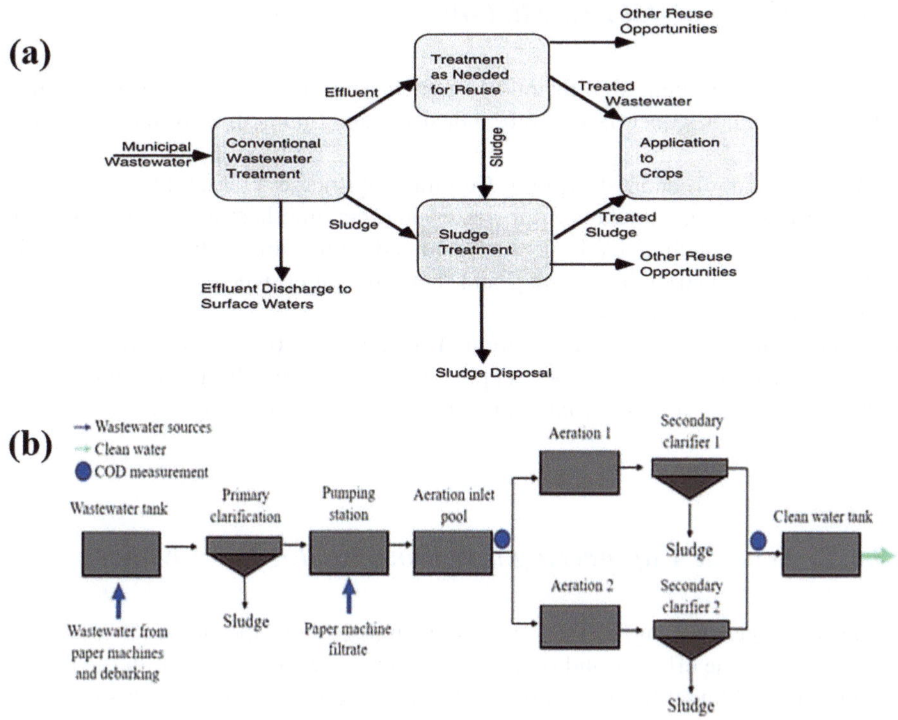

Fig. 13.10 (**a**) Municipal wastewater is discharged to surface waters or reused, or before discharge to surface waters (not illustrated). Before being reused, more treatment could be required. Sludge from wastewater treatment processes are treated and then disposed or reused in crop production or other application reproduced from Rahimi and Rezaii (2021) Copyright © 2021 Abolfazl Rahimi and Mohammad Rezaii. (**b**) A detailed diagram illustrating the wastewater treatment plant and the activated sludge process used for industrial wastewater analysis. The schematic highlights the various stages involved, including wastewater collection, preliminary treatment, aeration, and the biological degradation of pollutants through the activated sludge system. The layout emphasizes key components such as aeration tanks, clarifiers, and sludge recycling, which are essential for breaking down organic matter and ensuring the efficient treatment of industrial effluents. This process ensures the removal of contaminants before the treated water is safely discharged or reused reproduced from Pörhö et al. (2023) Copyright © 2023 by the authors. Licensee MDPI, Basel, Switzerland

of these persistent compounds, offering a more effective alternative for treating pharmaceutical waste.

Dyes and Pesticides: Fungi have demonstrated the ability to degrade industrial dyes and pesticides, pollutants that are difficult to treat through conventional methods. These capabilities make fungal treatment especially useful in industries where such compounds are prevalent, such as textile manufacturing and agriculture.

13.6.2 Heavy Metal Remediation

Fungal species can help significantly in remediating heavy metals in wastewater, which is a major concern in both industrial and municipal systems (Shakya et al., 2015).

Bioaccumulation and Adsorption: Certain fungi possess the ability to bioaccumulate or adsorb heavy metals from wastewater, making them an effective tool for removing metals such as lead, mercury, and cadmium. This feature is particularly important for industries like mining and metal finishing, where heavy metal contamination is a persistent issue.

Potential for Widespread Application: The capability of fungi to remove heavy metals expands the range of their application, positioning fungal treatment as a viable solution for treating contaminated water in both industrial and municipal settings.

13.6.3 Genetic Engineering and Biotechnological Advances

Advances in genetic engineering and biotechnology hold the potential to significantly enhance the efficacy and scope of fungal wastewater treatment in the future.

Optimizing Fungal Strains: Through genetic modification, researchers can optimize fungal strains to target specific pollutants with greater efficiency. Techniques like CRISPR-Cas9 and synthetic biology will enable the enhancement of fungal metabolic pathways, allowing for faster degradation of a broader range of contaminants (Stein et al., 2018). The scalability and economic viability of treating wastewater with fungi could be greatly enhanced by these optimized strains.

Enhanced Metabolic Pathways: By improving the metabolic pathways of fungi, researchers aim to create strains that are more robust and can function in diverse environmental conditions, making them more suited for industrial-scale applications.

13.6.4 Development of Hybrid Treatment Systems

Hybrid treatment systems, which combine fungi with other microorganisms, are an exciting area of research that promises to further improve the efficiency and adaptability of wastewater treatment.

Synergistic Relationships: In hybrid systems, fungi can work alongside bacteria, algae, or even higher plants to provide a more comprehensive treatment process. Fungi are highly effective at breaking down complex organic compounds, while bacteria can degrade the simpler byproducts. Algae contribute to nutrient removal and biomass production, improving the overall efficiency of the system.

Applications in Different Wastewater Streams: These hybrid systems can be tailored to treat a diversity of wastewater streams, from industrial effluents to municipal sewage. By leveraging the complementary capabilities of fungi and other organisms, hybrid systems offer a more integrated and adaptive solution for wastewater treatment.

13.6.5 Advancements in Bioreactor Design

Bioreactor technology will play a significant role in the future scalability and effectiveness of fungal wastewater treatment. Innovations in reactor design are necessary to optimize environmental conditions for fungal growth and pollutant degradation.

Optimization for Fungal Growth: Novel bioreactors are being developed that cater specifically to the environmental needs of fungi, such as maintaining optimal pH, temperature, and oxygen levels (Kumar et al., 2021). These reactors can significantly enhance the performance of fungal wastewater treatment systems, reducing treatment times and improving overall efficiency.

Energy Efficiency and Continuous Operation: Future bioreactors will focus on minimizing energy consumption while allowing for continuous operation. Continuous flow systems, which allow wastewater to pass through the reactor without interruption, can make fungal systems more scalable for large-scale applications.

Modular Reactor Systems: Modular reactor designs, which allow for easy expansion and adaptability, will enable fungal wastewater treatment systems to be integrated into existing infrastructure (Battilani et al., 2010). These modular systems can be scaled up or down depending on the volume and type of wastewater, offering flexibility for industrial and municipal applications.

13.6.6 Scalability and Integration with Existing Infrastructure

For fungal wastewater treatment to become widespread, it must be scalable and capable of integrating with existing wastewater treatment infrastructure.

Large-Scale Industrial and Municipal Applications: The development of scalable fungal treatment systems will make it possible for industries and municipalities to adopt these technologies on a larger scale. With continuous advancements in reactor design and genetic optimization, fungal systems are poised to handle high volumes of wastewater efficiently.

Integration with Conventional Systems: Fungal systems can also be integrated with traditional wastewater treatment processes, such as activated sludge systems, to complement bacterial treatment and address pollutants that bacteria struggle to degrade. This integration will provide a more comprehensive approach to pollutant removal, making wastewater treatment systems more effective and sustainable overall (Philip et al., n.d.).

13.7 Conclusion

The future of fungi in wastewater treatment is not only promising but also crucial for addressing the growing challenges of global water management and environmental sustainability. Fungal treatment systems are gaining recognition for their efficiency in degrading complex pollutants and their relatively low environmental impact, making them an attractive solution for both industrial and municipal applications. Their ability to effectively treat contaminants such as heavy metals and pharmaceuticals positions them as valuable alternatives to conventional methods, particularly in regions where traditional wastewater treatment infrastructure is lacking or insufficient. Future research is likely to focus on optimizing fungal consortia and hybrid systems, developing advanced bioreactors, and enhancing fungal strains through genetic engineering. These advancements will create more robust, efficient, and adaptable systems capable of tackling a wider range of contaminants while maximizing resource recovery. Real-time monitoring and automation will further streamline operations, making fungal wastewater treatment scalable for broader applications. As we move toward a more sustainable future, the versatility and resilience of fungal-based technologies will play a pivotal role in transforming wastewater treatment practices globally. Incorporating fungi into wastewater treatment systems could play a crucial role in advancing the circular economy by extracting valuable nutrients and energy from waste. As noted by renowned mycologist Paul Stamets, fungi might be instrumental in addressing the climate crisis. By leveraging their unique properties, we can create cutting-edge approaches that not only improve wastewater processing but also foster environmental equilibrium and long-term sustainability, thereby ensuring access to cleaner water for upcoming generations.

References

Al Mayyahi, A., & Ali Abed Al-Asadi, H. (n.d.). Advanced oxidation processes (AOPs) for wastewater treatment. *Asian Journal of Water, Environment and Pollution, 2*(3). Retrieved October 6, 2024, from www.ajast.net

Ayaz, M., Feizienė, D., Tilvikienė, V., Akhtar, K., Stulpinaitė, U., & Iqbal, R. (2021). Biochar role in the sustainability of agriculture and environment. *Sustainability, 13*(3), 1330. https://doi.org/10.3390/SU13031330

Battilani, A., Steiner, M., Andersen, M., Back, S. N., Lorenzen, J., Schweitzer, A., Dalsgaard, A., Forslund, A., Gola, S., Klopmann, W., Plauborg, F., & Andersen, M. N. (2010). Decentralised water and wastewater treatment technologies to produce functional water for irrigation. *Agricultural Water Management, 98*(3), 385–402. https://doi.org/10.1016/J.AGWAT.2010.10.010

Bhatnagar, S., & Kumari, R. (2013). Bioremediation: A sustainable tool for environmental management—A review. *Annual Review & Research in Biology, 3*, 974–993.

Bibbins-Martínez, M., Juárez-Hernández, J., López-Domínguez, J. Y., Nava-Galicia, S. B., Martínez-Tozcano, L. J., Juárez-Atonal, R., Cortés-Espinosa, D., Díaz-Godinez, G., & Jeb, J. (2023). Potential application of fungal biosorption and/or bioaccumulation for the bioremediation of wastewater contamination: A review. *Journal of Environmental Biology, 44*, 135–145. https://doi.org/10.22438/jeb/44/2/MRN-5093

Chandra, P., & Enespa. (2019). Fungal enzymes for bioremediation of contaminated soil. In *Recent advancement in white biotechnology through fungi* (pp. 189–215). Springer. https://doi.org/10.1007/978-3-030-25506-0_7

Chandran, P., Suresh, S., Balasubramain, B., Gangwar, J., Raj, A. S., Aarathy, U. L., Meyyazhagan, A., Pappuswamy, M., & Sebastian, J. K. (2023). Biological treatment solutions using bioreactors for environmental contaminants from industrial waste water. *Journal of Umm Al-Qura University for Applied Sciences, 2023*(1), 1–23. https://doi.org/10.1007/S43994-023-00071-4

Chatterjee, S., & Venkata Mohan, S. (2022). Fungal biorefinery for sustainable resource recovery from waste. *Bioresource Technology, 345*, 126443. https://doi.org/10.1016/J.BIORTECH.2021.126443

Chen, L., Zhang, X., Zhang, M., Zhu, Y., & Zhuo, R. (2022). Removal of heavy-metal pollutants by white rot fungi: Mechanisms, achievements, and perspectives. *Journal of Cleaner Production, 354*, 131681. https://doi.org/10.1016/J.JCLEPRO.2022.131681

Chen, S. H., Cheow, Y. L., Ng, S. L., & Ting, A. S. Y. (2020). Bioaccumulation and biosorption activities of indoor metal-tolerant Penicillium simplicissimum for removal of toxic metals. *International Journal of Environmental Research, 14*(2), 235–242. https://doi.org/10.1007/S41742-020-00253-6/METRICS

Chen, Z., Da Oh, W., & Yap, P. S. (2022). Recent advances in the utilization of immobilized laccase for the degradation of phenolic compounds in aqueous solutions: A review. *Chemosphere, 307*, 135824. https://doi.org/10.1016/J.CHEMOSPHERE.2022.135824

Cuellar-Bermudez, S. P., Aleman-Nava, G. S., Chandra, R., Garcia-Perez, J. S., Contreras-Angulo, J. R., Markou, G., Muylaert, K., Rittmann, B. E., & Parra-Saldivar, R. (2017). Nutrients utilization and contaminants removal. A review of two approaches of algae and cyanobacteria in wastewater. *Algal Research, 24*, 438–449. https://doi.org/10.1016/J.ALGAL.2016.08.018

Dave, S., & Das, J. (2021). Role of microbial enzymes for biodegradation and bioremediation of environmental pollutants: Challenges and future prospects. In *Bioremediation for environmental sustainability* (pp. 325–346). Elsevier. https://doi.org/10.1016/B978-0-12-820524-2.00013-4

de Mello, A. F. M., de Souza Vandenberghe, L. P., Herrmann, L. W., Letti, L. A. J., Burgos, W. J. M., Scapini, T., Manzoki, M. C., de Oliveira, P. Z., & Soccol, C. R. (2023). Strategies and engineering aspects on the scale-up of bioreactors for different bioprocesses. *Systems Microbiology and Biomanufacturing, 4*(2), 365–385. https://doi.org/10.1007/S43393-023-00205-Z

Demarche, P., Junghanns, C., Nair, R. R., & Agathos, S. N. (2012). Harnessing the power of enzymes for environmental stewardship. *Biotechnology Advances, 30*(5), 933–953. https://doi.org/10.1016/J.BIOTECHADV.2011.05.013

Dhankhar, R., & Hooda, A. (2011). Fungal biosorption—An alternative to meet the challenges of heavy metal pollution in aqueous solutions. *Environmental Technology, 32*(5), 467–491. https://doi.org/10.1080/09593330.2011.572922

Espinosa-Ortiz, E. J., Rene, E. R., & Gerlach, R. (2022). Potential use of fungal-bacterial co-cultures for the removal of organic pollutants. *Critical Reviews in Biotechnology, 42*(3), 361–383. https://doi.org/10.1080/07388551.2021.1940831

Fu, F., & Wang, Q. (2011). Removal of heavy metal ions from wastewaters: A review. *Journal of Environmental Management, 92*(3), 407–418. https://doi.org/10.1016/J.JENVMAN.2010.11.011

Gernaey, K. V., Van Loosdrecht, M. C. M., Henze, M., Lind, M., & Jørgensen, S. B. (2004). Activated sludge wastewater treatment plant modelling and simulation: State of the art. *Environmental Modelling & Software, 19*(9), 763–783. https://doi.org/10.1016/J.ENVSOFT.2003.03.005

Gherghel, A., Teodosiu, C., & De Gisi, S. (2019). A review on wastewater sludge valorisation and its challenges in the context of circular economy. *Journal of Cleaner Production, 228*, 244–263. https://doi.org/10.1016/J.JCLEPRO.2019.04.240

Ghosh Ray, S., & Ghangrekar, M. M. (2018). Comprehensive review on treatment of high-strength distillery wastewater in advanced physico-chemical and biological degradation pathways. *International Journal of Environmental Science and Technology, 16*(1), 527–546. https://doi.org/10.1007/S13762-018-1786-8

Ghosh, S., Rusyn, I., Dmytruk, O. V., Dmytruk, K. V., Onyeaka, H., Gryzenhout, M., & Gafforov, Y. (2023). Filamentous fungi for sustainable remediation of pharmaceutical compounds, heavy metal and oil hydrocarbons. *Frontiers in Bioengineering and Biotechnology, 11*, 1106973. https://doi.org/10.3389/FBIOE.2023.1106973/BIBTEX

Gomaa, H., Emran, M. Y., & El-Gammal, M. A. (2023). Biodegradation of azo dye pollutants using microorganisms. In *Handbook of biodegradable materials* (pp. 781–809). Springer. https://doi.org/10.1007/978-3-031-09710-2_33

Gouda, S. A., & Taha, A. (2023). Biosorption of heavy metals as a new alternative method for wastewater treatment: A review. *Egyptian Journal of Aquatic Biology and Fisheries, 27*(2), 135–153. https://doi.org/10.21608/EJABF.2023.291671

Hakawati, R., Smyth, B. M., McCullough, G., De Rosa, F., & Rooney, D. (2017). What is the most energy efficient route for biogas utilization: Heat, electricity or transport? *Applied Energy, 206*, 1076–1087. https://doi.org/10.1016/J.APENERGY.2017.08.068

Harms, H., Schlosser, D., & Wick, L. Y. (2011). Untapped potential: Exploiting fungi in bioremediation of hazardous chemicals. *Nature Reviews Microbiology, 9*(3), 177–192. https://doi.org/10.1038/nrmicro2519

Hayat, R., Ahmed, I., & Sheirdil, R. A. (2012). Enhancing plant growth promoting rhizobacterial activities through consortium exposure: A review. *Frontiers in Bioengineering and Biotechnology, 11*, 557–579. https://doi.org/10.3389/fbioe.2023.1099999

Horst, J., Drane, S., & Gattenby, J. (2020). Nature-based remediation: Growing opportunities in the harnessing of natural systems. *Groundwater Monitoring and Remediation, 40*(1), 14. https://doi.org/10.1111/GWMR.12360

Hu, Z., Qi, Y., Zhao, L., & Chen, G. (2019). Interactions between microalgae and microorganisms for wastewater remediation and biofuel production. *Waste and Biomass Valorization, 10*(12), 3907–3919. https://doi.org/10.1007/S12649-018-0325-7/FIGURES/2

Hyde, K. D., Jones, E. B. G., Leaño, E., Pointing, S. B., Poonyth, A. D., & Vrijmoed, L. L. P. (1998). Role of fungi in marine ecosystems. *Biodiversity and Conservation, 7*(9), 1147–1161. https://doi.org/10.1023/A:1008823515157/METRICS

Ibrahim, M., Kabir, J., Kwanashie, C., & Salawudeen, M. (2017). Identification of aspergillus species in feed fed to caged birds using morphological characteristics in Zaria, Nigeria. *Microbiology Research International, 5*(2), 16–24.

Imran, M., Anwar, Z., Irshad, M., Asad, M. J., & Ashfaq, H. (2016). Cellulase production from species of fungi and bacteria from agricultural wastes and its utilization in industry: A review. *Advances in Enzyme Research, 4*(2), 44–55. https://doi.org/10.4236/AER.2016.42005

Jain, M., Khan, S. A., Sharma, K., Jadhao, P. R., Pant, K. K., Ziora, Z. M., & Blaskovich, M. A. T. (2022). Current perspective of innovative strategies for bioremediation of organic pollutants from wastewater. *Bioresource Technology, 344*, 126305. https://doi.org/10.1016/J.BIORTECH.2021.126305

Kesari, K. K., Soni, R., Jamal, Q. M. S., Tripathi, P., Lal, J. A., Jha, N. K., Siddiqui, M. H., Kumar, P., Tripathi, V., & Ruokolainen, J. (2021). Wastewater treatment and reuse: A review of its applications and health implications. *Water, Air, and Soil Pollution, 232*(5), 1–28. https://doi.org/10.1007/S11270-021-05154-8/FIGURES/5

Khalil, H., Legin, E., Kurek, B., Perre, P., & Taidi, B. (2021). Morphological growth pattern of Phanerochaete chrysosporium cultivated on different Miscanthus x giganteus biomass fractions. *BMC Microbiology, 21*(1), 1–18. https://doi.org/10.1186/S12866-021-02350-8/PEER-REVIEW

Kookana, R. S., Drechsel, P., Jamwal, P., & Vanderzalm, J. (2020). Urbanisation and emerging economies: Issues and potential solutions for water and food security. *Science of the Total Environment, 732*, 139057. https://doi.org/10.1016/J.SCITOTENV.2020.139057

Kües, U. (2015). Fungal enzymes for environmental management. *Current Opinion in Biotechnology, 33*, 268–278. https://doi.org/10.1016/J.COPBIO.2015.03.006

Kumar Singh, N., Pandey, S., Singh, R. P., Muzamil Gani, K., Yadav, M., Thanki, A., & Kumar, T. (2020). Bioreactor and bioprocess technology for bioremediation of domestic and municipal wastewater. In *Bioremediation of pollutants* (pp. 251–273). Elsevier. https://doi.org/10.1016/B978-0-12-819025-8.00011-9

Kumar, V., Ahluwalia, V., Saran, S., Kumar, J., Patel, A. K., & Singhania, R. R. (2021). Recent developments on solid-state fermentation for production of microbial secondary metabolites: Challenges and solutions. *Bioresource Technology, 323*, 124566. https://doi.org/10.1016/J.BIORTECH.2020.124566

Larsen, T. A., Hoffmann, S., Lüthi, C., Truffer, B., & Maurer, M. (2016). Emerging solutions to the water challenges of an urbanizing world. *Science, 352*(6288), 928–933. https://doi.org/10.1126/SCIENCE.AAD8641

Leng, L., Li, W., Chen, J., Leng, S., Chen, J., Wei, L., Peng, H., Li, J., Zhou, W., & Huang, H. (2021). Co-culture of fungi-microalgae consortium for wastewater treatment: A review. *Bioresource Technology, 330*, 125008. https://doi.org/10.1016/J.BIORTECH.2021.125008

Levasseur, A., Lomascolo, A., Chabrol, O., Ruiz-Dueñas, F. J., Boukhris-Uzan, E., Piumi, F., Kües, U., Ram, A. F. J., Murat, C., Haon, M., Benoit, I., Arfi, Y., Chevret, D., Drula, E., Kwon, M. J., Gouret, P., Lesage-Meessen, L., Lombard, V., Mariette, J., & Record, E. (2014). The genome of the white-rot fungus Pycnoporus cinnabarinus: A basidiomycete model with a versatile arsenal for lignocellulosic biomass breakdown. *BMC Genomics, 15*(1), 1–24. https://doi.org/10.1186/1471-2164-15-486/FIGURES/8

Maqsood, Q., Hussain, N., Mumtaz, M., Bilal, M., & Iqbal, H. M. N. (2022). Novel strategies and advancement in reducing heavy metals from the contaminated environment. *Archives of Microbiology, 204*(8), 1–18. https://doi.org/10.1007/S00203-022-03087-2

Morin-Crini, N., Lichtfouse, E., Fourmentin, M., Ribeiro, A. R. L., Noutsopoulos, C., Mapelli, F., Fenyvesi, É., Vieira, M. G. A., Picos-Corrales, L. A., Moreno-Piraján, J. C., Giraldo, L., Sohajda, T., Huq, M. M., Soltan, J., Torri, G., Magureanu, M., Bradu, C., & Crini, G. (2022). Removal of emerging contaminants from wastewater using advanced treatments. A review. *Environmental Chemistry Letters, 20*(2), 1333–1375. https://doi.org/10.1007/S10311-021-01379-5

Naveena, J., & Latha, L. (2012). *Fungal cell walls as protective barriers for toxic metals*. https://www.researchgate.net/publication/322701166.

Nji, Q. N., Babalola, O. O., & Mwanza, M. (2023). Soil aspergillus species, pathogenicity and control perspectives. *Journal of Fungi, 9*(7), 766. https://doi.org/10.3390/JOF9070766

Ojha, N., Karn, R., Abbas, S., & Bhugra, S. (2021). Bioremediation of industrial wastewater: A review. *IOP Conference Series: Earth and Environmental Science, 796*(1), 012012. https://doi.org/10.1088/1755-1315/796/1/012012

Peng, X., Wilken, S. E., Lankiewicz, T. S., Gilmore, S. P., Brown, J. L., Henske, J. K., Swift, C. L., Salamov, A., Barry, K., Grigoriev, I. V., Theodorou, M. K., Valentine, D. L., & O'Malley, M. A. (2021). Genomic and functional analyses of fungal and bacterial consortia that enable lignocellulose breakdown in goat gut microbiomes. *Nature Microbiology, 6*(4), 499–511. https://doi.org/10.1038/s41564-020-00861-0

Pereira, C. P., Goldenstein, J. P. N., & Bassin, J. P. (2021). Industrial wastewater contaminants and their hazardous impacts. *Biosorption for Wastewater Contaminants*. https://doi.org/10.1002/9781119737629.CH1

Periasamy, D., Mani, S., & Ambikapathi, R. (2019). White rot fungi and their enzymes for the treatment of industrial dye effluents. *Biotechnology Advances, 22*, 161–187. https://doi.org/10.1007/978-3-030-25506-0_4

Philip, L., Pradeep, T., & Murty Bhallamudi, S. (n.d.). *Technological solutions for water sustainability: Challenges and prospects towards a water-secure India*. IWA Publishing.

Pörhö, H., Tomperi, J., Sorsa, A., Juuso, E., Ruuska, J., & Ruusunen, M. (2023). Data-based modelling of chemical oxygen demand for industrial wastewater treatment. *Applied Sciences, 13*, 7848. https://doi.org/10.3390/APP13137848

Priya, A. K., Gnanasekaran, L., Dutta, K., Rajendran, S., Balakrishnan, D., & Soto-Moscoso, M. (2022). Biosorption of heavy metals by microorganisms: Evaluation of different underlying mechanisms. *Chemosphere, 307*, 135957. https://doi.org/10.1016/J.CHEMOSPHERE.2022.135957

Pryce, D., Kapelan, Z., & Memon, F. A. (2022). Modelling the performance of an integrated fixed-film activated sludge (IFAS) system: A systematic approach to automated calibration. *Scientific Reports, 12*(1), 1–19. https://doi.org/10.1038/s41598-022-13779-w

Puyol, D., Batstone, D. J., Hülsen, T., Astals, S., Peces, M., & Krömer, J. O. (2017). Resource recovery from wastewater by biological technologies: Opportunities, challenges, and prospects. *Frontiers in Microbiology, 7*(JAN), 218193. https://doi.org/10.3389/FMICB.2016.02106/BIBTEX

Rahimi, A., & Rezaii, M. (2021). Seismic performance evaluation of buried sewage collection pipelines. *Shock and Vibration, 2021*(1), 6669505. https://doi.org/10.1155/2021/6669505

Rao, M. A., Scelza, R., Acevedo, F., Diez, M. C., & Gianfreda, L. (2014). Enzymes as useful tools for environmental purposes. *Chemosphere, 107*, 145–162. https://doi.org/10.1016/J.CHEMOSPHERE.2013.12.059

Robles, Á., Aguado, D., Barat, R., Borrás, L., Bouzas, A., Giménez, J. B., Martí, N., Ribes, J., Ruano, M. V., Serralta, J., Ferrer, J., & Seco, A. (2020). New frontiers from removal to recycling of nitrogen and phosphorus from wastewater in the circular economy. *Bioresource Technology, 300*, 122673. https://doi.org/10.1016/J.BIORTECH.2019.122673

Rollié, S., Mangold, M., & Sundmacher, K. (2012). Designing biological systems: Systems engineering meets synthetic biology. *Chemical Engineering Science, 69*(1), 1–29. https://doi.org/10.1016/J.CES.2011.10.068

Sağ, Y. (2001). Biosorption of heavy metals by fungal biomass and modeling of fungal biosorption: A review. *Separation and Purification Methods, 30*(1), 1–48. https://doi.org/10.1081/SPM-100102984

Sankaran, S., Khanal, S. K., Jasti, N., Jin, B., Pometto, A. L., & Van Leeuwen, J. H. (2010). Use of filamentous fungi for wastewater treatment and production of high value fungal byproducts: A review. *Critical Reviews in Environmental Science and Technology, 40*(5), 400–449. https://doi.org/10.1080/10643380802278943

Saravanan, A., Kumar, P. S., Vo, D. V. N., Jeevanantham, S., Karishma, S., & Yaashikaa, P. R. (2021). A review on catalytic-enzyme degradation of toxic environmental pollutants: Microbial enzymes. *Journal of Hazardous Materials, 419*, 126451. https://doi.org/10.1016/J.JHAZMAT.2021.126451

Sato, T., Qadir, M., Yamamoto, S., Endo, T., & Zahoor, A. (2013). Global, regional, and country level need for data on wastewater generation, treatment, and use. *Agricultural Water Management, 130*, 1–13. https://doi.org/10.1016/J.AGWAT.2013.08.007

Sellami, K., Couvert, A., Nasrallah, N., Maachi, R., Abouseoud, M., & Amrane, A. (2022). Peroxidase enzymes as green catalysts for bioremediation and biotechnological applications: A review. *Science of the Total Environment, 806*, 150500. https://doi.org/10.1016/J.SCITOTENV.2021.150500

Shakya, M., Sharma, P., Meryem, S. S., Mahmood, Q., & Kumar, A. (2015). Heavy metal removal from industrial wastewater using fungi: Uptake mechanism and biochemical aspects. *Journal of Environmental Engineering, 142*(9), C6015001. https://doi.org/10.1061/(ASCE)EE.1943-7870.0000983

Sharma, M., Agarwal, S., Agarwal Malik, R., Kumar, G., Pal, D. B., Mandal, M., Sarkar, A., Bantun, F., Haque, S., Singh, P., Srivastava, N., & Gupta, V. K. (2023). Recent advances in microbial engineering approaches for wastewater treatment: A review. *Bioengineered, 14*(1), 2184518. https://doi.org/10.1080/21655979.2023.2184518

Sharma, P., Rani, L., Grewal, A. S., & Srivastav, A. L. (2022). Impact of pharmaceuticals and antibiotics waste on the river ecosystem: A growing threat. In *Ecological significance of river ecosystems: Challenges and management strategies* (pp. 15–36). Elsevier. https://doi.org/10.1016/B978-0-323-85045-2.00015-7

Singh, S., & Tiwari, S. (2019). Climate change, water and wastewater treatment: Interrelationship and consequences. In *Water conservation, recycling and reuse: Issues and challenges* (pp. 203–214). Springer. https://doi.org/10.1007/978-981-13-3179-4_11

Sirisha, V. L., Jain, A., & Jain, A. (2016). Enzyme immobilization: An overview on methods, support material, and applications of immobilized enzymes. *Advances in Food and Nutrition Research, 79*, 179–211. https://doi.org/10.1016/BS.AFNR.2016.07.004

Smol, M., Adam, C., & Preisner, M. (2020). Circular economy model framework in the European water and wastewater sector. *Journal of Material Cycles and Waste Management, 22*(3), 682–697. https://doi.org/10.1007/S10163-019-00960-Z/FIGURES/4

Stein, H. P., Navajas-Pérez, R., & Aranda, E. (2018). Potential for CRISPR genetic engineering to increase xenobiotic degradation capacities in model fungi. In *Approaches in bioremediation* (pp. 61–78). Springer. https://doi.org/10.1007/978-3-030-02369-0_4

Stottmeister, U., Wießner, A., Kuschk, P., Kappelmeyer, U., Kästner, M., Bederski, O., Müller, R. A., & Moormann, H. (2003). Effects of plants and microorganisms in constructed wetlands for wastewater treatment. *Biotechnology Advances, 22*(1–2), 93–117. https://doi.org/10.1016/J.BIOTECHADV.2003.08.010

Tariq, A., & Mushtaq, A. (2023). Untreated wastewater reasons and causes: A review of most affected areas and cities. *IJCBS, 23*(1), 2023. www.iscientific.org/Journal.html

Thirumalaivasan, N., Gnanasekaran, L., Kumar, S., Durvasulu, R., Sundaram, T., Rajendran, S., Nangan, S., & Kanagaraj, K. (2024). Utilization of fungal and bacterial bioremediation techniques for the treatment of toxic waste and biowaste. *Frontiers in Materials, 11*, 1416445. https://doi.org/10.3389/FMATS.2024.1416445/BIBTEX

Vaksmaa, A., Guerrero-Cruz, S., Ghosh, P., Zeghal, E., Hernando-Morales, V., & Niemann, H. (2023). Role of fungi in bioremediation of emerging pollutants. *Frontiers in Marine Science, 10*, 1070905. https://doi.org/10.3389/FMARS.2023.1070905/BIBTEX

Visvanathan, C., Ben Aim, R., & Parameshwaran, K. (2000). Membrane separation bioreactors for wastewater treatment. *Critical Reviews in Environmental Science and Technology, 30*(1), 1–48. https://doi.org/10.1080/10643380091184165

Wesenberg, D., Kyriakides, I., & Agathos, S. N. (2003). White-rot fungi and their enzymes for the treatment of industrial dye effluents. *Biotechnology Advances, 22*(1-2), 161–187. https://doi.org/10.1016/J.BIOTECHADV.2003.08.011

Zabochnicka-Świątek, M., & Krzywonos, M. (2014). Potentials of biosorption and bioaccumulation processes for heavy metal removal. *Polish Journal of Environmental Studies, 23*(2), 551–561. https://www.pjoes.com/Potentials-of-Biosorption-and-Bioaccumulation-r-nProcesses-for-Heavy-Metal-Removal,89224,0,2.html

Correction to: Fungi in Waste Water Treatment

Chaudhery Mustansar Hussain, Neha Srivastava, and Ghazanfar Hussain

Correction to: Chapter 7, in C. M. Hussain et al. (eds.), Fungi in Waste Water Treatment, Springer Water, https://doi.org/10.1007/978-3-032-12693-1_7

In the original version of the chapter, the following correction has been incorporated: In Chap. 7, the authors' names have been changed from "S. Umamaheswari" and "Sudipta Choudhuy" to "Umamaheshwari S" and "Sudipta Choudhury". The correction chapter and the book have been updated with the change.

The updated version of this chapter can be found at
https://doi.org/10.1007/978-3-032-12693-1_7

C. M. Hussain et al. (eds.), *Fungi in Waste Water Treatment*, Springer Water,
https://doi.org/10.1007/978-3-032-12693-1_14

MIX
Papier aus verantwortungsvollen Quellen
Paper from responsible sources
FSC® C105338

If you have any concerns about our products,
you can contact us on
ProductSafety@springernature.com

In case Publisher is established outside the EU,
the EU authorized representative is:
Springer Nature Customer Service Center GmbH
Europaplatz 3, 69115 Heidelberg, Germany

Printed by Libri Plureos GmbH
in Hamburg, Germany